California
Earthquakes

California Earthquakes

Science, Risk &
the Politics of Hazard Mitigation

CARL-HENRY GESCHWIND

THE JOHNS HOPKINS UNIVERSITY PRESS
Baltimore & London

© 2001 The Johns Hopkins University Press
All rights reserved. Published 2001
Printed in the United States of America on acid-free paper
9 8 7 6 5 4 3 2 1

The Johns Hopkins University Press
2715 North Charles Street
Baltimore, Maryland 21218-4363
www.press.jhu.edu

Library of Congress Cataloging-in-Publication Data

Geschwind, Carl-Henry, 1965–
California earthquakes : science, risk, and the politics of hazard mitigation /
Carl-Henry Geschwind.
    p. cm.
Includes bibliographical references and index.
ISBN 0-8018-6596-4 (alk. paper)
    1. Earthquake hazard analysis—California—History.   2. Earthquakes—
California—Safety measures—History.   I. Title.
QE535.2.U6 G485   2001
363.34′95′09794—dc21

                                                    00-010627

A catalog record for this book is available from the British Library.

For Mom and Dad

Contents

Acknowledgments

The research and writing of this book were made possible through financial assistance from the National Science Foundation (Grant SBR-9411270), the Huntington Library, the Maurice Biot Archives Fund at the California Institute of Technology, the Johns Hopkins University Departments of Earth and Planetary Science and of the History of Science, Medicine, and Technology, the American Institute of Physics, the Sourisseau Academy of California, the American Philosophical Society (John Clarke Slater predoctoral fellowship), the Center for History of Recent Science at George Washington University (postdoctoral fellowship), and my parents, Herman and Mary Geschwind.

My deepest debt of gratitude goes to Sharon Kingsland, who first suggested that I look at the history of seismology in California. She was the perfect dissertation advisor, allowing me to follow my own path while never losing faith in me. I am also very grateful to Harry Marks, who first introduced me to the literature on regulatory science and the state, and to Bruce Marsh, who hosted me in the Department of Earth and Planetary Science while I was at Hopkins. My fellow students, including Keith Barbera, Jesse Bump, Hunter and Kathleen Crowther-Heyck, Sandy Gliboff, Shoshanna Green, Scott Knowles, Tom Lassman, Tanya Levin, Susan Morris, Buhm Soon Park, Dave Roberts, Karen Stupski, Xinye Zhang, and above all Melody Herr provided me with continuing support and constructive feedback during the writing of my dissertation and afterwards. At George Washington University, I thank Horace Judson for providing me with the opportunity to complete this manuscript and my colleagues Nathaniel Comfort, Richard Burian, Anne Fitzpatrick, Dave Grier, Mark Lesney, Patrick McCray, Mark Parascandola, Tomoko Steen, and Steve Weiss for their many insightful comments.

During my research I was fortunate to receive assistance from numerous helpful and dedicated archivists and librarians, including Judith Goodstein, Shelley Erwin, and Bonnie Ludt at the California Institute of Technology, Elizabeth Andrews at the Massachusetts Institute of Technology, Lisa Miller at the National Archives Regional Facility in San Bruno, California, Dave Butler at the Natural Hazards Center Library in Boulder, Colorado, Carla Bowen at the National Academy of Sciences, Rob Ritchie and Jennifer Martinez at the Huntington Library, the special collections staffs at Stanford University, the Bancroft Library, the Library of Congress, and the California State Archives, and reference librarians at the Earthquake Engineering Research Center Library, the Los Angeles Public Library, the Berkeley Public Library, the Pasadena Public Library, and the California State Library.

I am grateful to the many geologists, geophysicists, and other researchers who shared their insights into the history and current state of earthquake research, including Yash Aggarwal, Manuel Bonilla, Louise Comfort, Jerry Eaton, Jim Goltz, Hiroo Kanamori, Carl Kisslinger, Dennis Mileti, Amos Nur, Chris Scholz, Kerry Sieh, Lynn Sykes, Fred Turner, and Gilbert White. My thanks also go to the many individuals who commented on portions or all of my manuscript, including Duncan Agnew, Kai-Henrik Barth, Ron Doel, Suzette Hemberger, Bill Leslie, Chuck ReVelle, Margaret Rossiter, Seth Stein, Robert Yeats, and several anonymous reviewers. In addition, I benefited greatly from questions posed by seminar audiences at the California Institute of Technology, Pomona College, the University of Maryland, the National Museum of American History, the Potomac Geophysical Society, the History of Science Society, and the Geological Society of America. Finally, I thank Anne Whitmore, Bob Brugger, and others at the Johns Hopkins University Press for overseeing the final steps of publication.

California
Earthquakes

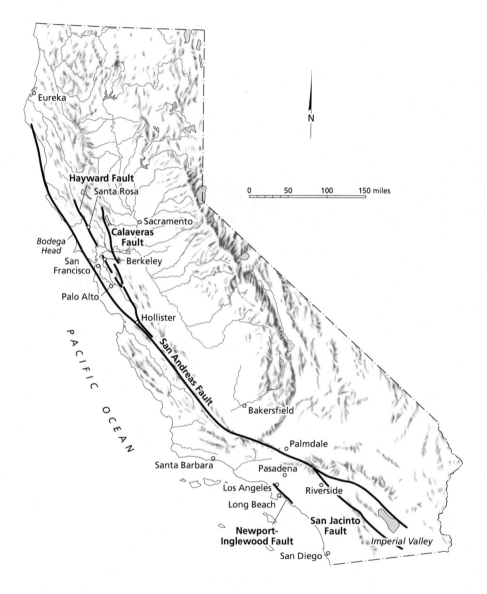

Eureka

Hayward Fault
Santa Rosa

Sacramento

Bodega Head

Calaveras Fault

San Francisco

Berkeley

Palo Alto

Hollister

San Andreas Fault

PACIFIC OCEAN

Bakersfield

Palmdale

Santa Barbara

Pasadena

Los Angeles

Riverside

Long Beach

San Jacinto Fault

Newport-Inglewood Fault

Imperial Valley

San Diego

N

0   50   100   150 miles

by Bill Nelson

Introduction

Earthquakes can be terrifying. The earth that had seemed so reassuringly solid suddenly lurches, disrupting both physical and emotional balance. Skyscrapers sway visibly and apparently sound buildings suddenly collapse. Familiar geographies are rearranged: bridges fall down, streets are torn apart, and landslides scar hillsides and alter the course of rivers. The terror remains even after the vibrations from the main shock, which last a minute or two. As the ground settles back into place following the shock, it continues to quiver in a series of small aftershocks. The aftershocks are sufficient to keep survivors of the main shock—be they in Turkey, Taiwan, Mexico, or California—from sleeping in their houses. The main earthquake has disrupted all expectations of normalcy, shattered confidence in buildings and other structures. How soon will another large and damaging shock strike, perhaps killing those spared by the first quake? Only after the passage of weeks, even months, does this fear subside and is trust in the solidity of the earth reestablished. But this universal emotional reaction is coupled with many different responses to the threat of future earthquakes. In some communities, earthquake victims react by increasing their religious fervor or by accepting the recurrence of earthquakes with fatalistic resignation. In other places, people have chosen to prepare themselves for future earthquakes by constructing their buildings so that they do not collapse when shaken and by developing contingency plans so that an earthquake does not entirely disrupt the life of a community.

Earthquake preparedness has become particularly sophisticated in California. Here, building codes enforced by local inspectors require that residential and commercial structures be designed and built to resist seismic vibrations. State law requires that critical structures, such as schools, hospitals, and fire stations, be designed to even more exacting standards. Many cities and counties have also insisted that old structures be retrofitted to comply with more recent building codes. A state law prohibits new housing developments from being placed across active earthquake faults; numerous local jurisdictions go even further in preventing human habitations from being built near other areas of particular seismic hazard. Meanwhile, the California Office of Emergency Services and local emergency preparedness agencies have developed emergency response plans, so that firefighters, police officers, medical teams, utility crews, and others will know what to do when an earthquake strikes. Watching over all of these government regulatory activities is the California Seismic Safety Commission, an independent government agency that serves as an advocate for greater awareness of earthquake safety in the state government. Aiding the regulatory apparatus are government-funded scientists and engineers who monitor the level of seismic hazard in the state, search out better ways for designing buildings to resist earthquakes, and educate the public on the earthquake threat.

This regulatory apparatus and the concern with reducing seismic hazards that underlies it are quite recent phenomena, even in California. In the aftermath of the disastrous 1906 San Francisco earthquake, California officials and newspapers insisted that their state was not particularly subject to seismic hazards. Looking to the future, they asserted that the state would not experience another significant earthquake for generations or even centuries to come. As a result of this general disparaging of seismic hazards, little was done in the rebuilding of San Francisco to mitigate damage from future earthquakes. Even as late as the 1920s, there were no effective building codes requiring earthquake-resistant design in California, and there were no state agencies charged with assuring seismic safety. The powerful regulatory apparatus that addresses earthquake hazards in present-day California is entirely a twentieth-century innovation. The central goal of this book is to trace and to explain its emergence. How did earthquakes come to be viewed as a threat in California? And how did an approach that relies on state regulation of human behavior come to be seen as the most appropriate response to this threat?

Answering such questions will provide insights into not only the history of earthquake hazard mitigation in California but also the evolution of attitudes toward natural hazards in general in the industrial world. A regulatory-state approach also governs, for example, the response of Japan and New Zealand to earthquakes, of the eastern United States to hurricanes, and of Italy and the Caribbean to volcanic eruptions. But California's apparatus for dealing with earthquakes is among the world's most advanced. Thus, understanding in detail its emergence will provide an important entry into the larger problem of explaining how natural hazards became the domain of science and the state.

Progressivism and the Regulatory State

At its most fundamental, the regulatory-state apparatus in California is an outgrowth of the ideology of Progressivism that took hold among scientists, engineers, and other professionals around the turn of the century. By this time, science had attained considerable cachet in American culture. Scientific advances underlay the rapidly growing electrical and chemical industries. In areas as diverse as home economics, epidemiology, and factory production, scientific studies were bringing about greater efficiency and effectiveness. Moreover, Darwin's theory of evolution by natural selection had shaken old religious assumptions to their very core. Science, in the minds of many, stood for progress; the scientists' values of objectivity, efficiency, and expert guidance pointed the way to a better future.

Many scientists, engineers, and other members of the new urban professional middle class sought to extend these values to ever more areas of public life. One area was the conservation of natural resources, such as timber, water, and wildlife. Progressive reformers were appalled at the enormous waste produced by the rapacious exploitation of resources that characterized the laissez-faire economy of the nineteenth century. To remedy this situation, reformers urged the application of scientific and technical expertise in order to produce a more efficient use of natural resources. To promote this goal, Progressives formed a number of advocacy groups, including the Sierra Club and the Save-the-Redwoods League in California. But they also sought the employment of state power, in the guise of such new resource management agencies as the U.S. Forest Service, the Bureau of Reclamation, and the National Park Service, to further their agenda. This marriage of science and the state, the combination of technical expertise with government power (as long as that power followed the dictates of disinterested expertise rather than of greedy developers and other corrupters) seemed to many Progressives the ideal solution for promoting efficiency and progress.[1]

The campaign for seismic safety in California arose from the same Progressive impulse. After the 1906 San Francisco earthquake, a few scientists recognized the threat of more earthquakes' shaking the state in the near future. They also became increasingly uneasy over the nonchalance of the state's boosters and its press regarding seismic hazards and the lack of care with which shoddy buildings were erected by developers. To them, such practices were highly inefficient for society. While corner-cutting might produce profits for contractors in the short run, it would be costly in the long term when carelessly built structures collapsed in the next earthquake. To warn the California public about seismic hazards and to urge more efficient construction practices, these scientists organized themselves into the Seismological Society of America, headquartered in the San Francisco Bay area.

At first, the society's leaders believed that an accumulation of scientific data on earthquake hazards, coupled with moral suasion, would be sufficient to bring about progress in seismic safety. Consequently, they established several networks for gathering information on earthquakes and earthquake faults in California. Society presidents John C. Branner and Bailey Willis exhorted numerous civic groups, newspaper readers, and professional associations to develop greater earthquake awareness. Willis even went so far as to predict a seismic catastrophe for southern California by the mid-1930s. These efforts, however, failed to stir either the general public or the building trades into preparing for future earthquakes. Thus, by the late 1920s, Seismological Society leaders were seeking help from the state, in the form of building codes that would require more earthquake-resistant construction. The first application of state power to the cause of seismic safety came in the wake of the moderate Long Beach earthquake of 1933, when the California legislature for the first time required new schools and other public buildings to be earthquake resistant and many cities and counties in southern California introduced earthquake provisions into their building codes.

The Progressive alliance of science and the state in the promotion of seismic safety grew ever stronger after the Second World War. In the Cold War years of the 1950s and early 1960s, as the federal government recognized the contributions that seismology and earthquake engineering might make to national security, federal funding of these fields grew sixtyfold. In the late 1960s and the 1970s, leading earthquake researchers focused this federal largesse specifically on the field of earthquake hazards research. With the establishment of the National Earthquake Hazards Reduction Program in 1977, the federal government committed itself to spending more than $50 million annually on seismological and earthquake engineering research. With this money, the government maintained a large corps of earthquake experts who could push the campaign for greater seismic safety. At the same time, scientists and engineers in California worked to extend the reach of state power over seismic hazard mitigation. They lobbied hard for stricter building codes, ordinances that prohibited construction on active faults, laws requiring old buildings to be retrofitted, and other measures that would force Californians to better prepare for the next earthquake. The partnership of science and the state culminated with the establishment in 1969 of the California legislature's Joint Committee on Seismic Safety, which provided a formal means for scientists and engineers to advise legislators on hazard mitigation measures, and the creation in 1975 of the California Seismic Safety Commission as a state agency responsible for promoting earthquake safety.

In continuing to seek an alliance between science and state, seismic safety advocates in the 1960s and '70s diverged from other environmental reformers. Environmentalists became less interested in promoting efficiency than in preserving the quality of life by protecting wilderness areas and preventing the proliferation of environmental risks and pollution. Moreover, environmentalists began to share in an increasingly radical critique of government as a reactionary force allied with antidemocratic elements in industry and the military. This critique grew out of the bitter experiences of both civil rights activists and anti-Vietnam protesters, who repeatedly saw their hopes for meaningful reform dashed by the inaction or opposition of the federal government. It found its purest expression in the activities of the so-called "New Left," typified by the Students for a Democratic Society. Disenchanted environmental reformers joined this critique as they saw bureaucrats friendly to industry approve a series of nuclear power plants, big dams, and other environmentally damaging projects. As a result, many environmentalist groups transformed themselves into mass-based protest groups that aggressively used letter-writing campaigns, lawsuits, and both legislative and administrative lobbying to confront industry on environmental issues. Environmentalists still sought the power of the state to protect the environment through, for example, the Clean Air, Clean Water, and Endangered Species acts. At the same time, though, they had also become distrustful of state power, and they resorted to confrontational tactics to ensure that it did not fall into the wrong hands.[2]

Seismic safety advocates in the 1960s and '70s, however, continued to prefer quiet negotiations and consensus building over public or legal confrontations. True to their Progressive roots, they valued objectivity, efficiency, and deference to expertise over grassroots organization and protest. In some cases, earthquake researchers even discouraged the expression of grassroots protest about inadequate seismic safety measures, fearing that such protests would be uninformed and insufficiently deferential to their expertise. In essence, earthquake researchers were happy to be part of the regulatory-state apparatus; the marriage of science and the state has been a contented one in this case. The story of earthquake hazard mitigation in twentieth-century California, then, is the story of the Progressive impulse among a small group of California scientists and engineers and the persistence of that impulse even after other reformers had become disenchanted with the state.

Science and Earthquakes before 1906

Scientists and engineers in twentieth-century California were by no means the first to believe that humans could influence the severity of an earthquake's impact. Earlier peoples had used religious rather than technological means to mitigate quakes. In ancient times, people attributed frightful manifestations of nature—earthquakes, volcanoes, thunderstorms, droughts—to the anger of divine beings and sought to appease the wrath of heaven through ritual offerings. A variant of this approach still dominated in Puritan New England. In 1727 and again in 1755, moderate earthquakes visited Massachusetts, striking down brick chimneys and stone walls over large areas but causing no recorded deaths. In sermons with titles such as "Earthquakes a Token of the Righteous Anger of God" and "The Lord's Voice in the Earthquake Crieth to Careless and Secure Sinners," Puritan ministers and professors described the ultimate cause of the earthquakes as God's wrath at human sinfulness. They called for repentance and moral renewal as ways for propitiating that wrath, although the object of such propitiation was other-wordly rather than this-wordly: sinners should be concerned far more about the utterly horrific catastrophe they might encounter on the Day of Judgment rather than the damage another earthquake might produce. Nevertheless, some sermonizers seem to have thought that proper religious behavior might also be effective in mitigating earthquakes in this world. They pointed out that Catholic cities, such as Lisbon in Portugal and Lima in Peru, suffered far worse seismic catastrophes than Puritan Boston; thus, they concluded, a steadfast adherence to Protestantism did seem to lessen earthquake hazards.[3]

Such religious interpretations of earthquakes persisted in nineteenth century and early twentieth century America. After a strong earthquake ripped through the San Francisco Bay area in October 1868, killing about 30 people, the Reverend A. L. Stone preached a sermon in San Francisco proclaiming the disaster a call from God that "He would be recognized; He would be feared; He would be held in reverence. He would lead back this community and all this people from godlessness to godliness." The sermon was published as a pamphlet, reverently excerpted in one of the city's leading newspapers, and lampooned in another—all signs that its message still spoke to important segments of the community.[4] After the earthquake and fire that devastated San Francisco in 1906, the Seventh-Day Adventist magazine *The Signs of the Times,* published just south of the city in San Mateo, asked: "Was it not the fruit of awful wickedness and the mercy of God which brought the awful catastrophe? Like Babylon and Tyre in their preliminary punishments, was not this modern proud and beautiful Babylon and modern thrifty Tyre meeting the fate that her open, flaunting, God-defying wickedness invited?" By this time, though, such views had begun to lose their hold on mainstream public opinion. The *San Francisco Examiner,* for example, argued that God could not have been so cruel and indifferent as to destroy churches and saloons alike; therefore, the disaster had to be seen as the result of "unchangeable LAW" rather than of God's anger.[5]

The decreasing importance of religious interpretations of earthquakes went hand-in-hand with the rise of scientific explanations. Puritan ministers themselves had encouraged scientific speculation on the natural laws underlying earthquakes. They were concerned with showing that God's grandeur and power manifested itself not only in miracles and supernatural revelations, but even in the every-day workings of nature. Consequently, many of the sermons preached on the eighteenth-century Massachusetts earthquakes began with elaborate discourses on the possible natural causes of earthquakes (chief among them the ignition of sulfurous vapors in cavities deep in the earth) before discussing how God used such natural processes to reveal his anger at human sin.[6] As the nineteenth century progressed, the attention of natural philosphers shifted away from the considerations of how Providence manifested itself in nature. Geology became an increasingly professionalized discipline. Beginning soon after 1800 in both Europe and the United States, scientists established surveys to explore the geology of their countries, formed geological societies to exchange the results of their research, and taught geology as an independent academic subject.[7] With these trends, inquiry into the natural causes of earthquakes became an end in itself, divorced from any consideration of God. The rise of science, in turn, would have profound consequences on how people approached earthquake hazards.

In the late eighteenth century and nineteenth century, European scientists developed three methods for studying earthquakes; these were all adopted by American researchers after 1850. The first method was the compilation of catalogues of earthquake occurrence. As early as the 1750s, antiquarians who had searched through old chronicles published lists of earthquakes for such areas as England and Switzerland that went back to the Middle Ages. By the 1840s, the German researcher Karl von Hoff had assembled a list of more than 2,000 earthquakes across the world going back to 1606 B.C. Von Hoff also compiled annual lists of earthquakes, culling accounts from newspapers and foreign scientific reports. This work was continued by Frenchman Alexis Perrey, whose annual lists of earthquakes for 1843 to 1871 included more than 21,000 events. Perrey also compiled historical catalogues of earthquakes for 21 of the world's regions, ranging from the British Isles to Alaska. Perrey's motivation for putting together these lists was not merely antiquarian; he was interested in statistical analyses to determine whether the occurrence of earthquakes varied with the month of the year or the position of the moon (he claimed to find such variations, although later work has tended not to confirm his conclusions). The compilation of historical catalogues culminated in the work of Irish engineer Robert Mallet, who in the 1850s published a list of nearly 7,000 earthquakes that struck between 1606 B.C. and A.D. 1843. Moreover, Mallet produced a map showing the areas disturbed by these earthquakes, thereby demonstrating conclusively that the incidence of tremors is quite uneven across the globe: earthquakes tend to occur much more frequently in mountainous areas than in the broad basins that characterize both the ocean floor and the interior of continents.[8]

Mallet also set the standard for another method of earthquake study, the investigation of individual earthquakes. The first seismic event to be subjected to a detailed scientific investigation was the series of six shocks that killed 35,000 people in the southern Italian region of Calabria in February and March of 1783. The Neapolitan Academy of Sciences and Fine Letters appointed a commission to report on the earthquakes, the smaller aftershocks that followed them, and the extent of the damage they caused. Several Neapolitan court officials, a British diplomat, and a French geologist also made observations in the devastated area. Their reports showed that the center of disturbance had gradually moved northward during the series of shocks; they also concluded that buildings constructed on clay soils fared far worse than those erected on bedrock.[9]

In 1857, Mallet investigated another destructive earthquake in the kingdom of Naples. By this time, scientists had concluded that the shaking of an earthquake was due to the passage of a wave traveling through the solid earth, just as sound is due to the passage of a wave traveling through air. Mallet set himself the goal of determining the focus of the earthquake, that is, the point from which the seismic wave originated. Throughout the damaged area, he noted the direction in which walls, pillars, and other tall objects had been thrown. All of these directions converged near the village of Caggiano, which thus appeared to be the earthquake's epicenter (the point on the earth's surface directly above the earthquake's focus). Mallet also constructed a map of isoseismal lines, that is, lines where the earthquake had been felt with equal intensity. In particular, he showed the respective outer boundaries of the areas in which the earthquake had been barely perceived, in which there had been some damage to buildings, in which some people had been killed, and in which towns were for the most part obliterated. As expected, Caggiano fell within the area of highest intensity, thereby lending greater credibility to the conclusion that the epicenter had been there. Mallet finally tried to determine the depth of the focus below the earth's surface, by using the orientation of cracks in the walls of buildings to judge the angle at which the seismic waves had emerged from the earth. This attempt proved less than satisfactory, as the angles were widely scattered even in nearby locations.[10]

A final method of earthquake study was the recording of earthquake waves with mechanical instruments. Soon after the Calabrian earthquakes of 1783, Italians began using suspended pendulums to scratch a record in a box of sand or move a pencil over a piece of paper whenever an earthquake struck. The records from such seismoscopes were useful for determining the direction in which the seismic waves moved, because the pendulum would be set to vibrating in that direction. By the mid-nineteenth century, new ways of suspending a stationary mass or pendulum relative to a moving earth had produced seismoscopes sensitive enough to record some earthquakes not felt by nearby people. A few were also rigged up to stop a clock (by closing an electrical circuit or mechanically stopping a clock pendulum) and thereby to indicate the time at which the seismic wave had passed. One such seismoscope recorded more than 500 earthquakes in Tokyo from 1875 to 1885.[11]

The 1880s saw the development of a more powerful earthquake recorder, the seismograph. It was developed primarily by John Milne, J. A. Ewing, and Thomas Gray, who were among the hundreds of Western scientists and engineers who in the 1870s had been invited by the Japanese government to impart knowledge of Western technology as teachers at Japanese universities. Disturbed by a moderately destructive earthquake in Yokohama in early 1880, many of the Westerners and some of their Japanese students formed the Seismological Society of Japan, the first association in the world devoted to the scientific study of earthquakes. Under its auspices, Milne and his colleagues developed instruments that would not only register the occurrence of an earthquake but also produce a full record of the passing seismic wave, which would allow the amplitude and period of the wave to be determined. They did so by having a suspended pendulum write on a revolving piece of glass covered with soot or on a revolving drum covered with paper. At the same time, a clock would make a periodic mark on the record, so that the duration of the seismic vibration as a whole and of individual waves within it could be measured. With this ability to produce an accurate representation of the earthquake wave, modern seismology was born.[12]

During the second half of the nineteenth century, American researchers used all of these methods of earthquake study—the compilation of earthquake catalogues, the investigation of individual earthquakes, and the instrumental recording of seismic waves. One of the first to compile an earthquake catalogue was John Boardman Trask, a physician who had studied geology as well as medicine during a brief stint at Yale College before moving to California in 1849. In the early 1850s, Trask investigated California's geology, with particular emphasis on the state's potential for both mining and agriculture. From 1856 to 1865, he collected newspaper and telegraphic reports of earthquakes and compiled them into annual earthquake lists published in the proceedings of the new California Academy of Sciences. He also interviewed old residents of the state to produce an earthquake catalogue reaching back to 1800. As his more recent biographers state, one of his main purposes was to produce evidence that other scientists could use "to test theories on the causes of earthquakes."[13]

In 1868, an opportunity arose for studying a significant earthquake in detail. On October 21, a temblor struck the San Francisco Bay area, causing considerable damage, especially on the east side of the bay, and leading to about 30 deaths. Within two weeks, the California Academy of Sciences, headquartered in San Francisco, considered appointing a scientific investigating committee; such a committee, with nine members, was finally established three weeks later under the auspices of the San Francisco Chamber of Commerce. The committee gathered considerable information on where the quake had been felt, where and what kind of damage it had caused, and in what direction local observers thought the seismic waves had moved. It did not publish this information, however, because it could not raise sufficient money to complete its investigation and because it suffered from internal dissension. The committee included not only Trask and a few other scientists but also local lawyers, merchants, and others who were interested less in gathering detailed observations than in speculating about the causes of earthquakes (variously attributed by them to the passage of a meteor shower, electrical disturbances, or cooling of the earth). Even though San

Francisco was a booming metropolis, with a population that reached nearly 150,000 in 1870, scientific institutions there were still too weak to support a sustained investigation: the Academy of Sciences comprised mostly amateur members; the newly chartered University of California had not yet opened its doors; and the state geologist, disappointed at the lack of financial support for his survey, had just returned to a teaching position at Harvard. Under these circumstances, the study of the 1868 earthquake descended into bickering among supporters of various speculative theories.[14]

In the 1870s and '80s, the institutional setting of science was far more propitious for the study of earthquakes on the East Coast of the United States. Here, established universities already had a long tradition of geological inquiry. At Yale University, James D. Dana, one of the country's leading geologists, edited the *American Journal of Science*. At his request, Charles G. Rockwood, an 1866 graduate of Yale who eventually joined the faculty at Princeton, in the early 1870s began to compile records of American earthquakes for the journal. Like Trask and Perrey before him, Rockwood relied primarily on newspaper reports, supplemented by correspondence with residents of areas that had experienced some of the larger earthquakes.[15]

The East Coast also benefited from the existence of several governmental science agencies with an interest in sustained seismological investigations. The United States Geological Survey, founded in 1879, counted among its members Clarence E. Dutton, Grove K. Gilbert, William M. Davis, and other geologists who saw the study of earthquakes as a key for understanding larger geological issues, such as the origin of mountains and continents.[16] Meanwhile, astronomer and meteorologist Cleveland Abbe, who played a key part in transforming the U.S. Army's Signal Service into a national weather service in the early 1870s, also desired to study earthquakes. As head of the service's research department, he instructed weather observers to report on earthquakes as well as meteorological phenomena. In 1884, the service's efforts were augmented by the arrival of Thomas C. Mendenhall, a physicist who earlier had been a member of the Seismological Society of Japan, during a two-year sojourn in that country. Over the next several years, Mendenhall, assisted by his former student C. F. Marvin, attempted to develop a simple seismograph that might be used by Signal Service observers for the routine recording of earthquakes.[17]

All of these scientists were ready when a destructive earthquake struck Charleston, South Carolina, on August 31, 1886. The earthquake, probably as large as magnitude 7 on the Richter scale, damaged about four-fifths of all buildings in Charleston and killed upwards of 80 people. Within a day of hearing about the earthquake, Rockwood had already sent inquiries to newspapers and special correspondents across the eastern half of the country asking for their observations. Meanwhile, survey geologist WJ McGee quickly traveled to Charleston to inspect both the damage and the geological changes (such as landslides and fissures) produced by the earthquake. Over the next few months, several other survey geologists as well as Mendenhall from the Signal Service also visited South Carolina. Moreover, McGee engaged a local engineer to make detailed observations of bent railroad tracks and other damage that required extensive surveying. To record the numerous aftershocks, the Geological Survey distributed a few seismoscopes to interested residents; these were often quite simple instruments, in one case consisting of a bowl filled to the brim with molasses that would spill whenever a shock occurred. Finally, the Signal Service collected records from many of its weather observers and made them available to the Geological Survey. Unfortunately, no seismograph was yet in operation at the time of the August earthquake; the Signal Service's seismograph did not record its first earthquake until October 22, 1886.[18]

The task of organizing all of these data fell to the Geological Survey's Dutton, who published his final report in 1889. For constructing a map of isoseismal lines, Dutton had available to him more than 4,000 observations from about 1,600 localities. The resulting map showed that some places nearer Charleston had felt the earthquake far less intensely than some places farther away. Dutton argued that these areas had been shielded from the earthquake waves, perhaps by geological structures such as mountains. A detailed examination of the distribution of damage near Charleston suggested that the earthquake had two epicenters about a dozen miles apart. Dutton also sought to determine how fast seismic waves had traveled from these epicenters. While he did not have any records from seismographs, he did have evidence from stopped clocks and the testimony of alert observers who had noted the time when they felt the earthquake. His task was made far easier by the introduction in 1883 of standard time zones, which had been developed to allow the coordination of railroad schedules. By being able to refer to standard time rather than a local time that was different for each city, Dutton could arrange his time observations on a common scale. His analysis showed that the earthquake wave had traveled at more than three miles per second, much faster than wave speeds observed by European seismologists but in keeping with other theoretical calculations of what seismic wave velocities should be, given the density and elasticity of the earth's crust. Dutton's attempts to determine the depth of the earthquake's focus, however, were as unsuccessful as Mallet's had been.[19]

While Dutton was investigating the Charleston earthquake, the facilities for studying earthquakes in California were improving markedly, thanks mostly to the efforts of astronomer Edward S. Holden. President of the University of California, Holden in 1885 was also appointed as first director of the newly established Lick Astronomical Observatory, atop Mount Hamilton, about 50 miles southeast of San Francisco. Under his leadership, seismographs were set up in 1887 both at Lick and at the Students' Astronomical Observatory on the university campus. Holden's ostensible reason for installing them was to produce a list of the shocks that, while too small to be noticed by humans, might nevertheless have thrown telescopes out of alignment and thereby necessitated adjustments to astronomical observations. But Holden was also interested in the scientific study of earthquakes as such. Like pioneer French seismologist Perrey before him, he sought to determine whether earthquake occurrences showed any kind of periodicity that might be correlated with astronomical or meteorological phenomena, such as the solstices or the onset of the rainy season. To further this endeavor, Holden in 1887 published a catalogue that listed all earthquakes known to have occurred in California, Oregon, or Washington since 1769. Over the next decade, he and his assistants supplemented the catalogue with annual lists published by the U.S. Geological Survey. Holden wished to ensure that these lists would be as complete as possible, so, to supplement the seismographs at Lick and at Berkeley, he distributed seismoscopes to a number of his acquaintances in the San Francisco Bay area and to the State Weather Bureau in Carson, Nevada, and collected the records made by these instruments.[20]

By 1900, then, the scientific approach to understanding earthquakes had become well established among American researchers. Astronomers on the Pacific Coast and meteorologists back east were recording earthquakes with instruments and compiling their observations into earthquake catalogues. Meanwhile, geologists and other scientists stood ready to investigate strong earthquakes in detail, in order to determine the quake's epicenter and the speed at which seismic waves traveled away from it. Such studies were designed to illuminate broader questions of geological theory, such as the physical properties of the earth's crust (which affected seismic wave velocities), the spatial relation of earthquakes to areas of mountain building, and the influence of astronomical and meteorological events (such as solstices and the seasons) on the forces that governed the occurrence of earthquakes.

These scientific studies contributed to a much broader trend of secularization that dominated intellectual discourse in the nineteenth century. As scientific research became increasingly sophisticated and professionalized, the educated classes in Western society saw God less and less as playing a role in determining or causing events in the physical world. Earthquakes and other natural catastrophes were no longer understood as an expression of God's wrath; their occurrence was seen as the result of purely natural processes that were susceptible to scientific understanding. This trend toward secularization, however, decreased the sense that humans had the power to affect earthquakes. Because scientists had removed God as a causative agent for earthquakes, the public no longer could hope to avoid earthquakes by behaving morally and appearing pleasing in God's eyes.

A few Californians did seek technological means for mitigating seismic hazards, to replace the moral actions upon which their forebears had relied. One was George Gordon, an energetic San Francisco merchant and real estate developer who advocated earthquake-resistant construction. Within days of the 1868 bay area earthquake, Gordon wrote a lengthy letter to the editor detailing how brick buildings might be reinforced and how the structural elements of houses might be tied together so that structures could ride out the waves of an earthquake as a ship rides waves at sea. Soon, Gordon had persuaded the San Francisco Chamber of Commerce to set up the Committee on Earthquake Topics, which would gather scientific data on the recent quake, solicit suggestions for earthquake-resistant construction, and conduct experiments on buildings and building materials. But Gordon was not able to raise the money necessary to complete this ambitious project. Over the next few months, several proposals for earthquake-resistant design—bolting wooden houses to their foundations, using diagonal sheathing with lumber rather than plaster to strengthen walls, relying on cast iron rather than bricks as a structural frame—appeared in the local press. The only action local builders seem to have taken, however, was to insert iron anchor ties into brick walls to hold them together better and tie them more closely to the internal timber frame that many structures still used. Even this rudimentary (and, as later experience would show, not very effective) measure does not seem to have been widely adopted. Most builders, it appears, simply blamed the damage from the 1868 earthquake on poor workmanship and assumed that merely building properly would take care of all risks from future earthquakes.[21]

Just as California in the late nineteenth century did not yet have the resources to support a detailed scientific investigation of an earthquake, it also did not yet have the capability to provide extensive engineering solutions for the earthquake problem. Although by the 1860s California was home to quite a few technically trained engineers, most were mining engineers concerned with setting up capital-intensive machinery in the state's gold mines. Building construction was still the province of contractors, laborers, and a few architects, who mostly operated according to tradition. Professional, college-trained civil engineers using experimental and theoretical data to back up their designs did not appear on the Pacific Coast until near the end of the century. Consequently, California did not have an organized group of people who could draw engineering lessons from the 1868 earthquake and incorporate them into future building designs.[22]

With faith in repentance as an effective means of mitigating earthquake hazards on the wane because of the rise of science but faith in technology not yet widespread, most Californians saw destruction by earthquakes as an inevitable event that could not be altered by human intervention. The proper response to such an event was dogged persistence and denial. Two newspaper editorials published shortly after an earthquake had devastated several rural towns near Sacramento in 1892 clearly demonstrate this attitude. One celebrated the "pluck and push" of local merchants and artisans who reopened their businesses in makeshift quarters only hours after their shops had been destroyed. Another gave "all honor to courageous business men" who, "though the earth quakes and shakes down their places, . . . recover their momentum, spit on their hands and take a fresh grip." A determination not to let a mere little shaking disrupt life and business—that was the proper attitude for these pioneers and sons of pioneers.[23]

Hand-in-hand with this refusal to be daunted went a stout assertion that earthquakes or seismic hazards were not all that bad in California. As one San Francisco newspaper editorialized in 1868, "earthquakes are trifles as compared with runaway horses, apothecaries' mistakes, accidents with firearms, and a hundred other little contingencies, which we all face without fear." The same newspaper pointed out that not only California, but Ireland, England, New York City, even the rock of Gibraltar experienced earthquakes—so California surely was not unique in the natural hazards it had to live with.[24] Meanwhile, the San Francisco Chamber of Commerce sent out a telegraph message across America and to Europe stating that the cost of property damage in the 1868 earthquake had amounted to less than $300,000—even though some local residents put the damage at 10 times that figure.[25] After the 1892 earthquake, another San Francisco paper editorialized, "It is folly to think or speak of California as an earthquake country because we have a shock occasionally, usually perfectly harmless. One Western cyclone will do more damage than all the earthquakes California has ever known." The *Sacramento Record-Union* similarly opined, "Not so much life or property has been injured or destroyed by all earthquakes in California territory in a hundred years as by a single cyclone in the Mississippi Valley."[26]

In many cases, these statements were made in newspapers that were mouthpieces for real estate developers and other economic boosters who feared that stories about earthquakes might lessen the inflow of new residents and capital into the state. But scientists also joined the chorus. John Trask, the first compiler of an earthquake catalogue for California, was motivated not only by scientific interests, but also the desire "to disprove stories circulating outside the state about how dangerous the California quakes were."[27] Edward Holden, in his lists of California earthquakes, concluded, "When we take into account the whole damage to life and property produced by all the California earthquakes, it is clear that the earthquakes of a whole century in California have been less destructive than the tornadoes or the floods of a single year in less favored regions."[28] Finally, as late as 1904, Andrew Lawson, who had been professor of geology at the University of California since 1890, dismissed the possibility of an imminent severe earthquake in the bay area. In a front-page story in the university's student newspaper at a time when there were more small tremors in the area than was usual, Lawson asserted that "earthquakes in this locality have never been of a very violent nature, and so far as I can judge from the nature of the recent disturbances and from acounts [sic] of past occurrences there is no occasion for alarm at the present time."[29]

Thus, the attitude of Californians toward earthquakes around 1900 was truly fatalistic. Earthquakes and the devastation they brought were indeed the result of "unchangeable LAW," as the San Francisco Examiner put it in 1906: they were the product of natural processes that scientists could understand, but humans could do little to alter. Given this sense of their powerlessness, Californians reacted by denying the severity of the problem and by doggedly rebuilding their structures exactly as they had been whenever they were shaken down. This approach to earthquakes would change only gradually during the twentieth century, as scientists and engineers slowly persuaded the public that, by marrying technical expertise to the power of the state, humans might actually mitigate seismic hazards.

CHAPTER 1

Reactions to the San Francisco Earthquake and Fire of 1906

At 5:12 A.M. on April 18, 1906, residents of coastal central California were awakened suddenly by a strong shaking, which increased in severity for half a minute before culminating in a violent twisting lurch. This earthquake, now estimated to have reached a magnitude of 8.3 on the Richter scale, damaged buildings and other structures in an area 350 miles long by 70 miles wide, stretching from Eureka in northern California to Salinas in the south. Damage was particularly heavy in the San Francisco Bay area. In San Jose, then a town of 20,000 inhabitants, nearly all the brick buildings in the central business district were severely damaged, and the nearby State Insane Asylum at Agnews collapsed, killing 112. At Stanford University, numerous walls as well as the tower of the Memorial Church fell down. At the other end of the bay area, the earthquake and a subsequent fire wiped out the business district of Santa Rosa (pop. 6,700). The largest amount of damage, though, was suffered by San Francisco, at that time still California's largest city, with 400,000 residents. Here, too, the destruction was wrought not only by the earthquake itself but also by numerous fires started by downed electrical wires, cracked gas lines, and damaged stoves. With water mains extensively broken by the earthquake, the San Francisco fire department was powerless to stop these fires, which by noon of April 18 had coalesced into a general conflagration. Over the next two-and-a-half days, the fire laid waste to four square miles, including the central business district and numerous densely populated residential sections. The total number of lives claimed by the earthquake and fire is unknown. At the time, about 700 deaths were attributed to the disaster, but more recent estimates place the number at several thousand. The earthquake and fire rendered more than 200,000 people homeless, and property losses estimated at $500 million in 1906 dollars (roughly $10 billion in 2001 dollars) made it the most expensive natural disaster to strike the United States until Hurricane Andrew ripped through Florida in 1992.[1]

Californians reacted to this unprecedented disaster by relying on patterns of behavior established in response to previous earthquakes. Newspapers and other boosters denied that seismic hazards were all that serious and urged the rapid rebuilding of San Francisco. As in 1868, a few engineers ventured suggestions for making buildings more earthquake-resistant, but the engineering community as a whole showed little interest in improving the seismic safety of California's built environment. Instead, engineers focused on how to increase fire resistance, a topic then at the forefront of national engineering concern. Meanwhile, scientists undertook a detailed investigation of the earthquake, describing the extent of the damage it caused and locating its epicenter. The initial reports from the investigation focused on scientific matters and had little to say about future seismic hazards in California; as a result, they had no immediate impact on how the public perceived earthquake risks. Yet these scientists laid the groundwork that would eventually lead to a concerted campaign for earthquake hazard mitigation in the state. They identified the San Andreas fault as a potent source of seismic hazard and established a seismological society to continue monitoring earthquake activity. Moreover, in 1909 one geologist who investigated the earthquake, Grove K. Gilbert, issued a resounding call to Californians to protect themselves against future earthquakes; this call would serve as the cornerstone for a campaign for seismic safety.

The Reaction of Boosters: Denial

Even the most optimistic boosters of California could not deny that their state had suffered a major calamity. But the California press generally reacted by minimizing the extent of earthquake damage and insisting that San Francisco had been destroyed by a fire rather than an earthquake.[2] Broadly speaking, this reaction simply followed the pattern that boosters had already established after previous earthquakes. But there were also specific reasons for why this response still dominated in 1906. Boosters were motivated in part by a justifiable desire to counter the many exaggerated accounts of earthquake damage published throughout the nation following the disaster.[3] There was also a direct economic reason: most fire insurance companies had included in their policies a "fallen building" clause, which absolved them of liability if a building had been damaged by other causes before being destroyed by fire. Many San Francisco residents therefore insisted that their property had not been injured by the earthquake, so they could collect on their fire insurance.[4]

The boosters' campaign got off to a quick start. Already on the day after the earthquake, while San Francisco was still burning, California governor George C. Pardee assured his Massachusetts counterpart that "destruction was wrought by fire far more than by earthquake." Two days later, the Sacramento Valley Development Association issued a press release emphasizing that "the earthquake was accompanied by serious results in no part of the State except on the peninsula, on which San Francisco is situated, and south to San Jose and at a few points near the Bay. . . . Of the one hundred and fifty-seven thousand square miles in the State the injury to person or property occurred on less than the small area of one hundred square miles."[5] By late April 1906, the Real Estate Board of San Francisco had passed resolutions "pledging its members to speak hereafter of the disaster as 'the great fire' instead of 'the great earthquake.' " Meanwhile, James Horsburgh, Jr., of the Southern Pacific railroad company, the state's dominant business organization, urged chambers of commerce throughout California to "call attention to the small area of the State that was affected by the earthquake and the relatively small results in the way of destruction, and point out the great buildings of the business section of San Francisco and the residence portion of the city that escaped burning as proof of the fact that San Francisco did not suffer greatly from the earthquake."[6] Following its own advice, the Southern Pacific Company in the early summer of 1906 issued several publications illustrated with panoramic shots of San Francisco taken on the morning of April 18 after the earthquake but before the fire to show "most emphatically that the destruction was due to fire and not to earthquake." The *San Francisco Examiner* went even further and in early June 1906 urged city authorities to remove "the twisted flagstaff surmounting the tower of the Ferry Building," which dominated the waterfront. "The sight of that crooked pole convinces [every incoming stranger] that nothing standing upright on the morning of April 18th could possibly have resisted the shock, which, as everybody knows, is a preposterous error."[7]

Besides arguing that little damage had been done by the earthquake of April 18th, California's boosters also dismissed the danger of future earthquakes. In the days immediately after the San Francisco catastrophe, California newspapers assured readers that "the earthquake of San Francisco is a thing of the past, and a closed incident," that "the danger from a heavy shock of a destructive character is gone," that "Central California is now safe from any destructive earthquake for centuries to come," and even that the earthquake was "not likely to be repeated in a thousand years."[8] Moreover, Californians disparaged their seismic hazards by pointing to larger hazards elsewhere. As the *Los Angeles Times* editorialized, "all the earthquakes that had been experienced on the Pacific Coast, up to the time of the San Francisco disaster, caused less loss of life and property than an ordinary tornado or 'cyclone' causes in the Middle West. There are more deaths annually by lightning in almost any State of the Union, outside of the Pacific Coast, than all the deaths from earthquake in California."[9] Similarly, when a killer heatwave struck the East Coast in July 1906, the *San Francisco Chronicle* asserted that "during the last half century there have been 100 deaths from sunstroke and tornadoes at the East for one death resulting from earthquakes on the Pacific Coast." Californians even insisted that earthquakes were more of a danger elsewhere; as they pointed out, the strongest one to strike the United States in recorded history had occurred in 1811 in Missouri, not California. Furthermore, so the *Chronicle* stated, "earthquakes have occurred within historic times . . . in many parts of the United States"; as a result, "the next severe shock is as likely to be in New York as in San Francisco."[10]

In addition to belittling the severity of the earthquake, the San Francisco press insisted that whatever damage had been done by it was the result of shoddy workmanship and "reckless neglect of duty" by builders. These comments had wider political connotations in a city whose government was about to be exposed and put on trial for widespread corruption. One commentator asserted, "The earthquake was a stern exposer of sham and it ruthlessly searched out the work of ignorance, cupidity and graft." Another observer commenting on "wretchedly built" structures focused on the destroyed City Hall as an "architectural fraud"; the building, he claimed, "collapsed miserably, because every stone of it was laid in putrid politics; it is a disgraceful ruin, the great dome is stripped of its veneer of stone as thoroughly as the iniquity of the builders stands plain to every beholder." The corollary of this position was that "all honest construction with proper materials endured the shock without serious injury."[11] Thus, the California press asserted, all that was needed to render earthquakes completely harmless in California was a focus on moral values among contractors, which would eliminate greed, corruption, and dishonesty; there was no need to devise special engineering measures to counteract earthquakes.

The Reaction of Engineers: Focus on Fire

Like California's boosters, engineers both in California and elsewhere who discussed San Francisco's destruction concentrated on the fire rather than the earthquake. Their reaction was shaped, however, not by a desire to protect California's reputation as a haven for immigrants and investors, but rather by an ongoing campaign for fire prevention and fire-resistant construction. This campaign was well-justified; fires posed a considerable threat to both lives and property in turn-of-the-century America. But as a result of their preoccupation with fire, engineers, like the boosters, did little to encourage earthquake-resistant construction in the rebuilding of San Francisco.

The engineers' campaign against fire hazards had developed in response to frequent urban conflagrations in the post–Civil War era. In October 1871, a fire had swept through much of Chicago, burning more than three square miles, leaving about 100,000 people homeless, and causing nearly $200 million in damage. This catastrophe had been followed a year later by a fire that destroyed Boston's business district, with property losses estimated at $75 million. Finally, in February 1904, a blaze had gutted Baltimore's downtown, causing $50 million in damage. Beyond these vast conflagrations, America frequently experienced smaller city fires. In 1904 alone, there were sizable fires in Toronto, New York City, Rochester, Sioux City, and Yazoo City (in Mississippi) in addition to Baltimore. As one fire-prevention activist pointed out in 1907, the overall fire loss in the United States in recent years had averaged 7,000 deaths and $200 million in property damage annually.[12]

In response to these immense fire losses, a campaign for fire prevention developed in the late nineteenth century, led primarily by the fire insurance industry. The National Board of Fire Underwriters, a trade association of fire insurance companies formed in 1866, first urged the need for fire-resistant construction in the early 1870s, after the Chicago and Boston fires. By the mid-1890s, the National Board was supporting the Underwriters' Laboratories, which tests various materials for their fire-resisting qualities. Meanwhile, a group of insurance executives formed the National Fire Protection Association, which spearheaded the public campaign for fire prevention and fire-resistant construction. Finally, in 1901 the National Board itself officially adopted as one of its main goals "to influence the introduction of improved and safe methods of building construction, encourage the adoption of fire-protective measures, secure efficient organization and equipment of Fire Departments, with adequate and improved water systems, and establish rules designed to regulate all hazards constituting a menace to the business." After the 1904 Baltimore fire, the National Board redoubled its efforts, investigating hazards and protection in New York City, Chicago, Detroit, and other cities.[13] The conclusions that National Board investigators reached in October of 1905 concerning San Francisco were particularly ominous: "In view of the excessively large areas, great heights, numerous unprotected openings, general absence of fire breaks and stops, [and] highly combustible nature of the buildings, . . . the potential hazard [for a conflagration] is very severe. . . . In fact, San Francisco has violated all underwriting traditions and precedents by not burning up; that it has not done so is largely due to the vigilance of the fire department, which can not be relied upon indefinitely to stave off the inevitable."[14] When San Francisco did burn up in April 1906, advocates of fire prevention could say "I told you so."

Influenced by the well-developed and ongoing campaign for better fire protection, the national engineering press from the beginning interpreted San Francisco's devastation as yet another example of the need for more fire prevention rather than as an opportunity for developing earthquake-resistant construction. In its issue for April 26, 1906, *Engineering News,* the leading weekly magazine for engineers across the country, asserted: "The fire is merely a repetition, unaltered in its main phases, of the great Baltimore fire of 1904. Its teachings are a confirmation and reinforcement of the Baltimore lessons." As a consequence, so the *Engineering News* editorialized, "the smoldering ruins of San Francisco should awaken . . . every city in the land . . . to look to their water supply and fire protection as never before." To reinforce this point, the engineering magazine gave lengthy summaries of the National Board of Fire Underwriters' reports on the conflagration hazard faced by San Francisco and New York City. In this context, the fact that the San Francisco fire had been started by an earthquake was considered of only secondary importance. As a St. Louis engineer later put it, earthquakes "concern most cities only remotely," while the "fire lessons . . . are of universal interest and application."[15]

Although broadly speaking the San Francisco fire merely confirmed the lesson of the Baltimore fire that the close crowding of flammable timber structures within a central business district produces a severe conflagration threat, numerous engineers nevertheless flocked to the California metropolis to inspect the fire damage for themselves. Their interest and concern was sparked by the fact that even supposedly fireproof buildings had not withstood the flames. Indeed, modern high-rise office buildings constructed of steel, terra cotta, and brick had been reduced to rubble just as much as old-fashioned brick-and-timber buildings. Close inspection of the fire ruins showed that most supposedly fireproof buildings had suffered from two main weaknesses. First, openings in exterior walls, such as doors, windows and sky lights, had not been adequately protected to prevent fire from entering the building. Second, the protective material covering steel columns had not been sufficient to prevent the columns from softening and buckling under the intense heat of the blaze. To correct these weaknesses, the investigating engineers recommended that fireproof buildings be provided with wire glass (which would not completely disintegrate during a fire), metal window frames, and metal shutters to protect window openings. They also insisted that steel columns in such buildings be covered with a thick and well-designed covering of protective material, preferably concrete. In this manner, the investigation of San Francisco's fire damage led to the formulation of several new guidelines applicable to fire protection in business districts across the country.[16]

The lessons to be drawn from the San Francisco fire were studied not only by visiting experts but also by the local engineers who were responsible for rebuilding the city. The meetings of the Structural Association of San Francisco, a group of local engineers, architects, and general contractors established in June 1906, revolved largely around methods for providing better fire protection. Members on numerous occasions discussed the relative merits of metal sashes, wire glass, and metal window shutters for protecting wall openings against fire.[17] They also strenuously debated the relative value of terra cotta, brick, and concrete for protecting steel columns, and they concerned themselves with the use of sprinkler systems and circuit breakers for fire prevention.[18] Leaders of the Structural Association even set out to establish a "Fire Testing and Experiment Station" at the University of California to test the fire-resisting qualities of various building materials; this effort fizzled out when the $10,000 necessary for such a station could not be raised.[19]

Beyond gathering data on how better to protect buildings against fire, California engineers also enthusiastically joined their colleagues nationwide in urging building owners to adopt more fire-resistant construction methods. Luther Wagoner, for example, proclaimed to his fellow members in the Structural Association of San Francisco, "It is incumbent upon this association to agitate this question thoroughly with the end in view of educating people about their duties and obligations as owners and designers." Similarly, W. J. Miller stressed that "the line along which our association should work is the education of owners" about proper methods for fireproofing.[20] The actual work of such public education was carried out less by the Structural Association of San Francisco, which faded out of existence in December 1906, than by *Architect and Engineer of California,* which by 1906 had become the leading trade magazine for the local building professions. In late 1906 *Architect and Engineer* retained F. W. Fitzpatrick, a Washington architect and prominent advocate of fire protection, to provide free advice on details of fireproofing to local building owners.[21] *Architect and Engineer* also printed a large number of exhortatory articles by Fitzpatrick urging continued vigilance against "the red monster" and pressing the need for "better fire-prevention," and it used strong editorials to reinforce his preaching.[22]

The focus on fire prevention did lead to at least one improvement in San Francisco's infrastructure that took the earthquake hazard into account. Several commentators had noted that the widespread cracking of San Francisco's water pipes by the earthquake had from the very beginning doomed the fire department's efforts to stem the conflagration. Water from a couple of fireboats along the shoreline and from some buried cisterns in the older portions of the city, as well as salt water pumped through short lengths of fire hose from the San Francisco Bay, had helped to save a few buildings along the waterfront, but these supplies had been far too meager to halt the burning of the city.[23] After San Francisco's devastation, many engineers and others interested in fire protection clamored for some system whereby large quantities of water could be delivered throughout the city even after an earthquake. The most popular solution was a network of pipes through which water could be pumped under high pressure from the bay; this salt water system would be completely separate from the more vulnerable pipes for drinking water.[24] The city's voters approved bonds for the salt water system in 1908, and the entire system was put in place by 1914.[25] Those laying out the system's main trunks sought the advice of geologist Andrew Lawson on which areas of the city to avoid because of their high susceptibility to earthquake damage.[26] Nevertheless, the drive for a salt-water firefighting system came less from an appreciation of earthquake hazards than from a desire for better fire protection, and fire-prevention activists also called for improvements in water supply systems for New York City and other urban areas not as prone to earthquakes.[27]

Although they were primarily interested in the effects of the fire, many of the engineers and architects examining devastated San Francisco did make remarks about earthquake damage. Most experts, for example, noted that earthquake damage had been concentrated on loose ground and particularly on the artificial fill along the waterfront on which much of the city's central business district had been built.[28] From this observation engineers did not, however, draw the conclusion that builders should avoid such areas; rather, they emphasized the value of strong, solid building foundations in preventing earthquake damage even in regions of such adverse soil conditions.[29] Engineers also noted the need for proper bracing and tying together of structural frames. They applied this recommendation particularly to high-rise office buildings of steel frame construction, in which floors, walls, and the roof were carried on a structural steel frame composed of horizontal and vertical steel bars riveted or bolted together, but they did not insist that such buildings be braced against the horizontal stresses produced by earthquakes. Rather, engineers simply urged that steel frame structures be properly braced against the horizontal forces produced by winds, blithely assuming that such bracing would be sufficient to take care of earthquakes as well. A new building code for San Francisco drawn up by engineers and architects and adopted in the summer of 1906, for example, required that buildings with steel frames be braced against wind loads as high as 30 pounds per square foot but made no mention of earthquake forces.[30]

Besides urging stronger foundations and the proper bracing and tying together of structural frames, a number of engineers and architects responded by recommending increased use of reinforced concrete. This recommendation arose as much from a longstanding vested interest in promoting reinforced concrete as a building material as it did from a desire to encourage earthquake-resistant construction. Concrete, a mixture of gravel and cement, had been used in construction since Roman times. While it possessed great strength under compression, it nevertheless offered little resistance to being pulled apart. In the middle of the nineteenth century, a few European engineers had begun inserting iron or steel rods into concrete to increase its tensional strength. This reinforced concrete was introduced in California in the 1880s, and by the turn of the century a number of engineers and architects were enthusiastic advocates of the new material. It was stronger than brick and therefore allowed the construction of taller structures at a time when land was becoming increasingly valuable in central business districts. Reinforced concrete structures were also far less expensive to build than steel-frame buildings and cost only slightly more than brick-and-timber structures. Furthermore, the complex behavior of this new material required extensive consultation of engineers during design and construction; by contrast, brick construction had become so routinized that many contractors dispensed entirely with the services of civil engineers. From some parts of the construction industry reinforced concrete faced considerable opposition in California. Numerous architects disliked the gray and dull material on esthetic grounds, while brick manufacturers and bricklayers saw in it a threat to their livelihoods. In large part because of the opposition of the politically powerful bricklayers' union, the San Francisco building code until 1906 prohibited structures built entirely of reinforced concrete.[31]

After San Francisco's devastation, those interested in advancing the cause of reinforced concrete pointed to its excellent performance in the earthquake as well as the fire. One architect asserted: "The advantage of re-enforced concrete over brick construction is clearly shown in two instances: In Alameda [near Oakland] a two-story re-enforced concrete building stands intact, while a brick building about 50 feet away is badly wrecked. In San Francisco . . . a warehouse was in process of construction with brick external walls and interior columns, girders and floors of re-enforced concrete. The brick walls are completely shattered, but the concrete work does not show a single crack."[32] Advocates also cited a number of other examples of superior performance. The California Academy of Sciences in San Francisco "went through the earthquake and was entirely gutted by the fire; the brick walls were completely shattered, but the columns and floors [of reinforced concrete] remain intact." At Stanford University, buildings constructed of reinforced concrete also fared much better than brick or stone buildings. And a house in Berkeley built of reinforced concrete weathered the earthquake practically uninjured.[33] John B. Leonard, one of the leading local proponents of reinforced concrete, summarized this evidence in a glowing testimonial in *Architect and Engineer of California:* "an inquiry among architects and engineers, together with my own observations, ha[s] failed to reveal any instance of failure on the part of reinforced concrete" in the San Francisco Bay area.[34] Engineer William Hammond Hall, who had long championed "the cause of reinforced concrete," similarly proclaimed, in a series of articles published in the *San Francisco Chronicle,* that "re-enforced concrete walls of buildings withstood the earthquake where brick walls and combined brick and stone walls, both good according to California standards, have failed in material degree, or were, in part, demolished."[35]

Using such evidence, advocates of reinforced concrete succeeded in having the ban on reinforced concrete construction in San Francisco lifted. A month after San Francisco's devastation, a new building code that explicitly allowed for structures constructed entirely of reinforced concrete was submitted to the city's Board of Supervisors by a committee of engineers, architects, and contractors headed by John D. Galloway, an engineer who had declared that during an earthquake "brick houses are a menace to life." The supporters of reinforced concrete mobilized considerable political support for the new building code, including the active endorsement of the popular ex-mayor and millionaire businessman James Phelan, who had lost three brick warehouses in the earthquake.[36] As a result, the new building code passed the Board of Supervisors in late June 1906, and within a year more than 70 reinforced concrete buildings, mostly office buildings and warehouses, had been erected or were under construction in San Francisco.[37]

The claims made by advocates of reinforced concrete did not, of course, remain unchallenged by backers of other building materials. Engineer A. L. A. Himmelwright, writing for one of the premier constructors of steel-frame buildings, proclaimed that "enthusiastic persons in favor of concrete construction, and interested contractors, have recently published and disseminated much reading matter advocating buildings constructed entirely of reinforced concrete as a type well adapted to resist earthquake shocks. Much of the information and many of the statements so published are misleading. . . . Reinforced concrete is not as well adapted as steel for columns, girders and beams [in tall buildings], and should not be used for this purpose where steel is available."[38] Representatives of brick interests also attacked the claims that brick buildings had fared poorly in San Francisco. They argued that, on the contrary, many properly built brick structures were still standing undamaged, while nearby buildings containing reinforced concrete had sustained considerable damage.[39] Taking up this argument, Edward J. Brandon of the San Francisco bricklayer's union at several meetings of the city's Board of Supervisors dismissed reinforced concrete as an experimental and as yet unproven material and urged the supervisors to reject the new building code.[40] Though he was unsuccessful, his warnings about reinforced concrete were proven true in November of 1906, when a reinforced-concrete hotel under construction in southern California collapsed under its own weight, killing ten workmen.[41] Reinforced concrete was not necessarily the panacea it had been made out to be by its supporters.

In general, those engineers and architects not completely engrossed in urging better fire prevention responded to the earthquake by promoting the greater use of reinforced concrete and by recommending strong foundations and the proper bracing and tying together of structural frames. Few ventured beyond these general recommendations to embark on a sustained campaign for more earthquake-resistant construction. Of those who did, the most vocal was Charles Derleth, Jr., a young engineering professor at the University of California. Unlike many of his nonacademic colleagues, whose time was taken up with the day-to-day business of reconstructing the California metropolis, Derleth was able to inspect numerous areas outside of San Francisco where the earthquake damage had not been masked by fire damage. Perhaps as a result of this research, Derleth had a greater appreciation than his fellow California engineers for the extensive damage wrought by the earthquake and the need to prevent its recurrence.

Derleth repeatedly emphasized the importance of "good materials, high grade workmanship, and intelligent design" and gave specific examples of what would constitute intelligent design in an earthquake-plagued region. Included in his list of 44 recommendations for better building, 27 of which concerned earthquake-resistant construction, were warnings against terra cotta and hollow tile as building materials and a recommendation to reinforce brick chimneys. Derleth also went against his fellow engineers' blind promotion of reinforced concrete. He pointed out that the weaknesses in the joints between vertical columns and horizontal girders commonly found in reinforced concrete buildings might compromise their earthquake resistance, and he therefore urged that great care be taken in the design of such structures.[42] Finally, Derleth alone among his peers assailed the boosters' campaign to disparage the earthquake's effects by asserting that, "rather than try to tell outsiders that San Francisco was visited by a conflagration, . . . it will do San Francisco and California in general more lasting good to admit that there was an earthquake and that with honest and intelligent construction and the avoidance of geologically weak locations our losses would not have been so great."[43]

Derleth vigorously pushed his campaign for more earthquake-resistant construction. He contributed numerous articles to the local and national engineering press and wrote a long, well-illustrated article for a popular volume on the earthquake put together by Stanford University's president. He also addressed the public, in August 1906 at the University of California, in a lecture on "the destructive extent of the San Francisco earthquake" during which he graphically illustrated the earthquake damage and emphasized the need for earthquake-resistant construction in California.[44]

While Derleth was the most vocal and persistent of the California engineers who campaigned for earthquake-resistant construction in 1906, a few others also made specific recommendations on how to render structures safe against earthquakes. Robert Gardner and F. C. Davis, for example, described methods by which brickwork could be reinforced using steel bars so as to withstand earthquakes, while Lewis A. Hicks advocated the use of concrete backing for the same purpose. W. J. Cuthbertson, the former city architect of San Francisco, even argued that isolated brick chimneys, unreinforced gable walls, and parapet walls—all of which had suffered extensive damage in the earthquake—should be outlawed by the city building code.[45]

This fledgling campaign for more earthquake-resistant construction did not, however, receive general backing from the engineering and architectural professions in California. On the contrary, a number of local engineers and architects chose to support the assertions of California's boosters that damage from the earthquake had been minimal. Civil engineer W. W. Breite, for example, proclaimed, "A conservative estimate of the loss due to the quake made by a board of experts competent to judge, has been placed at about three-quarters of one per cent of the loss due to the fire immediately following, and the statement has well been made that, were it not for the fire, business would have quickly assumed its normal aspect, and the 'shake' would have been a thing of the past, long ago."[46] Edwin Duryea, Jr., the locally prominent president of the San Francisco Association of the American Society of Civil Engineers, also agreed with California journalists that earthquakes in San Francisco "are no worse than in Charleston— and are no more to be dreaded than sea-floods at Galveston, cyclones at St. Louis and elsewhere, river-floods at Kansas City and down the Ohio and Mississippi, and fires at Baltimore, Boston and Chicago."[47] Finally, a committee of the San Francisco Association headed by Duryea's colleague H. L. Haehl reiterated local claims of immunity against future earthquakes by stating, "while the time of occurrence of such movements cannot be predicted, the relief of the accumulated stresses gives reasonable assurance of freedom from a recurrence of earthquakes of similar intensity, within the area affected, for some time—probably at least fifty years."[48] In this manner, many California engineers and architects followed local boosters in minimizing the severity of the earthquake and the danger of its recurrence, thereby implying that special measures for ensuring earthquake-resistant construction in the rebuilding of San Francisco were not really necessary.

These engineers and architects also undermined Derleth's campaign for earthquake-resistant construction by echoing the San Francisco press in blaming damage primarily on poor workmanship. The State Board of Architecture, for example, declared in an official statement issued two weeks after San Francisco's devastation that "inspections in the burned district show that safety is not a question of style of architecture, but of quality of workmanship."[49] Similarly, Los Angeles architect Octavius Morgan emphasized that "well-built buildings" were still standing in San Francisco, while engineer William Hammond Hall blamed most of the damage in the city on "sham construction," and architect Charles Peter Weeks indicted "short-sighted greed, carelessness and weakness" as the crimes that caused the city's devastation, proclaiming that "the contractor is to blame for not giving golden workmanship for golden recompense." [50] Out of this conviction that earthquake damage was due primarily to poor workmanship arose the belief that safer construction required only an increase in the quality of workmanship rather than thoroughgoing changes in building design. In particular, a number of architects insisted that cornices, the decorative and often elaborate projections along rooflines that were a beloved feature of turn-of-the-century architecture, need not be omitted from future construction despite Derleth's argument that such cornices were inherently too weak to withstand strong shaking. [51]

Just as in 1868, little was done in the rebuilding of San Francisco in 1906 to make the city safer during earthquakes, even though a few engineers and architects did propose ideas for more earthquake-resistant construction. The local architectural and engineering communities were far better organized in 1906 than they had been in 1868, with local societies, journals, and professional schools that might have served to spread a greater awareness of earthquake safety in the construction industry. But the attention of most local architects and engineers, influenced as they were by the national campaign for fire prevention, was directed toward the effects of the fire rather than the earthquake. Moreover, many of the engineers accepted wholeheartedly the boosters' assertion that seismic hazards were not a serious problem in California—an assertion that scientists did not challenge until several years after the earthquake. Thus, few engineers saw the need for greater seismic safety even after witnessing the devastation wrought by the 1906 earthquake.

The Reaction of Scientists: A Belated Call to Action

Like California's boosters and engineers, scientists reacted to San Francisco's destruction by relying on previously established patterns of behavior. As they had done after the 1868 bay area earthquake, they set out to investigate the quake, to determine the extent of damage it had produced, and to fix its epicenter and its cause. But in 1906, the scientific resources available for such an investigation in central California were far more plentiful than they had been in 1868. The bay area by 1906 boasted active geology departments at both the University of California and Stanford University that could supply enthusiastic faculty and students to conduct the investigation. The scientific investigative group commissioned to study the 1906 earthquake, unlike the one established in 1868, completed its work.

It published a final report that not only documented the devastation un-
leashed by the earthquake but also established the San Andreas fault as
the prime cause of the earthquake and set forth a general theory explaining fault-generated earthquakes. In addition, commission members, local
geologists, and others interested in the study of seismic phenomena created the Seismological Society of America to provide continuing support
for the study of earthquakes in California. Finally, scientists issued a general call to the public for action to reduce seismic hazards in California,
although they did not do so until 1909, too late to affect the attitudes of
the boosters, engineers, and others who were already busily reconstructing
San Francisco.

The severity of the San Francisco earthquake surprised California's scientists, who also had long maintained that earthquakes posed little threat.
Very soon after the quake, though, geologists rallied to initiate a formal
study of the shock. Within a couple of days of the shaking, Andrew C. Lawson, the head of the geology department at the University of California,
sent a telegram to Governor Pardee requesting "the appointment of a scientific commission to investigate the earthquake in this state." On April 21,
Pardee took enough time from his frantic efforts to organize relief ship-
ments for the devastated area to issue a letter appointing eight scientists as
a State Earthquake Investigation Commission. Among its members were
Lawson, soon elected its chairman, and John C. Branner, professor of geology at Stanford, who on his own initiative had already sent 12 assistants to scour the region from San Francisco south to Santa Cruz for evidence on the earthquake's origin. Joining them on the commission were
George C. Davidson, the octogenarian dean of San Francisco's scientific
community, who had been a member of the 1868 commission and later
had operated a seismograph in the city, and three astronomers who also
maintained seismographs: A. O. Leuschner of the University of California,
W. W. Campbell of the Lick Observatory, and Charles Burckhalter of Oakland's Chabot Observatory. Rounding out the commission were Grove K.
Gilbert, a member of the U.S. Geological Survey with a longstanding interest in earthquakes, who happened to be stationed in Berkeley at the time of
the shock, and Harry Fielding Reid, a professor at Johns Hopkins University in Baltimore, Maryland, who several years earlier had begun collecting
information on earthquakes for the survey.[52]

Early on, the commission's members decided not to seek funds for their
work from the local business community. As Gilbert noted at the time, "an
appeal to men of wealth would be useless because just now men of wealth
in California have in general suffered heavy losses and have had their re-

sources further drawn upon by contributions for the relief of the desti-
tute."[53] Instead, commission members turned to the Carnegie Institution of
Washington, which had been set up in 1901 with an endowment of $10 mil-
lion to support scientific research. As early as April 18, the day of the earthquake, Lawson telegraphed Robert S. Woodward, president of the Carnegie Institution, asking for $5,000 to fund an investigation of the earthquake
by Branner, Gilbert, and himself. Without knowing of this action, Branner
on April 24 also sent a telegram to Woodward requesting $1,000 to pay
for his earthquake investigation at Stanford. Over the next several weeks,
Lawson and other commission members repeatedly lobbied Woodward for
approval of the two requests. On April 27, while awaiting word from the
Carnegie Institution, Lawson obtained $500 from the Citizens' Committee
of Fifty, a relief organization headed by San Francisco's ex-mayor James
Phelan, to pay for the commission's immediate expenses; it was agreed that
this money would be repaid as soon as funding from the Carnegie Institution came through. In late May, the Carnegie Institution's executive committee finally approved a grant of $5,000 to the earthquake commission.
This grant, together with a later supplemental appropriation, eventually
covered the commission's total expenses of $5,640.89.[54] The Carnegie Insti-
tution's support ensured that the commission, unlike its 1868 predecessor,
could complete its work free from worries over funding.

The commission's study of the earthquake progressed rapidly. Within
two weeks of the quake, parties of geologists and students sent out from
Berkeley were noting earthquake damage in Sonoma, Marin, and Monterey counties and the East Bay area, thereby complementing the work of
Branner's assistants on the San Francisco peninsula.[55] These parties soon
found that damage was particularly concentrated along the San Andreas
Fault. In 1893, Lawson had first defined and mapped this feature as an
abrupt boundary between different geological strata that extended for 20
miles on the San Francisco peninsula. In later writings Lawson and his students had downplayed its importance, claiming it to be an insignificant geological feature.[56] But in 1906 Branner found abundant evidence that during
the recent earthquake the two sides of the San Andreas Fault had moved
past each other by about eight feet on the peninsula. Roads had been disrupted, and fenceposts that formerly had been aligned in a straight line
were now offset. Gilbert found similar evidence of up to 20 feet of movement along an extension of the fault in Marin County north of San Francisco. To the south, Lawson noted horizontal displacement near Salinas.[57]
After piecing together this information, members of the earthquake commission concluded that there had been horizontal slipping along one fault

extending for at least 100 miles along the California coast and that this slip-ping had produced the seismic waves of the San Francisco earthquake.[58] The San Andreas Fault thus was both much longer and much more impor-tant than geologists had realized.

Acting upon a suggestion made by Branner, the earthquake commis-sion on May 31 submitted a preliminary report embodying its early con-clusions to Governor Pardee, who quickly approved it for publication. The report, after describing the commission's origin and early actions, docu-mented the earthquake fault from Point Arena in Mendocino County all the way to Mt. Pinos in Ventura County, 375 miles to the south. The re-port also outlined the area in which damage had been severe and noted that the amount of damage was correlated both with distance from the fault and with the nature of the ground on which structures had been erected, with those on loose alluvial sediments or artificial fill along the waterfront experiencing much more damage than those on hard rock. Finally, the pre-liminary report provided some general recommendations for strengthening buildings against earthquakes, focusing mostly on the need for solid foun-dations. On the important question of the recurrence, that is, how soon (if at all) another destructive shock could be expected, the report was silent.[59]

As many as 3,000 copies of the commission's preliminary report were distributed in booklet form, primarily in California; the Southern Pacific Company alone received 100 copies. *Science,* America's leading weekly newsmagazine for scientists, and the *Mining and Scientific Press,* which was geared toward engineers on the Pacific Coast, both reprinted the report.[60] Numerous newspaper and journal articles as well as public addresses, in-cluding a keynote speech by Branner at a teachers' convention, gave further prominence to the earthquake commission's conclusions. Even when not referring directly to the commission's work, these public statements repeat-edly traced the earthquake fault along the California coast for the bene-fit not only of scientists and engineers but also of the general public. On May 8, for example, the *San Francisco Examiner* published a map of the San Francisco peninsula showing the "course of [the] old fault which rebroke and caused the earthquake of April 18th, according to expert Branner."[61]

Widespread dissemination of knowledge about the San Andreas Fault was not accompanied by a call for increased vigilance against seismic risks. This was because the commission did not address the issue of earthquake recurrence. In an article reviewing the earthquake commission's work in August 1906, Gilbert did stress the need for considering this question: "Must the citizens of San Francisco and the bay district face the danger of experiencing within a few generations a shock equal to or even greater than the one to which they have just been subjected? Or have they earned by their recent calamity a long immunity from violent disturbance? . . . If a broad and candid review of the facts shall give warrant for a forecast of practical immunity, the deep-rooted anxiety of the community will find therein a measure of relief. If a forecast of immunity shall not be warranted, the public should have the benefit of that information, to the end that it shall fully heed the counsel of those who maintain that the new city should be earthquake-proof."[62] Despite the importance of this question, both Gilbert and the earthquake commission as a whole remained silent on the outlook for San Francisco. Correspondence between Gilbert and Lawson suggests that, throughout 1906, Gilbert was unsure about this matter and several times changed his private opinion on San Francisco's immunity from further earthquakes.[63]

Others, however, did not keep so quiet. During the summer of 1906 numerous respected seismologists and other scientists publicly affirmed that San Francisco was now immune from large earthquakes for a long period, thereby reinforcing the efforts of California's boosters to disparage the earthquake hazard. The most prominent source of such reassuring statements was Fusakichi Omori, secretary of the Imperial Earthquake Investigation Committee of Japan (the successor of the Seismological Society of Japan) and an international leader in seismology. Omori arrived in San Francisco on May 18 to investigate the earthquake and remained in California for nearly three months. During this time he became an associate member of the California State Earthquake Investigation Commission.[64] On June 6, the *San Francisco Chronicle* published a lengthy interview with him. In its opening paragraph, this article proclaimed that, "according to Professor F. Omori of Japan, the world's greatest authority on seismology, who has made a careful and thorough study of the causes and effects of the disturbance of April 18th, the people of San Francisco may feel assured that there will be no more serious earthquakes in this vicinity in the near future."[65]

Three days later, Omori repeated his reassurance in a public lecture given before the Astronomical Society of the Pacific at the University of California. The text of this lecture, with its calming conclusion, was published in both the *Mining and Scientific Press* and the *San Francisco Chronicle,* which in addition ran an editorial emphasizing that "it is the deliberate judgment of Professor Omori—and this is the opinion of the man most competent of living men to speak on this subject—that the cycle of serious disturbances about the shores of the Pacific which have continued for some years is at an end. . . . There is perhaps not, at this time, any place on earth where one is so secure from the effects of earthquake shocks as San Francisco."[66] Over the next two months, Omori repeatedly assured California newspaper readers that "the city of San Francisco and the surrounding country will be free from these great earthquakes for fifty years or more" (a statement that has proven to be correct) and, even more dramatically, that "in all probability . . . there will never be an earthquake in the State as severe as the one in April" (a statement that turned out to be incorrect).[67]

While certainly the most vocal, Omori was not the only scientist who, during the summer and fall of 1906, suggested that the state was now immune from severe earthquakes. Alexander McAdie, head of the local Weather Bureau station in San Francisco and an avid seismographer, who was busily updating Holden's catalogue of Pacific Coast earthquakes, asserted, "Judging from the history of disturbances in this kindly land, the present generation will probably not undergo such experiences as those of April 18th." Similarly, astronomer T. J. J. See, in an address at Stanford University, claimed, "There is no danger of another serious earthquake at San Francisco for a long time, say at least fifty or seventy-five years."[68]

Only one geologist, as far as I have been able to determine, unambiguously warned San Franciscans in the summer of 1906 to expect severe earthquakes at any time. In an article in *Popular Science Monthly,* George Ashley of the U.S. Geological Survey reviewed evidence for movements of several hundred feet on California faults in geologically recent times and concluded that "a comparison of the few feet of motion in April with the hundreds of feet of movement that have taken place in very recent time suggests that fault adjustments of equal or greater violence are liable to occur at any time in the future. And since similar conditions are known to occur all over the Pacific Coast region, no place in that whole district can claim immunity."[69] Coming as it did from a specialist in coal geology writing from Washington, D.C., though, this foreboding statement had little chance of being heard among the many reassurances voiced by Omori and other seismologists in California.

After issuing their preliminary report, members of the earthquake commission pressed on with their investigation of the 1906 earthquake. During the spring of 1907, Lawson prepared the commission's final report by synthesizing the sometimes conflicting reports submitted by different field parties into a consistent account of the damage produced by the earthquake. By August 1907, the first volume of the final report had been submitted to the printer, and in October 1908 it was finally ready for public distribution.[70] This report included much valuable information on the causes and effects of the 1906 earthquake. After briefly reviewing the geology of the California Coast Ranges, it presented in great detail evidence for the extension of the San Andreas Fault zone from Humboldt County in northern California to the Colorado Desert in the south, a distance of over 600 miles. Among the many landscape features described and profusely illustrated as evidence for this fault zone, which had repeatedly been the site of earthquakes, were linear ridges and valleys, aligned ponds, and linear sags in the sides of hills. The report then turned to the evidence for significant horizontal movement along the fault zone in 1906. Such evidence included not only offset fences, roads, bridges, and pipelines, but also new geodetic measurements that showed displacements of up to 15 feet between points on either side of the fault. This mass of observational evidence, supplemented by 65 plates of photographs, added great weight to the conclusions reached by the earthquake commission in its preliminary report.[71]

The final report went much beyond the preliminary one by also including detailed descriptions of the damage to buildings and other structures throughout California. This portion of the report, which was culled from the notes of about 300 observers, covered more than 200 pages and included 58 plates of often dramatic photographs. Included was occasional advice on how buildings might be strengthened against earthquakes. For example, Robert Anderson, one of Branner's assistants at Stanford, provided detailed suggestions on the proper construction of wooden houses to resist earthquakes. The first volume of the earthquake commission's final report closed with a review of what was known about the damage produced by the severe California earthquakes of 1857, 1865, and 1868.[72]

In the final report's second volume, whose publication was delayed until 1910, commission member Harry Fielding Reid added a theoretical discussion of the earthquake. Not only did he present a number of seismographic records of the earthquake taken around the world, together with an account of the theory of seismographs, but he also propounded what came to be known as the elastic rebound theory of earthquakes.[73] This theory emerged from a discussion of the results of measurements carried out by the U.S. Coast and Geodetic Survey in California in the periods 1851–65, 1874–92, and 1906–7. The geodetic measurements showed progressive northward displacements of points west of the San Andreas Fault. The survey officers attributed these relative displacements to two sudden fault movements, one during the 1868 earthquake, the other in 1906.[74] This interpretation, however, could not readily account for the fact that, in the half-century from the 1850s to 1907, points within a few miles of the fault experienced a relative displacement of about 15 feet, while those scores of

miles away moved only about half that amount. Reid resolved this problem by arguing that the displacement of points distant from the fault had
not occurred suddenly during earthquakes, but rather was the result of a
slow creeping motion of the earth's crust that had been going on for much
longer than 50 years. The elastic strain produced over time by this gradual
movement, according to Reid, had finally exceeded the breaking point of
rocks along the San Andreas Fault, where the rapid displacement during
the earthquake in 1906 released the elastic strain built up over the previous
hundred years. While Reid was not able to offer a satisfactory explanation
for the forces causing the gradual movement of the earth's crust in California (this had to await the development of plate tectonics in the 1960s),
the elastic rebound theory did account quite well for the observed displacements before and after the 1906 earthquake. Moreover, it also offered a
convenient means for predicting future earthquakes. As Reid suggested,
repeated geodetic measurements in the future, by detecting the relative displacement of points on opposite sides of and far away from the San Andreas
Fault, would be able to ascertain whether strain in the region was once
again building up to dangerous levels.[75]

Although the careful documentation of damage produced by the 1906
earthquake and the development of the elastic rebound theory provided
powerful new tools for a campaign for seismic safety, the earthquake commission's report was still mainly silent on the crucial matter of earthquake
recurrence in California. Lawson did point out that the San Andreas Fault
was just one of many faults in California and that the presence of these
faults "suggests seismic possibilities."[76] Reid's assertion in 1910 that the
1906 earthquake had relieved a century's worth of strain buildup implied,
however, that at least central California might be immune from a similar
earthquake for another century.

In addition to assembling a vast amount of information on the San
Francisco earthquake, members of the earthquake commission played an
important role in setting up an organization for promoting future seismological research. The Seismological Society of America originated at a conference of those "interested in seismological equipment on the Pacific
Coast," which was held in Berkeley on July 11, 1906, under the auspices
of the earthquake commission. Among those in attendance were Lawson,
Reid, and Leuschner of the commission, as well as Alexander McAdie,
the head of the local Weather Bureau station, who had been interested in
recording earthquakes since the 1880s, and seismologist C. F. Marvin from
the Weather Bureau's national headquarters, who was in California examining seismographic records from the San Francisco quake. As a result

of this meeting, the earthquake commission passed a resolution encouraging the formation of a seismological society "for the furtherance of research work and investigation and the dissemination of useful knowledge to the community." The resolution also promoted the establishment of a government-supported seismological service for the United States.[77]

Four weeks later, a committee composed of Reid, Leuschner, and Marvin submitted a report to the earthquake commission on the shape that seismological research should take on the Pacific Coast. This committee envisioned a network of stations across the state equipped with instruments for recording both local and distant earthquakes; these stations would report to a central bureau that would also be charged with improving seismographs. Affiliated with this network would be the Pacific Coast Seismological Association, composed of 200 to 300 cooperating observers who, every time they felt an earthquake, would report the quake's time of occurrence, duration, intensity, and other pertinent information to the central bureau. In this way, members of the earthquake commission hoped, a complete catalogue of earthquakes on the Pacific Coast might be assembled.[78]

The task of actually organizing this seismological association was left to McAdie. On August 22, 1906, he issued an invitation for a preliminary meeting in San Francisco to consider the formation of the "Seismological Society of America." Thirteen people attended this meeting, and McAdie received letters of encouragement from about 50 more. At a general meeting in San Francisco two months later, the organization of the Seismological Society of America was finalized with the adoption of a constitution, which broadly defined the society's aims as being the "acquisition and diffusion of knowledge concerning earthquakes and allied phenomena and [enlistment of] the support of the people and the government in the attainment of these ends." By mid-1907, the Seismological Society, which had set dues at the low rate of $2 a year, had managed to attract between 150 and 200 members, of whom somewhat less than half were scientists and the rest engineers, architects, and other professionals.[79]

The Seismological Society generally drew its membership from the same circles of Progressive urban professionals as did other California environmental groups, such as the Sierra Club and (somewhat later) the Save-the-Redwoods League.[80] In fact, there was considerable overlap in the membership and leadership of the Seismological Society and the Sierra Club, which had been founded in 1892 to foster greater appreciation of the Sierra Nevada's scenic beauty and to advocate measures for protecting this beauty. Sierra Club founder and long-time president John Muir and San Francisco lawyer Warren Olney, in whose office the Sierra Club had been formally organized, were also members of the Seismological Society, while Seismological Society founder McAdie served as the Sierra Club's vice president from 1905 until at least 1914 and George Davidson, the Seismological Society's first president, chaired the Sierra Club's Committee on Parks and Recreation from 1900 to 1905.[81] Joseph N. LeConte, professor of mechanical engineering at the University of California, served both the Seismological Society and the Sierra Club as treasurer before 1911.[82]

Finally, in early 1909, one of the earthquake commission's members made a public statement urging Californians to guard against seismic hazards. This plea came from Gilbert, who among the commission's members was probably most closely aligned with the Progressive reform movement. Early in the twentieth century, he was an associate of numerous Progressive reformers and participated vigorously in the Sierra Club's activities. Beginning in 1905, he was stationed at the University of California to investigate the environmental impact on the state's rivers of hydraulic mining in the Sierra Nevada. He firmly believed that applied science and expertise could be used to improve the interaction of humans with their environment. Thus, it is not surprising that he would also inject a concern for applying lessons to society into his work for the earthquake commission.[83]

Gilbert made his plea for seismic safety in a presidential address to the American Association of Geographers, published in *Science* in January 1909. In this disquisition on "earthquake forecasts," Gilbert gave a critical review of various methods proposed by seismologists for predicting the time of future earthquakes. After discussing at length the opinion of Omori and others that San Francisco was safe from severe earthquakes for a long time to come, Gilbert concluded that the reasoning underlying such forecasts "has no sure support from observation, and is not in working order for either large or small areas. Its corollary of local immunity after local disaster is more alluring than safe." While Gilbert dismissed the ability of scientists to predict the time of future earthquakes, he did assert that the places subject to severe earthquake damage were already well known, both through the delineation of fault zones and through the mapping of areas of soft ground and artificial fill where damage tended to be concentrated. In Gilbert's words, "the determination of danger districts and danger spots belongs to the past, the present and the near future; the determination of times of danger belongs to the indefinite future." This state of affairs Gilbert considered fortunate, "for knowledge of place has far more practical value than knowledge of time"; but he cautioned that even if forecasting "were definite, timely and infallible, so that peril of life could be altogether avoided, property peril would still remain unless construction had been earthquake-proof. If, on the other hand, the places of peril are definitely known, even though the dates are indefinite, wise construction will take all necessary precautions, and the earthquake-proof house not only will insure itself but will practically insure its inmates."[84] From this conclusion, Gilbert drew a series of policy imperatives: "It is the duty of investigators—of seismologists, geologists and scientific engineers—to develop the theory of local danger spots, to discover the foci of recurrent shocks, to develop the theory of earthquake-proof construction. It is the duty of engineers and architects so to adjust construction to the character of the ground that safety shall be secured. It should be the policy of communities in the earthquake district to recognize the danger and make provision against it."[85]

In keeping with this call for a general effort to ensure earthquake safety in California, Gilbert ridiculed the public relations campaign that had dismissed California's earthquake risks. While he agreed with the boosters that the risk of death by earthquake was quite low, he pointed out that the property damage done by earthquakes in California was of the same order as that done by fires and thus "equally worthy of serious consideration." Public leaders, he noted, did not conceal fire risks but rather worked to reduce those risks by promoting fireproof construction. Because "earthquake damage is at least as preventable as fire damage," he insisted that public leaders should also squarely face the earthquake hazard and embrace earthquake-resistant construction methods "in the building regulations of all cities as well as in the entire building practi[c]e of the state."[86]

This appeal for greater seismic safety, coming as it did nearly three years after San Francisco's devastation, had little impact on the reconstruction of the city. The state's boosters had already convinced the public that California did not really face any serious seismic hazards, and the engineers responsible for rebuilding San Francisco, swayed by these assertions and by their own preoccupation with fire prevention, had already decided that there was no need to modify construction methods to take earthquake hazards into account. While Gilbert's appeal had little immediate effect, it did, however, become the kernel around which a campaign for seismic safety crystallized in California. Convinced that they had a duty to warn the public about seismic hazards, California scientists began to devote their efforts to mapping the state's earthquake risks and to urging improved construction methods.

CHAPTER 2

Setting Up a Scientific Infrastructure
Seismology California Style, 1910–1925

With the publication of the State Earthquake Investigation Commission's final report in 1908 and the issuance in January 1909 of G. K. Gilbert's call for mitigating earthquake hazards, the groundwork had been laid for a campaign for seismic safety in California. Yet the scientific and organizational resources necessary to sustain such a campaign were still lacking. In March 1909, Gilbert suffered a debilitating stroke, which forced him to curtail his activities; as a result, he could not follow up on his plea for greater earthquake awareness.[1] The young Seismological Society of America also was in no state to take the lead in a campaign for seismic safety. Despite some early plans to issue a journal that might inform the public about earthquakes and seismological research, the society had not taken any sustained action toward making this journal a reality. The society's Scientific Committee, which was supposed to coordinate the gathering of earthquake observations from members across the state, showed few signs of activity. The society also failed to solicit new members vigorously and consequently fell into dire financial straits. In December of 1910, the group's secretary, George D. Louderback, had to admit that the society had become dormant; other observers similarly noted its "moribund condition."[2]

But during the next two decades, two California scientists succeeded in creating an institutional framework for earthquake research and seismic safety advocacy in California. Stanford University professor John Branner, who took over as the Seismological Society's president in late 1910, revitalized the society, transforming it into an effective tool for gathering and disseminating information on earthquakes in the state. Then, in the early

1920s, Harry O. Wood, a former University of California instructor, obtained the backing of the Carnegie Institution of Washington to set up a network of seismographs in southern California, which allowed the recording of earthquakes too weak to be felt by humans. Their combined efforts resulted in the creation of a permanent community of researchers who stood ready to investigate California's seismicity and to warn the public about the state's seismic hazards.

John C. Branner and the Seismological Society of America

Branner resuscitated the Seismological Society by embarking on an aggressive membership drive and by establishing a quarterly publication for disseminating the results of seismological research. Throughout the 1910s, Branner drew on his many contacts among scientific colleagues, students, and alumni to obtain further financial and logistical support for the society's efforts. Under Branner's leadership, members of the society began to collect information on earthquake occurrences in California and to map a number of dangerous faults. In this fashion, Branner established a solid organizational and scientific foundation for promoting greater seismic safety in California.

A native of Tennessee, John Branner received his bachelor's degree in geology from Cornell University in 1882 and had been head of geology at Stanford since the university's opening in 1891. At Stanford, he trained numerous geologists and mining engineers, the most famous of whom, Herbert Hoover, would go on to become president of the United States. While his main professional interest was in the geology of Brazil, he also did considerable work on California's geology during his decades at Stanford; this work included not only purely scientific research but also substantial consulting for mining and water companies. Like many other scientists of this era, Branner was very interested in the conservation and efficient use of natural resources, and he was a charter member of the Sierra Club.[3]

Branner became involved in the study of earthquakes after the 1906 San Francisco earthquake. As a member of the State Earthquake Investigation Commission, he directed a corps of Stanford students who examined the damage wrought by the earthquake on the San Francisco peninsula. In 1908, at the end of a research trip to Brazil, Branner visited Valparaiso in Chile and Kingston in Jamaica, both of which had been devastated by earthquakes shortly after the San Francisco shock, as well as Mendoza in Argentina, which had suffered a strong earthquake in 1861. These visits apparently strengthened his desire to contribute to earthquake hazard mitigation.[4] As early as January 1907, Branner became a life member of the Seismological Society of America. A few months later, Branner was already soliciting new members for the society. While the society languished in early 1909, Branner was busy collecting earthquake reports from his fellow Stanford professors and transmitting them to Alexander G. McAdie, who was almost single-handedly trying to maintain an up-to-date catalogue of Pacific Coast earthquakes.[5]

In January 1909, a few weeks before Gilbert published his call for applied research on the location of seismic hazard zones, Branner addressed a meeting of the Seismological Society in San Francisco with his own appeal for practical research. According to Branner, seismology's most pressing need was for accurate knowledge of active faults. As he put it, "geologists know of a great many faults which are not now centres of seismic disturbances. . . . On the other hand there are probably faults so obscured by soil or by recent deposits, especially along valley floors, that we are unable to say just where they are; and there are still others that are but ill defined at the surface." Branner put little stock in seismographic instruments for locating faults and distinguishing active from inactive ones, but, he noted, "there is one excellent kind of a seismograph . . . that we can all use to great advantage if we will only set about it, and will take the trouble to put the records on paper and send them in. I refer to our own bodies." Branner therefore urged the society to implement its original intention of setting up a large corps of observers who would "note the time and intensity of all earthquake shocks felt by them and . . . send these notes to the compiler with the least possible delay."[6]

In later years, Branner repeatedly expanded, along the lines already laid down by Gilbert, on the uses to which such applied research should be put. In 1913 he explained, "There are plenty of 'active faults' in California, and we should be working now to locate them. When we know where they are, we can keep our houses, bridges, dams, pipe lines and other structures off them, or, we can do our engineering so that, when the next earth-slip comes, the damage will be negligible."[7] According to Branner, acquisition of this essential knowledge was hindered by a pervasive "conspiracy of silence," particularly among the business community.[8] "The idea back of this false position," he wrote, "is that earthquakes are detrimental to the good repute of the west coast, and that they are likely to keep away business and capital, and therefore the less said about them the better. This theory has led to the deliberate suppression of news about earthquakes, and even of the simple mention of them." Branner considered this attitude to be "most unfortunate, inexcusable, untenable. . . . The only way we know of to deal successfully with any natural phenomenon is to get acquainted with it, to find out all we can about it, and thus to meet it on its own grounds. . . . The terrors that earthquakes have for mankind are largely attributable to our own ignorance. The more we know about them the less harm they can do us, and the less reason we shall have to fear them."[9]

Consequently Branner, like Gilbert, considered applied seismological research, particularly on the location of dangerous fault zones, and public education necessary to reduce the fearfulness of earthquakes and to render California truly safe from seismic catastrophe. In issuing their call for a campaign for greater seismic safety, both Gilbert and Branner drew upon the rhetoric of the Progressive conservation movement in which they were also involved. Both suggested that short-sighted businessmen were producing inefficiency by belittling seismic hazards in California and thereby hindering efforts to prevent future earthquake damage. Both also argued that a more efficient and wiser stance toward earthquake hazards required the adoption of technical remedies suggested by scientific experts, including geologists and engineers.

Branner received the opportunity he needed to implement his program in late 1910, when he was elected president of the Seismological Society of America to replace Andrew C. Lawson. When Branner took over, the Seismological Society had 143 members, of whom 86 (60%) lived in California. The non-Californian members were predominantly scientists with a professional interest in furthering seismological research. California members tended to be concentrated in large cities and university towns and included scientists, engineers, and reformist businessmen like James D. Phelan, the millionaire ex-mayor of San Francisco who spearheaded many of California's Progressive reform movements.[10] Within a month of being elected the society's president, Branner embarked on a vigorous campaign to add to this membership. By one count, he wrote about 2,500 letters over the next year and a half soliciting new members. As a result of this intense effort, the society nearly tripled in membership by April 1912 to 422 members and institutional subscribers. Branner's ambitious long-term goal was a thousand members. In keeping with his intention of building up a corps of earthquake observers, Branner directed a large part of his effort toward recruiting new members in California, especially in the southern part of the state. During 1911, he managed to increase society membership in southern California from 4 to 18; he also recruited three new members in California's Central Valley. Meanwhile, in the bay area Branner added a number of professionals and Progressive businessmen, such as banker J. K. Moffitt, insurance executive Francis V. Keesling, and petroleum tycoon Joseph D. Grant, to the Seismological Society's rolls.[11]

While embarked on his campaign to increase the Seismological Society's membership and thereby improve its financial situation, Branner also moved rapidly to inaugurate the society's new journal, the *Bulletin of the Seismological Society of America*. In this work, Branner had the valuable assistance of Sidney D. Townley, a professor of astronomy at Stanford University who late in 1910 was made secretary of the society on Branner's recommendation and a few months later also became its treasurer. During 1911, Branner and Townley managed to put out four issues of the *Bulletin*. The *Bulletin* received a significant financial boost in late 1911, when Harvard's J. B. Woodworth, a seismographer who also shared Branner's professional interest in the geology of Brazil, secured a donation of $5,000 from a former student.[12] This gift enabled Branner to continue publishing the *Bulletin* four times a year and thereby disseminate the results of research on seismology and earthquake hazards.

In April 1912, Branner declined to be reelected as Seismological Society president. Over the next nine years, the society was headed in succession by six different seismologists, of whom five were not residents of California. They were, however, for the most part mere figureheads. Actual control of the society remained in the hands of Branner, who chaired the *Bulletin*'s editorial committee until illness incapacitated him in 1921, and of his assistant Townley, who for two decades retained his position as the society's secretary-treasurer, an office for which he received a modest salary beginning in 1912.[13] In 1918, for the first time, the majority of the Seismological Society's 12 directors resided outside of California, yet Townley insisted on continuing to have the board of directors meet at Stanford, with the predictable result that these meetings were frequently attended only by Branner, Townley, and one or two other local directors.[14] In this manner, Branner and Townley retained control over decisionmaking in the Seismological Society and used it to implement Branner's practical approach to earthquake research.

Through the *Bulletin*, Branner advanced his program for locating hazardous fault zones through noninstrumental observations of local earthquakes. As early as 1911, in the *Bulletin*'s second issue, he published a lengthy article by Harry O. Wood, an assistant to Lawson at the University of California who supported Branner's goal of setting up a corps of earthquake observers throughout the state. In this article, Wood gave detailed instructions for how local residents should observe and note the time of occurrence, duration, and intensity of earthquakes and emphasized that prompt and reliable reporting of such observations was necessary for locating areas subject to earthquakes. Branner himself also repeatedly reminded the *Bulletin*'s readers that all felt earthquakes should be reported to the Seismological Society's secretary.[15] Over the years, a number of Seismological Society members did contribute such reports, and Branner occasionally published them in the *Bulletin* under the heading "Seismological Notes," thereby disseminating knowledge about California's seismicity.

The Central California earthquake of July 1, 1911, which was felt throughout the San Francisco Bay area, offered Branner his first chance to test his program for locating active faults. The earthquake occurred while Branner was away in Brazil for the summer, and, to his dismay, no one took the trouble during his absence to collect observations on it. Upon his return from Brazil in late August, Branner immediately solicited data from Seismological Society members around California, and in the September issue of the *Bulletin* he published reports from 23 observers. Branner also dispatched his graduate student, E. C. Templeton, to the area southeast of San Jose that had reported the most shaking; here, Templeton interviewed local residents to gather more data on the earthquake's origin and also examined its effects in the field. As a result of this work, Templeton was able to pinpoint the epicenter of the earthquake, which, as Wood subsequently noted, was on a previously unmapped southern extension of the Hayward Fault.[16]

While Branner, with the help of his student's field work, succeeded in his investigation of the 1911 bay area earthquake, he had to concede by early 1913 that he simply could not rely on the volunteer observers he recruited for the Seismological Society to report on all the earthquakes they felt. Frustrated by the inability of his nongovernmental organization to produce a reliable listing of earthquakes, he took to publicly urging the U.S. Weather Bureau to have its numerous paid weather observers across the country collect information on earthquakes in addition to their other duties.[17] In the summer of 1914, the federal agency, after several failed attempts, finally received congressional authorization to engage in seismological research. With this authorization secured, the Weather Bureau's director instructed his weather observers to gather observations on local earthquakes, a move that drew Branner's hearty approval. In California, Andrew H. Palmer, now head of the Weather Bureau's San Francisco office, recruited several hundred weather observers for additional duty as earthquake observers. He also placed the data they gathered at the service of the Seismological Society and, beginning in 1916, published an annual report in the society's

Bulletin listing all earthquakes felt in California. While these reports listed the places where earthquakes had been felt rather than their places of origin, they nevertheless contributed toward outlining the areas of high seismicity in the state. In 1917, for example, Palmer pointed out a concentration of earthquakes in the Monterey Bay region of central California.[18]

Meanwhile, Branner refined his methods for using field investigations of earthquake damage in conjunction with noninstrumental observations to pinpoint the faults responsible for earthquakes. On November 8, 1914, a sharp earthquake struck the Santa Cruz mountains south of Stanford. The Seismological Society received only about 20 reports of this earthquake, so Branner dispatched his current graduate student, Carl H. Beal, into the area to interview local residents and inspect the damage. On the basis of his investigation, Beal attributed the earthquake to movement on the San Andreas fault at a location near the town of Laurel.[19] In early 1915, Beal, again at Branner's suggestion, traveled to a thinly populated area of Santa Barbara County in southern California to note the damage done there by a moderate earthquake. On the basis of his observations as well as conversations with local ranchers, he was able to attribute the shock to a fault near the town of Los Alamos. Beal's work on this earthquake confirmed Branner's conviction that prompt field observations were "the easiest, quickest, and surest way to locate an epicenter."[20]

Beal's final contribution to the study of California earthquakes came in the summer of 1915. The Seismological Society commissioned him to investigate a severe earthquake in southern California's Imperial Valley and neighboring parts of Mexico that had killed 6 people and caused about $900,000 worth of damage. After spending several weeks in southern California investigating the earthquake damage and studying the regional geology, Beal blamed the earthquake on the San Jacinto Fault, which he traced through the length of Imperial Valley. In his report, Beal also included graphic illustrations of buildings devastated by the shock and urged residents of Imperial Valley to use reinforced concrete to make structures more earthquake resistant. Branner contributed to Beal's investigation by donating $403 to pay for Beal's field expenses and for the photos accompanying Beal's report in the society's *Bulletin*.[21]

Branner's program for studying California's seismicity received another boost in the fall of 1916, when several well-connected southern Californians decided to aid the Seismological Society's efforts. The first of these was Homer Hamlin, a well-known hydraulic engineer who from 1906 to 1917 headed the engineering department of the city of Los Angeles. Drawing on his many contacts among southern California engineers, Hamlin energetically collected information on the region's felt earthquakes, which turned out to be far more numerous than he had expected. Hamlin's work enabled Branner to discuss an earthquake along the San Andreas Fault in the Tejon Pass region north of Los Angeles. Hamlin also helped Branner's student A. C. Mattei to gather data on several earthquakes that had originated in the Santa Barbara Channel off the coast of southern California.[22] Finally, Hamlin interested others in the society's work. Most important among them was Ralph Arnold, a former student of Branner's who had made his fortune as a petroleum geologist based in Los Angeles. Arnold offered to pay for part of Hamlin's expenses in gathering earthquake observations; he also arranged to have Branner address a joint meeting of several Los Angeles engineering societies on the Seismological Society's work.[23]

These resources allowed Branner to launch a vigorous investigation after the southern California towns of San Jacinto and Hemet were struck, on April 21, 1918, by an earthquake strong enough to be recorded by seismographs around the world. Branner almost immediately sent Townley to southern California, where he spent three days together with Hamlin investigating the damage. Hamlin also collected numerous reports of non-instrumental observations on the earthquake from his large network of acquaintances; these reports were checked against those collected by the Weather Bureau. Based on this work, Townley attributed the earthquake to the San Jacinto Fault, which Beal had first mapped as an active fault three years earlier. In his published report, graphically illustrated with numerous photos of destroyed buildings, Townley described at length the damage to brick buildings. He also pointed out that another earthquake of the same intensity had struck the same location in 1899, thereby indicating that this area was subject to frequent strong earthquakes.[24] Soon after Townley returned to Stanford, Arnold in Los Angeles, with Branner's blessing, sent two of his business associates, both of whom had recently graduated from Branner's geology program at Stanford, into the field to trace the San Jacinto Fault in greater detail. Drawing on the data collected by them and by Townley and Hamlin, as well as on unpublished data he obtained from the U.S. Geological Survey, Arnold published a report describing in detail the course of the San Jacinto and adjacent faults.[25]

The crowning achievement of the Seismological Society under Branner's leadership came in 1920, when seismologists were able to locate several faults directly threatening the booming southern California metropolis of Los Angeles. In the evening of June 21, 1920, a moderately strong earthquake did considerable damage in the business district of Inglewood, a suburb about 10 miles southwest of Los Angeles, and also caused some chimneys and plaster to fall in the city. It so happened that Stephen Taber, a former student of Branner, had just arrived at Stanford on the eve of the earthquake to spend the summer there as a visiting professor. As professor of geology at the University of South Carolina, Taber had used Branner's methods to locate the origins of earthquakes in South Carolina and in Puerto Rico. Branner immediately asked Taber to go to southern California, where Taber spent three days investigating the earthquake. Besides describing and illustrating the damage, Taber also located and reported the earthquake's epicenter on a fault he named the Inglewood–Newport–San Onofre Fault. Using some subtle geological indicators that had been overlooked by previous researchers, he traced this fault through the Los Angeles basin and along the coast of southern California. Taber concluded "that earthquakes are to be expected in [the Los Angeles coastal plain] from time to time. But if the inhabitants will heed nature's warnings and take proper precautions in the location and construction of buildings and other structures, such earthquakes are not likely to do much damage."[26]

Less than a month after the Inglewood earthquake, Los Angeles experienced a series of weak shocks that did slight damage and caused a few injuries from falling brick and glass. Taber once again traveled to southern California to investigate on behalf of the Seismological Society. Besides locating a series of short but apparently active faults within the city limits of Los Angeles, he noted that even geologically very young sedimentary deposits in the Los Angeles area had already experienced significant deformation, indicating a high degree of dislocation caused by earthquakes in recent times. From this Taber concluded, "There is every reason for believing that earthquakes will be felt in the Los Angeles district as frequently in the future as in the past." He then asserted that it was "of the utmost importance that all active faults [in the Los Angeles area] be located so that engineers and architects can take precautions in the location and construction of aqueducts, buildings, and other structures." He pointed out that "earthquakes can not be prevented, but with proper precautions their destructive effects can be reduced to a minimum." He therefore concluded his report by imploring "the people living in California" to "frankly admit that this is a region of relatively high earthquake frequency."[27]

After the Inglewood and Los Angeles earthquakes, Branner, who had long been frustrated because people in Los Angeles were "afraid to say earthquake out loud," along with his colleagues in the Seismological Society moved to promote greater earthquake awareness in southern California. During August 1920, Branner wrote to various geologists, engineers, and businessmen in the Los Angeles area, commenting on the "queer things" said by Los Angeles newspapers about earthquakes and sending copies of his earlier addresses urging Californians to support the Seismological Society's research. Meanwhile, Arnold, who had been elected chairman of the Southwestern Section of the American Institute of Mining Engineers two years earlier, arranged to have this organization host a symposium in Los Angeles on earthquakes.[28] The symposium, which drew an audience of at least 150, was addressed by Arnold, Taber, an engineer named William Mulholland, and Robert T. Hill, a petroleum geologist who had done considerable mapping in southern California. Not all statements made by these speakers countered the denial of seismic hazards still voiced by southern California boosters. Hill, for example, while describing in general terms the system of faults in southern California, asserted that earthquakes in the region were rather harmless, especially when compared to storms and floods in other areas of the country. But Taber took the opportunity to display a map of the active faults he had recently located in the Los Angeles area and to repeat his claim that earthquakes as severe as the one that had done widespread damage in 1812 would certainly occur again. Taber also urged his audience to support the work of the Seismological Society in locating active faults, so that engineers could avoid building on them. In closing the symposium, Arnold suggested the organization of a southern section of the Seismological Society. By January 1921, a committee of ten prominent southern California geologists, engineers, and architects headed by Arnold had organized such a section and recruited 45 new members for the society in the Los Angeles area.[29]

Despite these successes, the Seismological Society in 1920 still faced considerable obstacles to its program of research and public education on seismic hazards. After 1912, membership in the society, despite occasional efforts by Branner to solicit new members, had stagnated at around 400. This lack of growth in membership led to financial problems, because the rising costs of publishing the *Bulletin* could not be offset by increased revenues. By 1920, the situation had become so critical that Branner had to donate $451 of his own money to keep the *Bulletin* afloat.[30] The society also suffered from a decided lack of impact in its public education activities. Even though a number of architects and engineers were members of the society, there is little evidence that engineers or architects in California paid much attention to seismic hazards in designing new buildings during the 1910s. Newspapers and the general public also continued to ignore the earthquake threat.[31] Nevertheless, under Branner's leadership the Seismological Society did succeed in stimulating earthquake research in California. Trained observers now stood ready to record data whenever a sizable earthquake hit and to disseminate their findings through the society's *Bulletin*. And as a result of a decade worth of work, Seismological Society members had increased the list of fault zones known to be dangerous.

Harry O. Wood and the Carnegie Institution of Washington

In the 1910s, earthquake research in California was supported largely by local resources: the volunteer labor of observers organized through the Seismological Society, together with the funding, contacts, and organizational skills provided by local professors and their associates. In the early 1920s, a new and very different source of patronage appeared in support of applied seismological research in California. This new patron was the Carnegie Institution of Washington, a large philanthropic foundation that had been established in 1901 to sponsor scientific research across the nation. The Carnegie Institution had already been involved with seismological research when it paid for the expenses of the California State Earthquake Investigation Commission and published its lengthy report on the 1906 San Francisco earthquake, but this involvement had not led to continuing support for seismological research. In 1921, the Carnegie Institution agreed to fund a long-term program of applied seismological research in southern California proposed by Harry O. Wood, a former instructor at the University of California. As a Carnegie Institution research associate, Wood went about devising a seismograph capable of recording local earthquakes; he then set up a regional network of stations equipped with this instrument. He thereby added seismography to geological fieldwork as methods available to applied seismology for locating active fault zones. Following in his footsteps, other California scientists also took advantage of the Carnegie Institution's funds and political connections to advance seismological research.

Wood first became involved in earthquake research while at the University of California. After obtaining bachelor's and master's degrees at Harvard University, Wood, a native of Maine, came to Berkeley in 1904 as an instructor in the University of California's geology department, which was then headed by Lawson. Wood at first continued his earlier research in mineralogy, but his scientific interests changed after the 1906 San Francisco earthquake. Under Lawson's direction, he conducted a detailed investigation of damage in the city of San Francisco for the State Earthquake Investigation Commission. He also engaged in extended discussions with Gilbert, who was then beginning to formulate his call for applied earthquake research and public education on earthquake hazards in California. In 1908,

Gilbert gave Wood $1,000 to map possibly hazardous earthquake faults in northern California.[32] Several years later, Lawson put Wood in charge of the university's seismographs. Unlike most other American seismologists at the time, Wood did not confine his research to the study of seismic waves from distant earthquakes (used to study the density and other characteristics of the earth's interior through which the waves had passed) but was interested in recording local earthquakes and using them to locate potentially dangerous faults. As a result, Wood enthusiastically supported Branner's plans for revitalizing the Seismological Society as an institution for conducting applied seismological research. Wood's seismological studies at Berkeley culminated in the publication of a long paper assigning a large number of recorded earthquakes to known or inferred faults in California and using the resulting statistics to discuss the importance of various faults as generators of earthquakes. In the preface to this paper, Wood emphasized the importance of delineating "the districts of greatest danger" in this manner, so that "protection against the effects of strong shocks" might be more effectively provided. Wood thereby echoed the pronouncements of both Gilbert and Branner on the need for practical earthquake research.[33]

Although the study of California earthquakes gave a clear focus to Wood's research, career considerations led him to seek a position elsewhere. In 1912, after eight years at Berkeley, he was still a low-paid instructor with no hope for promotion. That year, he abruptly resigned his post. Soon, he found a position as an assistant to his former teacher T. A. Jaggar, Jr., who was then building a volcano observatory at Kilauea Volcano on Hawaii. Over the next five years, Wood devoted himself to recording the local earthquakes given off by Kilauea and interpreting them in terms of the movement of magma within the volcano.[34] In the fall of 1917, as the spreading world war diverted financial resources away from the volcano observatory and when his personal relations with Jaggar had deteriorated, Wood was fired from the observatory. But during his interlude in Hawaii, besides honing his skills in recording local earthquakes, he had made an important friend in Arthur L. Day, the influential director of the Carnegie Institution of Washington's Geophysical Laboratory and Home Secretary of the National Academy of Sciences, who occasionally visited Hawaii in order to conduct volcanological research at Kilauea.[35] In subsequent years, Day would often serve as Wood's mentor.

With no prospects in sight for a research position, Wood in late 1917 decided to take Day's advice and join other scientists in contributing to the American war effort. From January 1918 to April 1919, Wood, with a commission in the U.S. Army Engineer Reserve Corps, worked at the Bureau of Standards in Washington searching for a method to detect the position of enemy artillery through the ground vibrations emitted when guns were fired. Other scientists had already sought to apply standard seismographic techniques to the problem but had had to concede that acoustic methods, based on detecting the arrival of air waves from fired guns, worked far better than methods using ground vibrations. Yet Wood continued to experiment with seismic methods. As he later admitted, this work did not produce anything of military value, but it did give him valuable experience in designing seismographic equipment.[36]

While in Washington, Wood met many of the influential scientists of the day who were shaping the direction of scientific research. Among them was George Ellery Hale, the energetic director of the Carnegie Institution's Mount Wilson Observatory in Pasadena, just northeast of Los Angeles. In the uneasy months preceding American entry into the war, Hale had formed the National Research Council as an adjunct to the National Academy of Sciences, to encourage and coordinate American scientists' contributions to the war effort. Shortly before the war ended, Hale obtained an executive order from President Wilson turning the council into a permanent organization for coordinating public and private research. Wood also reestablished contact with John C. Merriam, who had been professor of paleontology at the University of California during Wood's years there. During the war, Merriam helped his friend Hale mobilize American scientists for the war effort; after the war, he would serve for a year as chairman of the National Research Council before, in 1921, becoming director of the Carnegie Institution of Washington. Wood, after being discharged from military work in the spring of 1919, maintained his contacts with these influential men by joining the National Research Council as assistant to the executive secretary. Somewhat later, he also became secretary of the newly formed American Geophysical Union, initially a branch of the National Research Council. Although Wood found this secretarial work to be quite dull, his position near the center of postwar planning for scientific research gave him an excellent base from which to promote his long-held dream of establishing a comprehensive seismological research program in California.[37]

In his first proposal for such a program, published in the *Bulletin of the Seismological Society* in 1916, Wood had advocated the establishment of an Institute of Seismology and Geophysics in California and suggested that this institute focus on a "thorough search for living faults." In justifying the need for this institute, Wood restated at length Gilbert and Branner's appeal for applied research on seismic hazards. As Wood put it, "if we look to their importance to human life alone the problems raised by earthquake occurrence in this region [California and neighboring states] demand systematic and sustained investigation in the open field, in seismological laboratories, and in engineering science. . . . Sound building practice, and sound, enforceable legislation in regard to building and building-sites, for the protection of the public, must depend primarily upon knowledge gained by such work. And such a body of experience and of concrete facts, *if heeded*, will save in money in a short term of years many, many times its money cost. Obviously it cannot be *heeded* before the knowledge is acquired." Unlike Branner, Wood argued that the search for active fault zones should be carried out using more than just geological mapping of faults. He suggested an interdisciplinary research program that combined geological mapping with geodetic measurements and seismographic recording of weak local earthquakes. The geodetic measurements might reveal distortions in the earth's surface due to fault movements, while fixing the epicenters of small earthquakes might point to the location of dangerous faults.[38]

Wood himself was primarily interested in the seismographic aspects of this research program. Pursuing them, however, would require him to design a new type of seismograph. The seismograph had undergone considerable development since its invention in Japan in the early 1880s, but virtually all of these developments had been directed toward improving the ability of seismographs to register the waves from far-distant earthquakes. Seismic waves emanating from an earthquake's epicenter generally have a short period, 0.5 to 2 seconds. As they travel through the earth's interior, though, these waves gradually become elongated until they have a period of 5 to 10 seconds. The seismographs that were commercially available in the 1910s were all tuned to pick up these long-period waves, which had traveled through the depths of the earth, and were generally incapable of recording the short-period waves from nearby earthquakes.[39]

Wood, one of the few seismographers actually interested in local earthquakes, had experienced considerable frustration in trying to register such earthquakes. Consequently, he proposed focusing his energies on devising a seismograph that would be suitable for recording local earthquakes. At first Wood envisioned that, once such a seismograph had been developed, a dense network of instrument stations stretching all the way from Seattle to Albuquerque would be established. But he soon limited his ambition to setting up a network for recording local earthquakes in southern California, with headquarters near Los Angeles. Because this region had repeatedly experienced strong quakes before 1857 but only weak or moderate ones since then, he argued that it was overdue for another large earthquake and therefore presented a particularly suitable area for research on seismic hazards.[40]

Wood's initial appeal, aimed as it was at the financially strapped members of the Seismological Society, produced no response from any potential sponsor who might fund his proposed institute. In 1919, with his war work coming to an end in Washington, Wood tried again to garner support for seismological research in California, this time making his pitch to Hale, Merriam, and other influential members of the National Research Council. After a year of lobbying, Wood in April 1920 obtained the approval of the National Research Council's Division of Geology and Geography. While this was a powerful recommendation for his proposal, it still did not carry with it any funds for implementing the plan, but merely the suggestion that $25,000 a year be sought from some foundation. Thus, during the summer and fall of 1920 Wood continued to use his contacts with the scientific elite in Washington to search out potential sponsors for his program.[41]

He finally achieved success when Merriam became president of the Carnegie Institution of Washington in January 1921. In common with other leaders of the National Research Council, Merriam firmly believed in the value of cooperative research programs, ones that spanned more than one discipline and involved the efforts of various scientific organizations.[42] Therefore, Wood's proposed program for seismological research in California, which called for the coordination of geodetic measurements, geologic mapping, and seismographic recording, particularly appealed to Merriam. In early March 1921, he obtained permission from the Executive Committee of the Carnegie Institution's Board of Trustees to appoint Wood as a research associate in seismology at an annual salary of $3,500, thereby giving Wood the means to begin putting his research program into action. The Carnegie Institution's trustees also gave Merriam permission to appoint an Advisory Committee in Seismology, with a budget of another $3,500, to stimulate "the initiation and development of research in the field of seismology" by securing the cooperation of all scientific agencies and organizations that might be interested in seismological research.[43] By the summer of 1921, the Advisory Committee, under the chairmanship of Wood's mentor, Day, had among its members: Andrew Lawson; Ralph Arnold; seismologist Harry Fielding Reid; Bailey Willis, who had replaced Branner as both head professor of geology at Stanford and as the chief driving force behind the Seismological Society of America; Robert Millikan, a prominent physicist who had expressed an interest in setting up a seismograph at the California Institute of Technology in Pasadena; and John Anderson, an astrophysicist and instrument designer at Hale's Mount Wilson Observatory, who was appointed primarily to balance the influence of the geologists on the committee.[44]

In the summer of 1921, Wood arrived in Pasadena, where he occupied an office provided by Hale at the Mount Wilson Observatory's headquarters. From this base, Wood concentrated over the next several years on developing a seismograph capable of recording short-period waves. He first tried to exploit the phenomenon of piezo-electricity, which he had encountered in his war-time research: when subjected to pressure variations, such as those produced by a passing seismic wave, certain crystals will emit a small, transient electrical current. Wood also pursued an idea for a more traditional seismograph, in which the vibrations of a suspended pendulum would be converted into electrical currents and recorded by a sensitive galvanometer. But he failed to make much headway using either of these approaches.[45]

In the fall of 1922, he received word from both Day and Merriam that continued support from the Carnegie Institution for his program depended on definite progress in developing a local seismograph. In desperation, Wood, collaborating with John Anderson, turned to yet another principle for recording earth vibrations, the torsion method. In a torsion seismograph, a weight is not suspended at one end and left to swing as a pendulum but rather is strung between two or more wires. As a vibration moves the weight relative to the earth, the supporting wires are twisted, and the resulting torsion produces a restoring force that moves the weight back to its original position. In the summer and fall of 1922, after working through the fundamental equations governing the theory of seismography, Wood and Anderson concluded that using torsion as a restoring force rather than the gravity method employed in a standard pendulum seismograph would provide a more accurate record of earth vibrations, particularly those of short period. In November of that year, Anderson built several crude torsion seismographs in his laboratory; these models demonstrated that the torsion principle indeed showed great promise. With a proper torsion seismograph built by a pair of local instrument builders, Wood in the spring of 1923 recorded a number of weak local earthquakes, thereby showing that this instrument was in fact suitable for routinely registering local seismicity. Records from the torsion instrument soon evoked considerable admiration from eastern seismologists as well as from Day and Merriam.[46]

In September 1923, the torsion seismograph having proved its value, Wood asked the Carnegie Institution to go ahead with setting up a seismographic network for recording local earthquakes. At this point, Wood envisioned a network of five stations: a central station in Pasadena and outlying stations on nearby Mount Wilson, at Riverside 50 miles to the east, on Santa Catalina Island 50 miles to the south, and at Fallbrook in San Diego County 70 miles to the southeast. Both Day and Merriam, impressed with Wood's progress, concurred in this recommendation; and at the end of the year the Carnegie Institution's trustees approved an increase in the seismology program's annual budget, from $8,000 to $20,000, enough to cover the cost of equipment and maintenance in the new seismograph stations, with buildings and grounds to be provided by cooperating institutions.[47]

In early 1924, Wood turned to the politically connected Arnold to generate the necessary local support for establishing the seismograph stations. Soon after New Year's Day, both Wood and Arnold addressed the Riverside Rotary Club on the ways in which the recording of local earthquakes at Riverside and elsewhere might help to reduce seismic hazards. As a result of these talks and subsequent prodding by Arnold, the mayor of Riverside agreed to provide funds for building a seismograph station there. Arnold's other attempts, however, brought few results. He several times addressed groups of San Diego County businessmen but failed to arouse any enthusiasm for funding a station at Fallbrook. He was also unable to persuade fellow Republican activist William Wrigley, Jr., the chewing gum magnate who owned Santa Catalina Island, to provide for a station on his property. Finally, Arnold made little headway in finding a suitable site for the central station in Pasadena.[48]

Wood and Day relied on other contacts to provide sites for their recording instruments. T. W. Vaughan, a geologist and oceanographer who earlier had shown some interest in recording earthquakes while working in the Caribbean and who also cooperated with the Carnegie Institution in other research endeavors, proved willing to house a seismograph at the Scripps Oceanographic Institution in La Jolla, just outside San Diego. Another offer for cooperation came through the efforts of Henry S. Pritchett, an influential Carnegie Institution trustee and friend of Merriam who often summered in the wealthy resort community of Santa Barbara, 90 miles northwest of Los Angeles. After a strong earthquake in June of 1925 caused extensive damage and killed 12 people in Santa Barbara, Pritchett interested a local landowner in providing a site for a seismograph station. This site turned out to be unsuitable, but Wood, after consulting with Pritchett and other concerned Santa Barbara residents, accepted an alternative site offered by the Santa Barbara Museum of Natural History.[49]

For the central station in Pasadena, the Carnegie Institution turned to the California Institute of Technology. In seeking Caltech's cooperation,

Day initially approached Hale, still the director of the Carnegie Institution's Mount Wilson Observatory and a principal organizer in the drive to bring Caltech to the forefront of science. In May 1924, Day asked Hale whether he might be able to help find a suitable site in Pasadena for the central station. In reply, Hale proposed that the Carnegie Institution's seismological work be made an integral part of a new department of geology and geophysics at Caltech. Wood expressed some concern over losing control of his program to Caltech, but after Hale clarified that he envisioned a cooperative agreeement under which the Carnegie Institution would retain control of the seismological work carried out in conjunction with the Caltech department, Day and Merriam gave their approval. Through Hale's influence a suitable site for the central station was acquired in western Pasadena, away from Caltech's main campus; Hale also got the head of Caltech's board of trustees to put up $75,000 for a building to house the station.[50]

Gradually, in the course of four years of hard work, the seismographic network was constructed. Wood was able to occupy the new central station, soon called the Seismological Laboratory, in December 1926. New seismographs were installed at the outlying stations in Riverside in October 1926, Santa Barbara in April 1927, La Jolla in May 1927, and Mount Wilson in December 1927. By the end of 1927 Wood and the Carnegie Institution had succeeded in putting together a system for the routine recording of local earthquakes in southern California.[51]

The Carnegie Institution also played a critical role in enlisting geodetics in the study of seismicity in California. In his original proposal, Harry Wood had called for using geodetic measurements and geological fieldwork to help locate active fault zones. This call was taken up by his former boss Andrew Lawson, who used the Carnegie Institution's political resources to obtain new geodetic measurements that might reveal distortions of the earth's surface near faults in California.

Lawson's interest in geodetic measurements in California stemmed from his work as chairman of the State Earthquake Investigation Commission. In its final report on the 1906 San Francisco earthquake, the commission had included a theoretical explanation of earthquakes, the so-called elastic rebound theory. According to this theory, earthquakes were due to the sudden release of strain that had been gradually built up by the constant creeping of the earth's surface near a fault. In his contribution to the commission's final report, Harry Fielding Reid had argued that there had indeed been a gradual distortion of the earth's surface near the San Andreas Fault during the late nineteenth century, just as the elastic rebound theory called for. Reid made this argument by comparing measurements taken by the U.S. Coast and Geodetic Survey in California at different times between 1851 and 1907. These surveys showed that points west of the San Andreas Fault had moved northward as much as 11 feet relative to points east of the fault in the 50 years prior to the 1906 earthquake.[52] According to the elastic rebound theory, this gradual creeping motion should have continued even after 1906. Lawson, who sometimes claimed credit for having originally suggested the elastic rebound theory, sought to demonstrate that this creep was indeed still taking place in the early 1920s.[53]

In September 1920, Lawson turned to data from the International Latitude Station at Ukiah (north of San Francisco), which had been set up several decades earlier to determine repeatedly the latitude of a fixed point using astronomical methods, in order to detect subtle changes in latitude due to a wobble in the earth's rotation around the pole. Lawson concluded that the Ukiah latitude measurements showed evidence of a northward shift in latitude due not only to the earth's wobble but also to a creep of several feet along the nearby San Andreas Fault. Over the next several weeks, Lawson obtained from the Lick Astronomical Observatory on Mount Hamilton, southeast of San Francisco, data that seemed to confirm a progressive shift in latitude near the San Andreas Fault. Lawson quickly publicized these data and his interpretation in a talk at the annual meeting of the Geological Society of America in December 1920 and in a paper published shortly thereafter.[54]

The astronomical determinations could fix latitude only to an accuracy of about ten feet. Thus, Lawson next turned to a more accurate technique for determining surface distortions—the triangulation method routinely used by the U.S. Coast and Geodetic Survey. In this geodetic method, the distance between two fixed points, usually located on mountain tops, is first very carefully measured, thereby producing a base line. Then the angles between this base line and the lines connecting the two points with a third point are painstakingly measured. Using simple geometrical considerations, the distance of the third point from the first two points can then be deduced. The lines connecting the two initial points to the third point are then used as base lines for establishing the distance to yet more points. By repeating this process over and over, a triangulation network covering a wide area and fixing the exact location of prominent points within that area can be produced. The Coast and Geodetic Survey had first established a triangulation network in California in the 1850s and 1860s. Some of these measurements were repeated in the 1880s and 1890s and once again in 1906 and 1907, after the earthquake.[55]

Soon after publishing his paper on the Ukiah and Lick Observatory latitude observations in early 1921, Lawson began to lobby for a return of the Coast and Geodetic Survey to California. In this endeavor he received the strong support of his former assistant Wood. While still in Washington, in the spring of 1921, Wood brought Lawson's article to the attention of William Bowie, chief of the Coast and Geodetic Survey's Division of Geodesy, with whom Wood had struck up a close friendship during the war. Bowie soon agreed that the agency should indeed retriangulate California, in order to check Lawson's hypothesis of continuing surface dislocation. Together with Wood's mentor, Arthur Day, chairman of the Carnegie Institution's Advisory Committee in Seismology, Bowie then obtained the enthusiastic support of the director of the Coast and Geodetic Survey for this project. Lawson turned to convincing Secretary of Commerce Herbert Hoover, in whose department the Coast and Geodetic Survey was located, to endorse the project. In August 1921, Hoover, a former student of Branner and a member of the Seismological Society since 1911, agreed to provide $15,000 in the survey's budget for 1922–23 to initiate retriangulation, with the understanding that this level of funding would be continued until the job had been completed. In March 1922, after considerable lobbying by Lawson and fellow members of the Carnegie Institution's Advisory Committee in Seismology, Congress approved this budget item.[56]

With funding for the retriangulation in California thus assured, the Coast and Geodetic Survey turned to doing the actual work of triangulation in the state. During the first season of work, in the summer of 1922, the retriangulation was carried out from the Lake Tahoe region near the California-Nevada border (which presumably was outside the seismic belt affected by surface deformation) to the San Francisco Bay area. In the following year, the triangulation network was extended south along the coast to Santa Barbara County, about 100 miles northwest of Los Angeles. A few observations were also made farther south in the state, but not connected to the main retriangulation. At the end of 1923, although the resurveying in California was still incomplete, the Coast and Geodetic Survey began to circulate preliminary results from the first two seasons of fieldwork. They were distributed among seismologists in October 1923 and published early the next year. These preliminary results were spectacular. While the measured movement of the earth's surface near Mount Hamilton was only four feet, not much larger than the observational error of geodetic measurements, the retriangulation revealed evidence of much larger movements to the south. A few points west of the San Andreas Fault in Santa Barbara County were found to have moved as much as 24 feet northward since last being surveyed in the 1880s.[57] Since no large earthquake had occurred in the Santa Barbara area during that interval, the measured surface deformation meant that an immense amount of strain had built up along the San Andreas Fault in that region and was ready to be released in a catastrophic quake.

Another California scientist who took advantage of the Carnegie Institution's entry into seismology was Bailey Willis, who had replaced the retiring Branner as head professor of geology at Stanford in 1915. At Stanford, Willis enthusiastically followed Branner's program for applied research and public education on seismic hazards. In May 1921, Willis was elected president of the Seismological Society of America.[58] By actively involving himself in the society's scientific and educational efforts, he became the true successor of Branner as driving force of the society. In 1921, Willis also was named a member of the Carnegie Institution's Advisory Committee in Seismology.

Soon after becoming president of the Seismological Society, he conceived the idea of issuing a special earthquake engineering issue of the society's *Bulletin*. This issue would discuss engineering and architectural provisions against earthquakes and would have "Safety First—Build for Security" as its motto. Willis also planned to compile and publish a map of all known faults in California as part of this special issue. The special issue did not materialize, but Willis did induce Sumner Hunt, a leading southern California architect, to contribute to the *Bulletin* a short article entitled "Building for Earthquake Resistance," which emphasized the need for solidity and honest workmanship in construction.[59] With help from the Carnegie Institution, Willis was able to pursue his plans for a fault map of California. It so happened that Wood, in his program for applied seismological research in southern California, had included a plan for compiling information on faults known or suspected to be active. Consequently, the production of a fault map for California became a joint venture involving both the Carnegie Institution and the Seismological Society.

Once Wood had arrived in Pasadena in early June 1921, he had busily started collecting data on local faults. In this work, he received considerable assistance from Los Angeles–based Ralph Arnold. Drawing on his extensive business connections, Arnold put Wood in contact with dozens of southern California petroleum geologists, who generously shared the data on local faults which they had obtained while exploring the oil-rich Los Angeles basin and neighboring areas of southern California. Arnold also released a press notice about Wood's seismological research program and arranged for Wood to address several gatherings of local scientists, engineers, and civic leaders, thereby advertising Wood's search for fault data. Wood transcribed the information gathered through these channels onto topographic maps supplied by the U.S. Geological Survey. He also spent a lot of time in the field inspecting faults for signs of recent activity and mapping faults in areas outside the reach of the Los Angeles oil fraternity. By mid-1922, Wood had nearly completed his compilation of data about southern California faults.[60] Meanwhile, Willis had added data on faults in the bay area which had been accumulated over the years by members of the Seismological Society, and he had managed to obtain some data on faults in central California from members of the U.S. Geological Survey. In the summer of 1922, Willis's son Robin, then a graduate student at Stanford, did a rapid geological survey of central California faults, for which he was paid $470.69 by the Seismological Society. As a result of this cooperative work, compilation of data for the statewide fault map had been completed to Willis's satisfaction by September 1922.[61]

Putting out a fault map involved much more than just gathering data, however. Willis also had to raise sufficient money to cover the cost of printing the location of the faults on a base map and distributing the resulting fault map to the public. This was particularly important because of the chronic shortage of funds faced by the Seismological Society. In order to remedy this situation, Willis sought to substantially increase the society's membership. Using his considerable charm, he succeeded in raising membership from 307 at the end of 1920 to 538 four years later. Among the new members were a number of California's leading businessmen, including railroad executive Paul Shoup, utilities magnate C. O. G. Miller, and banker William H. Crocker, who has been called "perhaps the most important single figure in [San Francisco's] business elite during the early twentieth century." In the fall of 1921, Willis also embarked on an ambitious campaign to raise $10,000 for the society, half of it in northern California and the other half in the southern part of the state. Five directors of the society issued a letter soliciting funds for publishing the fault map and stressing that "such a map will be of great service to engineers and capitalists engaged in the development of public and corporate works throughout the State" because it showed the faults that might pose a threat to such structures. Over the next year, Willis personally solicited a number of wealthy San Francisco businessmen and managed to obtain nearly $1200 for the Seismological Society.[62] In the south, Arnold also tried to convince local industrialists to support the society's work by arguing that knowledge of faults was essential for the companies then building the powerlines, water conduits, and other infrastructure on which Los Angeles would have to depend for its growth. But because he became preoccupied with other campaigns, Arnold did not follow up on these solicitations, and as a result no monetary contributions were received from southern California businessmen. The Carnegie Institution eventually had to come to the Seismological Society's rescue, by helping Willis to obtain $5,000 from the Carnegie Corporation of New York, a separate philanthropic foundation. That money was used to keep the society's *Bulletin* afloat.[63]

With compilation of data completed and at least some money for the work raised, Willis in September 1922 was ready to press for publication of the fault map. A manuscript version of the map was displayed at the annual meeting of the American Association for the Advancement of Science, in December 1922, and the printed version was finally issued in the fall of 1923. The map was distributed to all members of the Seismological Society as part of the society's *Bulletin*. In an accompanying article, Willis drew attention to some of the more prominent fault lines displayed on the map. He insisted that earthquakes emanating from these faults were "as certain as thunderstorms in New York in early summer," and he expressed the hope that greater public knowledge of the faults' locations would finally induce Californians to take steps to guard against earthquake damage.[64] Willis further promoted wide distribution of the fault map by donating a copy to his local public library, by prominently featuring it in numerous talks he gave to local civic organizations, and by announcing its availability in bay area newspapers. Ads for the fault map were also placed in *Science* and in the *Bulletin of the American Association of Petroleum Geologists*. By the spring of 1925, about a hundred copies of the fault map had been sold outside the Seismological Society, and more copies had been given away, in hopes of soliciting new members for the society.[65]

By the mid-1920s, then, much scientific progress had been made: a substantial infrastructure had been established for studying earthquakes and seismic hazards in California; the first fault map for the state had been published; and the U.S. Coast and Geodetic Survey had repeated its triangulation of California, providing evidence of a continuing buildup of seismic strain in the state. All of these scientific activities were made possible by the confluence of many different resources: the numerous friends and former students of John C. Branner, along with other California professionals who were willing to contribute to the Seismological Society's work; the deep pockets and political influence of the Carnegie Institution, which sat at the center of the emerging scientific national establishment; and the manpower provided by the federal government through the Weather Bureau's observers and the Coast and Geodetic Survey's surveyors. Branner, Wood, and Willis all combined these resources in pursuit of a single well-defined mission: the collection of scientific information about the seismic hazards threatening California. Through their persistence, these scientific entrepreneurs succeeded in setting up an infrastructure capable of achieving this mission.

Their efforts did not, however, noticeably improve seismic safety in California. Branner, Wood, and Lawson were preoccupied with gathering scientific information, and they naïvely assumed that disseminating this information among the Seismological Society's members would be enough to impress Californians of the need for earthquake preparedness. Willis alone among them recognized that scientific information was not enough, that engineers, architects, businessmen, and the general public had to be convinced, through a concerted public relations campaign, to act on the information. Willis would embark on just such a campaign in the mid-1920s.

CHAPTER 3

Bailey Willis and the Promotion of Earthquake Safety in the Mid-1920s

When Bailey Willis assumed the presidency of the Seismological Society in 1921, he was a long-experienced scientist and a veteran of the Progressive reform movement. Born in 1857, he had trained as a mining and civil engineer at the Columbia School of Mines in the late 1870s. In 1884, he had joined the U.S. Geological Survey, where he remained (with some interruptions) for three decades before moving to Stanford University to replace John C. Branner as head professor of geology. While at the survey he embraced the campaign for the conservation and rational use of natural resources that was so popular among engineers and scientists of the Progressive Era. He urged government involvement in preventing soil erosion and protecting watersheds in the Appalachian region, and he personally supported Gifford Pinchot in his efforts to preserve the national forest system as a showcase for responsible logging practices. Moreover, Willis contributed to the efforts to establish Mount Rainier and Glacier national parks in the Pacific Northwest. While on leave from the U.S. Geological Survey to work for the Argentine government in the early 1910s, he surveyed the southern part of Argentina and made numerous suggestions for the efficient and rational development of its natural resources. Finally, as professor emeritus at Stanford after 1922, he continued to advise municipal groups in California on the proper development of their water resources.[1] Given his involvement in these causes, it is no surprise that Willis took on the campaign for earthquake preparedness initiated by fellow Progressives Branner and Gilbert.

In pushing seismic safety, Willis went far beyond his fellow seismologists, who had been content with giving an occasional speech at an engineering convention. In the mid-1920s, Willis launched a concerted public-relations campaign to draw attention to the need for earthquake-resistant construction. He gave dozens of speeches and newspaper interviews, and he cultivated allies in the state's engineering, architectural, and business communities. One of his goals was to have building codes enacted that would make earthquake-resistant construction mandatory; in this manner, he sought the aid of state power for the cause of seismic safety. For the most part, though, Willis hoped Californians would embrace earthquake-resistant construction voluntarily, because it was in their best interest, and thus he employed a sermonizing tone that would appeal to the public's moral sensibilities.

Willis's campaign achieved a fair degree of success after a moderately destructive earthquake shook the southern California resort city of Santa Barbara in 1925. California boosters, as they had done after previous earthquakes, once again sought to downplay the seismic hazards in their state. But Willis, using the damage from the earthquake to great effect, convinced the city councils of Palo Alto and Santa Barbara to incorporate a requirement for earthquake-resistant construction in their building codes. In the late 1920s, though, Willis overreached when, dissatisfied with the slow pace with which earthquake consciousness was spreading in California, he resorted to predicting that a catastrophic earthquake would strike southern California within the next ten years. A geologist in the employ of southern California boosters soon found a crucial weakness in the scientific arguments underpinning this rash prediction, and the boosters used this discovery to discredit Willis and support their own contention that southern California was safe from earthquakes. Despite this failure, Willis permanently transformed the campaign for seismic safety, by adding organized lobbying and public-relations efforts to the purely scientific work.

Launching a Campaign for Seismic Safety

Willis most clearly articulated the Progressive philosophy underlying his public-relations efforts in a five-part series of articles on "Earthquake Risk in California," published beginning in 1924 in the Seismological Society's *Bulletin*. In these articles, he emphasized the need for understanding earthquake dangers and taking reasoned action against them. As he put it in his opening paragraph, "there are a good many ways of regarding the danger from earthquakes. It may be ignored or ridiculed, prayed against or fled from, or examined and guarded against. It is, however, useless to ignore a blow which is sure to fall and foolish to ridicule that which has the power to destroy us. Prayer, though long and devoutly practiced, has not averted shocks, even from sacred edifices and their congregations; and he who flees ignorantly is quite as likely to run into danger as out of it. The only safeguard against the forces of nature, whether they be lightning strokes or earth tremors, is *understanding,* by virtue of which we may be forearmed because forewarned."[2] Willis insisted that, in the light of such understanding, Californians should provide themselves with the "assurance that we shall not suffer material losses in property or lives through the falling of buildings or the setting of fires by earthquake, because we shall build structures that will stand in spite of trembling foundations, and we shall adequately guard against conflagration."[3] Willis sought to make Californians realize that they could actually protect themselves, if only they would take proper action. He urged "a profound respect for [the] power" of earthquakes, but he coupled this warning with the reassurance that "it is well within the capacity of engineering so to construct buildings . . . that they shall be earthquake proof."[4]

After outlining his philosophy, Willis set out in detail his understanding of the earthquake risks facing California and the means that could be taken to reduce them. He began by describing the major fault systems of California, thereby summarizing the information graphically recorded in the just-released fault map.[5] Based on the historical record of seismicity, as set down in the earthquake catalogs for California, he then asserted that, "taking the State as a whole, a severe earthquake must be expected every five years somewhere," while "great disasters have occurred in the coast ranges [of California] on the average once in thirty years."[6] In his third article, Willis described in detail both the faults and the historical record of earthquakes for several subsections of the state. He suggested that the 1906 San Francisco earthquake had afforded an "unusual degree of relief" of strain in the bay area, so that the next great earthquake there might be more than three decades away. For the Los Angeles area, though, Willis was far less sanguine. As he noted, "the earthquake record [in this area] is characterized by the long period of quiescence typical of all Southern California since 1857." Given the large number of faults in the area and the high degree of seismicity before the great earthquake of 1857, Willis considered this quiescence to be ominous: "A great shock may come soon, or within a decade, or not till after more than a decade. But it will come."[7]

Willis next discussed the factors affecting the intensity with which an earthquake was felt at a given point. As many commentators had done before him, he pointed out that, for any particular earthquake, ground-shaking tended to be much more severe on loose alluvial soil or mud than

on bedrock. But Willis also sought to quantify the horizontal acceleration produced by earthquake waves in order to specify the horizontal force
which engineers and architects would have to take into account in designing their structures. Based on admittedly qualitative data gathered by the
State Earthquake Investigation Commission and by Japanese investigators
of the 1923 Tokyo earthquake, Willis estimated that lateral, or horizontal,
acceleration in a strong earthquake might range between a tenth and a third
of the vertical acceleration produced by gravity (abbreviated as 0.10g to
0.33g).[8]

In his last article, Willis discussed the requirements for earthquake-
resistant construction. He stressed the need for "good materials and good
workmanship." Beyond this, there was also a need for structural "unity,"
which "should be achieved by tying the parts of a building together, superstructure to foundation, wall to wall, floors and roof trusses to supporting
columns or walls. The ties should be so firm that the inertia of the structure, developed in resisting the impulse given by the shock, shall not tear
the structure apart."[9] Unity also required a homogeneity in structural materials, he insisted; if materials with different elasticities, such as wood and
brick, were used together in a structure, their different responses to vibration would tear the structure apart in an earthquake.[10] In this series on seismic risk, Willis combined the scientific data accumulated by seismologists
over the previous 15 years with a number of Progressive themes, foremost
among them the assurance that progress in human adaptation to the natural
environment would come through scientific understanding and guidance by
technical experts.

Willis's campaign of lectures and newspaper interviews ranged widely.
In October 1923, he addressed the Commonwealth Club in San Francisco
and the San Jose Chamber of Commerce and issued a press release urging
Californians to take measures against earthquakes. Over the next half-year,
Willis addressed another eight or so organizations, including the Los Angeles Chamber of Commerce and the Palo Alto Rotary Club as well as groups
of architects and contractors.[11] In finding channels through which to convey
his message, he showed considerable dedication. While vacationing in the
resort community of Carmel in March 1924, for example, he used a slight
earthquake there as an occasion to urge earthquake-resistant building practices in an article for the local newspaper.[12] Willis's many addresses, usually
delivered with considerable wit and charm, certainly had an effect on at
least part of their audience. A Los Angeles architect, for example, was inspired by one of Willis's speeches to apply for membership in the Seismo-

logical Society and to order a copy of the fault map and of anything the society might have published on earthquake-resistant construction.[13]

As part of his public education campaign, Willis worked to involve engineers and architects more directly in the effort to improve building practices. In the summer of 1924, Willis organized the Committee on Building for Safety Against Earthquakes of the Seismological Society. This committee counted among its members engineer Henry D. Dewell of San Francisco and architect Sumner Hunt of Los Angeles, as well as representatives from the San Francisco City Engineer's Office, the Los Angeles Committee of Safety and Fire Prevention, and the Board of Fire Underwriters of the Pacific. Willis urged this committee to investigate "current methods of building in California, including all materials and all classes of structures with a view to ascertaining to what degree and in what particulars various types of structures fall short of being reasonably safe against earthquakes and in what manner and at what cost the current practices may reasonably be modified in the direction of safety."[14] At several meetings of the committee, Willis received the help of Harry Wood in elucidating the nature of the earthquake hazard in California.[15]

Other members of the committee focused on earthquake-resistant construction, and in 1925 and 1926 they published several reports in the Seismological Society's *Bulletin*. Dewell produced a technical treatise addressed to his fellow engineers on how properly to design a structure against the lateral stresses produced by earthquakes or heavy wind. An engineer working for the insurance industry emphasized the need to design water delivery systems against earthquakes, so that sufficient water would be available after a shock to fight any possible fire. In his report, he concentrated on recommendations for the proper design of pipelines, shutoff valves, and water tanks. A Los Angeles insurance executive focused on guarding against panic in the aftermath of a severe earthquake. He recommended both general education about earthquakes and training in earthquake response procedures, as well as development of efficient crowd control measures by the police and fire departments and other official bodies. Finally, architect Hunt provided easily understandable advice on the correct construction of walls and floors to withstand earthquakes. He recommended the use of reinforced concrete; if bricks had to be used, the mortar should by all means contain cement rather than just lime, so as to produce better bonding. Hunt also suggested using a simple floor plan, ample bracing, and proper connections between the foundation and the superstructure.[16]

Thus, by 1925 Willis had expanded the Seismological Society's campaign for greater earthquake awareness in California well beyond the foundation laid by Branner. Willis had not only reiterated the reasons for believing that great earthquakes would strike California again in the future but had initiated a conversation with engineers and architects on the details of earthquake-resistant construction, thereby refining the Seismological Society's message on what Californians should do against earthquakes. Moreover, Willis had begun a drive to bring his pleas for seismic safety before the general public. Meanwhile, the engineers and architects he had recruited to the Committee on Building for Safety introduced a greater concern for earthquake-resistant construction in California engineering and architectural circles. As a result of this considerable organizational work, Willis and his colleagues were ready to respond promptly and energetically when a moderately severe earthquake struck southern California in 1925.

Responses to the Santa Barbara Earthquake of 1925

On June 29, 1925, an earthquake caused substantial damage in Santa Barbara, a resort city about 90 miles northwest of Los Angeles. Numerous brick and reinforced concrete structures in the central business district were destroyed; 12 people were killed and scores more injured. The casualty rate could have been much higher, but since the quake struck at 6:45 A.M., few people were in the central business district and hence in the path of crumbling buildings.[17] Certain segments of the California business community reacted in the same way they had after the San Francisco earthquake of 1906: they once again insisted that earthquakes were not a real menace in California and could be safely ignored. In 1925, they put forward this soothing interpretation of California's seismic hazards in an even more organized fashion than before. This time, however, Bailey Willis and his fellow seismologists were ready to challenge them.

The organized boosting of California, and of southern California in particular, was driven by those who stood to profit enormously from the area's rapid population growth. Since the late 1880s, the Los Angeles Chamber of Commerce had vigorously promoted southern California as an ideal place for new settlers. Attracted by the promise of a mild climate and of escape from the harsh toils of farm life, hundreds of thousands of settlers migrated to the Los Angeles area, primarily from the Midwest. From 1890 to 1920, the city of Los Angeles increased in population more than tenfold, from 50,000 to 577,000. During the 1920s, another 660,000 moved to the city, while the rest of Los Angeles County gained 610,000 new residents.

This enormous boom in population brought large profit to those who could subdivide land for new neighborhoods and housing. Land speculators had to have utility and road connections to their land already installed by the time new settlers arrived. Because the development of new subdivisions required a considerable outlay of capital, speculators had to make sure that the stream of new residents would continue to swell without interruption if large profits were to be realized. To sustain the population boom, promoters formed several new promotional organizations in the early 1920s, including the All-Year Club of Southern California. Together with the long-established Chamber of Commerce and the statewide California Development Association, these booster organizations intensified the public relations efforts on behalf of southern California, sending out tens of thousands of brochures, press releases, and photos celebrating the climate, scenery, easy living, and other wonders of the Los Angeles area.[18]

Almost immediately after the Santa Barbara earthquake, this finely honed public relations machinery geared up to contain the damage to southern California's image. The first group to act was Californians, Inc., a San Francisco-based "non-profit organization of California citizens and institutions interested in the sound development of the state." Within hours of the quake, the organization's publicity department "was in conference with executives and editors of the national press associations and as a result . . . the extent and effects of the quake were definitely localized to Santa Barbara." The organization also telegraphed "a score of railroad traffic managers and tourist travel bureau executives," advising them that damage had been very limited and southern California was still safe for travelers. This message was reinforced by a special bulletin mailed to tourist bureaus and large newspapers across the country. The publicity department even asked newsreel producers to refrain from showing pictures of devastated Santa Barbara; at least one newsreel company complied. As the manager of Californians, Inc., later bragged, "the fact that we had a permanent functioning publicity organization in the Santa Barbara crisis proved of inestimable value . . . in an emergency which cast a shadow of serious consequences to the business interests of California."[19]

Not to be outdone, southern California boosters also rapidly mobilized. A couple of days after the earthquake, the Los Angeles Chamber of Commerce sent out telegrams to its counterparts across the country asserting that "stories concerning Santa Barbara earthquake greatly exaggerated. . . . City returning to normal rapidly and work of reconstruction already under way. There need be no anxiety as to safety of relatives and friends in that city. Earthquake shock purely local in character and in a region approximately 125 [*sic*] miles north of Los Angeles, and effect confined closely to small area."[20] In addition, the Chamber of Commerce sent more than a thousand pictures via air mail to newspapers nationwide ostensibly showing that damage to Santa Barbara had been minimal. The president of the Los Angeles Clearing House Association, a group of bankers, also sent telegrams to about 20 eastern cities emphasizing that damage had been limited to Santa Barbara and that reconstruction did not require assistance from non-Californians. Meanwhile, George I. Cochran, a life insurance executive who headed the Southern California Forward Movement, issued a widely circulated press release asserting the relative safety of southern California. As Cochran claimed, "there are instances . . . where elsewhere in the United States more people have been killed in a single flood or tornado than have been killed in California by quakes since the white man first came to these shores." Moreover, the less than 1,000 earthquake deaths in California's recorded history paled in comparison to the 65,232 fatalities in "ordinary accidents" across the United States in the year 1922 alone. Cochran concluded by urging Californians to contribute $1 million to a "truthtelling campaign" by the Los Angeles Chamber of Commerce and the All-Year Club of Southern California. He exclaimed, "With adequate advertising, the evil effect in the mind of people outside of California created by the Santa Barbara quake can be set right."[21]

On the morning of July 1, 51 hours after the earthquake, representatives from Californians, Inc., the All-Year Club of Southern California, the California Development Association, the chambers of commerce of San Diego and San Francisco, and several other organizations met in the offices of the Los Angeles Chamber of Commerce to coordinate their response to the earthquake. These groups agreed to stress that the devastation was so small there was no need for charitable contributions from outside the state. This particular interpretation of the earthquake was promptly conveyed to local and national newspaper editors. The manager of the California Development Association also convinced Arthur Brisbane, the nationally syndicated columnist of the Hearst newspaper chain, to repeat the statement that the Santa Barbara earthquake was "a local isolated slip with no recurrence probable for many years. Reconstruction has already started and local situation will return to normal promptly." Brisbane added: "Go to California, which is the world's finest summer resort, and don't let earthquakes worry you."[22]

California's newspapers by and large echoed the boosters' assertions. The *Los Angeles Times* noted that the fatalities in the quake were "about equal to the toll of [those] killed by automobiles every week in Southern California." Meanwhile, the *Los Angeles Examiner* stressed the speed with which Santa Barbara was returning to normalcy. The *San Francisco Chronicle* stated that Maine and New York were as much subject to earthquakes as California and that no place on earth could claim to be safe from natural or man-made calamities. The *Chronicle* reinforced this point in an editorial cartoon that shows a frightened rube clinging to a church steeple in the Midwest as a tornado swishes by and devastates a farm; the caption incredulously notes: "And he's afraid of earthquakes!" In Pasadena, home of the California Institute of Technology, the *Star-News* urged local residents to broadcast to their friends back east that, aside from Santa Barbara, "no other city of Southern California has suffered a great earthquake disaster, at any time," to refute the notion "that violent earthquakes are of common occurrence here and that life and property here are constantly imperiled thereby." A day later, this newspaper, asserting that the Santa Barbara shock was "a closed incident," categorically told its readers, "Forget earthquakes!" Even Santa Barbara's newspapers participated in this campaign. The *Santa Barbara Morning Press* reprinted a statement by the executive secretary of the All-Year Club of Southern California asserting that "in scores of places more have been killed in a single blow by cyclone or flood than have been killed by natural violence in California from earthquakes and all other natural causes in its entire history" and thus that "California is one of the most secure, if not the most secure place to live in the wide world." The *Santa Barbara Daily News,* endorsing what it called a "gospel of cheerfulness," also insisted that tornadoes, heat, and cold waves elsewhere took more lives than California's earthquakes.[23]

This concerted effort had its intended effect. The general manager of the California Development Association noted with relief that "the passenger and tourist travel [to California] has apparently not been affected" by the earthquake.[24] Santa Barbara residents also remarked, albeit critically, on the boosters' success. Seven weeks after the earthquake, a member of the Santa Barbara City Council pointed out that damage in the city had amounted to about $15 million (about $150 million in 2001 dollars), but that "the campaign of belittlement was so effective that the public as a whole, even in California, does not realize the extent of our damage." As a result, local residents found it very difficult to convince outside charitable organizations of their need for funds with which to reconstruct the city. Even the California Development Association, which had pledged itself to raise funds for reconstruction within the state, barely met half of its already modest fundraising goal. As a few angry Santa Barbara residents charged, the association had been far more eager to protect the interests of real estate speculators by downplaying the earthquake than to help the devastated city in its time of need.[25]

In asserting that damage from the Santa Barbara earthquake was inconsequential, California's boosters were reacting in much the same way as they had to the devastation of San Francisco in 1906. In the latter case, the minimization of the earthquake hazard had remained essentially unchallenged, because that earthquake had taken California's scientists and engineers by surprise. It was not until 1909, three years after disaster had befallen San Francisco, that Gilbert published his statement warning Californians to guard against future earthquakes. In 1925, however, both the scientific data and the organizational framework for a campaign to urge greater earthquake preparedness were already in place, and Bailey Willis took full advantage of these resources to counter the boosters' efforts with a public relations campaign of his own.

The Seismological Society was, of course, an important channel for Willis's efforts. The December 1925 issue of the society's *Bulletin* was devoted entirely to a discussion of the Santa Barbara earthquake. Willis and engineer Henry Dewell gave a detailed and profusely illustrated description of damaged buildings and emphasized the need for good workmanship, deep foundations, and a high degree of rigidity and homogeneity in the structural design, to provide resistance against earthquake vibrations. Willis also reproduced the portion of the Seismological Society's fault map showing the faults underlying Santa Barbara and discussed the geological effects produced by the earthquake.[26]

Willis further distilled these pieces of information into an emotionally charged sermon on the need for earthquake preparedness, written for the lay reader. After evoking Santa Barbara in poetic terms as "the city of sunlight, basked by the sea between the Mesa and the mountain foot," Willis pointed out that the resort owed its idyllic surroundings to the repeated visit of "the Earthquake . . . the architect of the mountains," whose sustained activity had raised the peaks towering over Santa Barbara. After presenting in nontechnical terms the geological evidence for faults in the area, Willis chastised the builders of Santa Barbara for constructing the city "in ignorance or in disregard of the fact that earthquakes had occurred [there] and must inevitably occur again from time to time." Driven "by the domination of self-interest," Willis charged:

> the city grew carelessly, heedless of the elements which seemed to treat it so generously. Neither violent winds nor heavy rains, nor heat nor fearful cold, not even floods threatened it. So they built thoughtlessly. . . .

Here and there a stately pleasure house was built regardless of cost, perhaps mindful of fire because of the treasures it was to guard; but in ignorance or utter disregard of the certainty of earthquake. Its beauty, nay, even its excellence of design and construction, availed little where the materials were ill-chosen or the structure was not framed to withstand the shock.

The hotels presented everything that a guest could seek—except safety. As was natural in that environment, they were luxurious to a degree, but behind the pleasing effects of tinted walls, behind even the appearance of substantial massiveness there was the weakness of sham construction and scant design.

Willis emphasized that the destruction resulting from this thoughtlessness had been entirely unnecessary: "Here and there among the residences as among the business buildings good designing and good construction had combined to produce an exceptional result which stood out after the earthquake, demonstrating how foolish the past practice, how needless the disaster." In these words he preached his message of earthquake preparedness to the readers of the Seismological Society's *Bulletin*, including the 119 new members who had joined the society in 1925.[27]

Meanwhile, Willis used additional means to counteract the boosters' campaign. He was aided by the fact that he happened to have been visiting Santa Barbara on business on the day of the earthquake and had personally experienced the tremor. This coincidence, together with Willis's earlier pronouncements on the imminence of a strong earthquake in southern California, led a few newspaper writers to assert erroneously that he had predicted the quake.[28] Taking advantage of the expert status thus conferred on him, Willis issued a press release that was widely circulated by the Associated Press. In it, he described the specific design flaws that had caused several of the more spectacular building failures, and he drew as his lesson the need for solid foundations and structural frames that are well tied together. The publicity department of Californians, Inc., managed to insert an introductory paragraph into the version distributed by the Associated Press which claimed that Willis was pronouncing Santa Barbara safe from earthquakes for many years. Even with this intentional editorial distortion, though, Willis was able to place his call for better building practices before the readers of many California newspapers.[29]

He also pushed his campaign for greater earthquake preparedness by personally addressing numerous groups after the Santa Barbara earthquake. On July 2, three days after the shock, Willis, together with Arthur Day of the Carnegie Institution, addressed a gathering of 150 bankers, insurance underwriters, engineers, architects, and industrialists at the San Francisco Chamber of Commerce. As Day later reported, he and Willis "spoke quite plainly" in urging a revision of the San Francisco building code, to take earthquakes into account.[30] A month later, Willis once again described in great detail the bay area's earthquake faults, in an address to the influential Commonwealth Club of California in San Francisco. He assured his audience that he was not opposed to San Francisco's growth; like them, he wished to foster the "*city-that-is-to-be,* the city that will spread around all the lands about the bay, covering the low land, and ascending the high land as far as water can be piped." But he argued that the future stability of this metropolis required a strict control on "that frail humanity which seeks to squeeze the last dollar out of its rents by saving on the investment." In earthquake matters, so he asserted, such control on greedy practices could come only through rigid building codes backed by well-trained building inspectors. Willis's plea was seconded by Dewell, who showed numerous slides of earthquake-damaged structures and discussed in some detail the requirements of earthquake-resistant construction.[31] In the months after the Santa Barbara shock, Willis also spoke to the Down Town Association of San Francisco, the Public Spirit Club of Oakland, the Berkeley Kiwanis Club, and various groups of insurance executives.[32]

Willis also proselytized among California's engineers and architects. At a meeting of the San Francisco Section of the American Society of Civil Engineers on July 28, 1925, attended by an exceptionally large audience of 400, Willis recounted his impressions of the Santa Barbara earthquake, and his ally Dewell once again urged a tightening of the local building codes to take earthquakes into account. Several of the other prominent San Francisco engineers who addressed this meeting echoed their sentiments. J. D. Galloway and L. H. Nishkian, for example, provided detailed suggestions on how to make structural frames more earthquake-resistant. In addition, Dewell reiterated the need for special design provisions against earthquakes in several articles that drew favorable comment in both the local and the national engineering press.[33]

Meanwhile, Willis gained the ear of architects through Sumner Hunt, the Los Angeles architect who, like Dewell, was a member of the Seismological Society's Committee on Building for Safety. In the month after the Santa Barbara shock, Hunt put together a special earthquake issue of the *Bulletin* of the Allied Architects Association of Los Angeles, and he invited both Willis and Harry Wood to contribute to it.[34] Willis, in an article placed at the very beginning, discussed both the natural geologic faults in the Santa Barbara area as well as the "unnatural faults" in building construction that had contributed to the large amount of damage. He concluded by urging, "It will be well for architects to respect His Majesty, the Earthquake, in making their designs. He will not spare them if they do not."[35] Wood, in turn, used the historical record of earthquakes in southern California and the results of the U.S. Coast and Geodetic Survey, which had discovered an unrelieved strain of 24 feet along the southern San Andreas Fault, to emphasize that strong shocks would recur in the area and needed to be provided against. In the remainder of the special issue, 35 architects, engineers, general contractors, and others discussed their observations and impressions of the earthquake damage. While a few continued to argue that competent workmanship was sufficient to render ordinary structures earthquake-proof, many others agreed that special precautions in design had to be taken for earthquake resistance. Architect Roy C. Mitchell, for example, made specific recommendations for designing staircases, elevator shafts, and reinforced concrete columns against the stresses produced by earthquake vibrations.[36] The various suggestions for earthquake-resistant construction proffered in this special issue were widely reprinted in architectural trade journals and even in the general press.[37]

Willis's considerable promotional efforts produced real results, in the form of stricter building codes requiring earthquake-resistant construction in the cities of Palo Alto and Santa Barbara. Willis had a direct hand in introducing an earthquake clause into the building code of Palo Alto, a college town whose affairs were dominated by professors from nearby Stanford University. Even before the Santa Barbara earthquake, Willis, with the help of Stanford engineering professors C. D. Marx and C. B. Wing, had drawn up an amendment to the Palo Alto building code that required structures to be designed to withstand a lateral acceleration amounting to 0.10–0.20g, depending on the foundation material.[38] Soon after the Santa Barbara earthquake, Willis submitted his proposed amendment to the Safety and Fire Prevention Committee of the Palo Alto Chamber of Commerce, of which he was chairman. Under his guidance, the committee as well as the chamber's Board of Directors endorsed the amendment. Willis also sought support among contractors, architects, and businessmen in Palo Alto, and he gained the enthusiastic backing of the *Palo Alto Times*. In early August 1925, the amendment was submitted to the Palo Alto City Council, of which Wing, one of its coauthors, was an influential member. The council referred it for consideration to the city engineer and the city Board of Public Works, which was headed by Marx, the other coauthor of the amendment.[39] Here, however, a snag developed. As Willis later recalled, it was discovered that the amendment "was not written in a language which the contractors could understand." In the spring of 1926, a new draft of the amendment was submitted that expressed the lateral forces buildings should be able to resist in terms of units commonly used in the construction industry. This revised amendment then sailed through the City Council with little comment, receiving final passage a day after the twentieth anniversary of the great San Francisco earthquake.[40]

Because of the delay necessary for revising the code's language, Palo Alto was not the first California municipality to require earthquake-resistant construction. That distinction goes to Santa Barbara, which beat Palo Alto to that achievement by four months. Several months after the earthquake had devastated large portions of Santa Barbara's business district, a draft of an entirely new building code was submitted for the consideration of its residents. Besides the usual provisions regarding the vertical strength of structures made of wood, brick, reinforced concrete, or steel, the proposed new code also included a clause requiring buildings to be designed against horizontal forces produced by either earthquake or wind. Furthermore, the new code required that structures be designed by accredited architects and thoroughly inspected during construction, to prevent shoddy workmanship or the use of defective materials.[41] The proposed new building code was part of a much larger program advanced by what historian Kevin Starr has called "a group of affluent, genteel, preservation-minded citizens." In the earthquake's aftermath, these upper-crust reformers sought to implement "a vision . . . of Santa Barbara as a Spanish dream city, beyond the gritty realities of American life." As part of their program, they pushed not only a building code that would outlaw shoddy construction but also a zoning ordinance that would restrict the kinds of buildings that could be erected in the city; furthermore, they supported the inauguration of an architectural board of review that would ensure that all new construction adhered to a Spanish-inspired style of architecture. This agenda of course evoked considerable opposition from both property owners and building contractors, who saw in the proposed new ordinances a significant infringement on their inalienable right to build as they pleased.[42] But the reformers, possessing greater wealth, political acumen, and public relations skill, prevailed. The new building code, requiring earthquake-resistant design, was enacted by the Santa Barbara City Council on December 17, 1925.[43]

From the surviving records it cannot be ascertained whether Willis played any direct role in inserting the earthquake clause into the new Santa Barbara building code. It is quite clear, however, that he vigorously supported the attempts of Santa Barbara's ruling elite to remake the city as a Spanish dream place. For example, he welcomed articles for the Seismological Society's *Bulletin* from Bernhard Hoffmann and Vern Hedden, both of whom were leading advocates of architectural renewal in Santa Barbara. Willis also gave his endorsement to the program of architectural beautification at a well-attended talk given on October 6, 1925, at the Santa Barbara Recreation Center. In return, the Hispanicizers gave their support to Willis's campaign for greater earthquake awareness. For example, Herbert Nunn, the Santa Barbara city manager, who was part of the reform group, contributed an account of his earthquake experiences to the Seismological Society's *Bulletin* and warned the League of California Municipalities at its annual meeting in the fall of 1925, "we must prepare and build and organize for future earthquakes in the state of California." The *Santa Barbara Morning Press,* which supported the reformist efforts, also made sure to discuss at length Willis's views on earthquake preparedness. Because of the close ties between Willis and the elite reform group, he was rapidly informed of the passage of the new building code, and just as rapidly he sent a congratulatory telegram to the Santa Barbara reformers.[44]

By drawing on his political connections, Willis was able to have earthquake preparedness included in the building codes of Palo Alto and Santa Barbara. Through his vigorous campaigning, he also made some progress toward inducing greater earthquake awareness in the much larger city of San Francisco. He succeeded, for example, in altering the editorial stance of the *San Francisco Chronicle.* Immediately after the Santa Barbara earthquake, this influential newspaper had repeated the claim that earthquakes were not really of much concern in the state and could thus be safely ignored.[45] After Willis addressed a large gathering at the San Francisco Chamber of Commerce on July 2, though, the *Chronicle* published several editorials explicitly endorsing Willis's call for a stricter building code that would require earthquake-resistant construction in the city.[46] After being addressed by Willis and Dewell, the San Francisco Section of the American Society of Civil Engineers also decided to back the move for greater earthquake preparedness. In August 1925, it appointed a committee of five, including Dewell, to investigate the state of building construction in San Francisco. While acknowledging that revision of the building code would be beneficial, this committee decided to concentrate first on upgrading the quality of building inspection in San Francisco. Together with committees from local groups of architects and general contractors, the engineers petitioned the San Francisco Board of Supervisors to step up inspection on construction jobs, so that faults in design, materials, and workmanship could be prevented. As a result of this petition, the staff of the city building department was increased, albeit only slightly.[47]

Immediately after the Santa Barbara earthquake, Willis also made sure to generate support for two new institutions that would in turn produce more data for his campaign: a local seismograph network in the San Francisco Bay area and an earthquake engineering laboratory at Stanford University. As early as January 1925, Willis advocated establishing a network of several stations equipped with Wood-Anderson seismographs in the bay area, along the lines of the Carnegie Institution's local network in southern California. Through such a network, Willis argued, weak local earthquakes could be pinpointed and used to locate active faults; this knowledge, in turn, could be used to raise earthquake awareness in the area.[48] After the Santa Barbara earthquake, Willis vigorously pushed his idea in talks to such groups as the San Francisco Chamber of Commerce, the Down Town Association of San Francisco, the Public Spirit Club of Oakland, and other businessmen's groups in the bay area. In October 1925, he received the backing of several influential members of the San Francisco Chamber of Commerce who promised to aid Willis in raising $22,000 for a four-station seismograph network. With their support, Willis was able to collect most of the required money over the next year, with local insurance companies, public utility companies, and banks each contributing about $4,500.[49] Bureaucratic difficulties significantly delayed the installation of instruments at the University of California at Berkeley and at the Golden Gate Park in San Francisco, but at Stanford University and at the Lick Astronomical Observatory on Mount Hamilton, seismographs went into operation in November 1927 and January 1928, respectively.[50] In succeeding years, this network, like the one in southern California, would contribute greatly toward delineating the state's active faults.

Willis also initiated an earthquake engineering laboratory at Stanford. As a result of his contacts with engineers and architects through the Seismological Society's Committee on Building for Safety, Willis had concluded that much more needed to be known about the precise behavior of wood, brick, reinforced concrete, and other building materials during an earthquake. In a series of memoranda to the Dean of Engineering at Stanford in late 1925 and early 1926, he suggested that the university's Department of Mechanical Engineering undertake experiments on realistic models of structures made of the various building materials, in order "to determine what are the limits of strength of the structures." The head professor of mechanical engineering objected to having this additional work foisted on his department, but the dean appeared willing at least to consider what might be undertaken in terms of "research in earthquake matters."[51]

Willis realized that he would have to raise funds for an earthquake engineering laboratory before he could enlist assistance from the engineering faculty in this work. In April 1926 he wrote a fundraising letter seeking $6,000 from architects, engineers, bankers, and manufacturers of building materials to "construct a vibrating machine which shall impart such impacts and vibrations to a movable platform as would be suffered by the foundations of a building in an earthquake." Willis explained, "The reaction of a structure fastened to the platform [would] be observed and recorded from initial minor stresses up to failure"; by repeating such experiments on different kinds of structures, it would be possible to ascertain "what the modern constructions will stand and what bracing, tying, or other improvement in design will afford greater security at reasonable cost." The inspiration for this experimental approach came from earthquake-plagued Japan, where seismologists had already conducted such research; a couple of engineers at Caltech were also experimenting with "vibrating models," although they did not pursue this line of research very far. In August 1926, Willis secured the entire $6,000 from the president of a cement manufacturing company who had turned to Willis for advice on seismic hazards.[52] With this money, a large shaking table was built at Stanford and put under the charge of Lydik S. Jacobsen, a young associate professor of mechanical engineering. Initial tests in November 1927 showed that this machine could indeed be used to test how models of buildings responded to repeated horizontal vibration. In the years to come, the shaking table at Stanford became the centerpiece of a flourishing earthquake engineering research program.[53]

In his efforts to promote seismic safety and combat boosterism after the Santa Barbara quake, Willis was able to draw on resources that had not been available after previous earthquakes: a well-rehearsed rhetorical stance on earthquake risks in California, the scientific knowledge embodied in the fault map and other publications of the Seismological Society, the detailed ideas on earthquake-resistant construction brought together by members of the society's Committee on Building for Safety, the cooperation of engineer Dewell and architect Hunt in pushing greater earthquake awareness among their colleagues, and the many contacts Willis had made with leading businessmen and public officials throughout the state. By employing these resources with great vigor and enthusiasm, Willis succeeded in scoring some victories for his campaign for greater earthquake preparedness in California: two municipalities had incorporated a requirement for earthquake-resistant construction into their building code, San Francisco engineers had begun to pay more attention to earthquakes in their work, and a number of bay area businessmen were giving their financial support to further scientific and engineering research on earthquakes. To be sure, many newspaper editorialists, public officials, businessmen, and others, particularly in southern California, still followed the boosters in believing that earthquakes were not a true menace to the state. Yet, because of the organizational work of seismologists between 1906 and 1925, after the Santa Barbara earthquake some Californians did take steps to prepare themselves against future earthquakes.

Earthquake Prediction and the Fall of Bailey Willis

Despite the gains made after the Santa Barbara earthquake, Willis was still dissatisfied with the general level of earthquake preparedness in California. So, he decided to embark on a new strategy: he would scare Californians by not only pointing backward at the recent seismic destruction but also predicting that a catastrophic earthquake, much larger than the Santa Barbara shock, would soon strike southern California. Seduced by a desire to prod Californians into action, he went beyond the limits of what scientific evidence could support, by asserting that this catastrophe would occur within the next ten years. This risky strategy eventually backfired on him. While it brought about greater attention to seismic risks among some California businessmen, it also exposed Willis to a concerted attack from Los Angeles real estate speculators. These speculators engaged the services of a geologist, who soon uncovered weaknesses in the scientific underpinnings of Willis's prediction, and they used these findings in a well-orchestrated campaign that succeeded in discrediting him.

As early as 1920, seismologist Harry Wood had argued that, as southern California had experienced severe earthquakes at roughly 45-year intervals in 1769, 1812, and 1857 but none since then, the area was overdue for another strong earthquake. Four years later, Willis used the same reasoning in his series of articles on earthquake risk, concluding that a great shock would strike southern California sooner or later. This analysis was supported by the U.S. Coast and Geodetic Survey's discovery, first announced in late 1923, that 24 feet of unrelieved strain had accumulated along the San Andreas Fault in southern California.[54] Neither Wood nor Willis believed that the Santa Barbara shock, a moderate earthquake that had originated along a local fault rather than the San Andreas, had done anything to relieve the regional buildup of strain.[55] Therefore, Willis continued to warn that southern California was due for a much larger earthquake.

Willis's initial notoriety for predicting imminent doom for Los Angeles came from a misrepresentation of a statement he made to the *Daily Palo Alto* in early November 1925. After discussing the history of seismicity in southern California and the results of the Coast and Geodetic Survey, he concluded, "No one knows whether it will be one year or ten before a severe earthquake comes, but when it does come it will come suddenly, and those who are not prepared will suffer." The editor of the *Daily Palo Alto* understood this to mean that the strain along the southern San Andreas Fault would "probably result in an earthquake shock in from one to ten years." National news media soon picked up this interpretation. The *New York Times* depicted Willis as predicting that "Los Angeles, or its immediate vicinity," would "experience a severe earthquake, probably more violent than that at San Francisco in 1906, in from one to ten years." Several weeks later *Time* magazine made the prediction appear even more certain: it claimed that, according to Willis, "within the next ten years Los Angeles" would be "wrenched by a tremor worse than that of San Francisco." By February 1926, accounts of the supposed prediction had reached even far-off Australia.[56]

Willis did not, however, refute this misinterpretation. In fact, in May 1926 he himself positively asserted that the expected great earthquake in southern California would come within the next decade. The occasion was the annual meeting of the National Board of Fire Underwriters in New York City. In his keynote address, Willis explained the behavior of earthquakes and the principles of earthquake-resistant construction. He touched upon the matter of future earthquakes in California by stating that "we have to expect a deep-seated . . . earthquake in southern California . . . likely to do much widespread damage." As evidence, he pointed once again to the absence of a strong earthquake in southern California since 1857; he also suggested that the moderate shocks that had occurred within the previous eight years (at San Jacinto in 1918, at Inglewood and Los Angeles in 1920, and at Santa Barbara in 1925) should be regarded as premonitory quakes. During his speech Willis was careful not to put any time frame on his prediction, but in response to a query during the question-and-answer period following the speech he boldly proclaimed: "I regard as probable that in Southern California there will be a severe shock which is more likely to come in three years than in ten and more likely to come in five years than in three."[57]

This prediction went beyond what the available scientific evidence could support. Most of the arguments advanced by Willis were qualitative: the earthquakes of 1769, 1812, and 1857, for example, had occurred when southern California was still very sparsely inhabited, and so their true severity was hard to gauge. There was also no truly compelling reason, within either California's seismic history or seismological theory, to regard the moderately strong earthquakes of the late 1910s and 1920s as premonitory events. The quantitative measurement of the earth's surface deformation produced by the Coast and Geodetic Survey, an organization with a reputation for highly precise work, was a persuasive piece of evidence that significant strain had built up in southern California, but no conclusion could be drawn from it as to when exactly this strain would be released. Willis himself later admitted that his prediction was based less on scientific data than on a desire to stir complacent Californians into greater earthquake preparedness. In October 1927 he wrote, "Inasmuch as we are ignorant of the gathering forces" leading up to an earthquake, "it is impossible to predict the time at which the [fault] may yield. But it has seemed wise to me to warn business interests that such a yielding may reasonably be expected within ten years and may not be postponed beyond three." [58]

Although Willis garnered much publicity for his prediction, it was at first generally ignored by southern Californians. As an East Coast engineer interested in earthquake matters reported after a trip to California in the summer of 1926, California businessmen were still "apathetic" with regard to earthquake matters. [59] One group, however, took Willis's prediction quite seriously: the insurance industry. Before 1925, there had been little demand for earthquake insurance in the United States. In California for the entire period from 1906 to 1924 the total amount of premiums paid for earthquake insurance came to only $1,250,000. After the Santa Barbara earthquake, insurance companies aggressively pushed earthquake insurance in California. As part of their sales strategy, they convinced a number of banks to require earthquake insurance as a precondition for mortgages on commercial buildings. Many building owners decided to purchase such protection, which was being offered at very low rates. As a result, the premiums collected on earthquake insurance in 1925 alone jumped to nearly $2,000,000, a sixfold increase over the previous year, and in 1926 premiums reached $2,500,000. Many insurance executives, trusting the assertions of California's boosters, did not really expect a destructive earthquake to occur again in the near future and therefore regarded the increased premium income as pure profit. Soon, however, some insurance companies realized that they had exposed themselves to heavy liability if an earthquake should indeed strike. They began to pay close attention to Willis's pronouncements.[60]

The anxiety of California insurance executives increased in early 1927, after a moderate earthquake struck the Imperial Valley town of Calexico near the Mexican border. Although the damage caused by this shock was not particularly severe, a number of insurance companies lost substantial money on buildings they had insured. Moreover, Wood was careful to point out that the Calexico quake had by no means relieved the strain on the San Andreas Fault and that therefore a severe earthquake, as predicted by Willis, was still to be expected in southern California. As a result, the Board of Fire Underwriters of the Pacific, whose member companies wrote about 85 percent of the earthquake insurance in California, decided to examine Willis's earthquake prediction more closely and solicited comments on it from leading geologists and engineers.[61] A few of the evaluations were negative. University of California geologist Andrew C. Lawson, for example, used a complex statistical manipulation of the historical record of southern California earthquakes to argue that a destructive earthquake in the Los Angeles area, while possible, was not very probable. Similarly, William Mulholland, the eminent engineer who had built the Los Angeles aqueduct, asserted that the city was immune from sharp earthquakes because he had never experienced one in his more than 50 years of residence there. Wood, however, endorsed the general notion that a destructive earthquake would soon strike southern California, although he was careful to point out that, unlike Willis, he was not specifying a definite time frame. John P. Buwalda, Caltech's professor of geology, also stated that he had "no doubt that a severe shock" would "occur in Southern California within a fairly limited period," an opinion in which he was joined by R. R. Martel, professor of civil engineering at Caltech. The engineer who compiled these comments for the Board of Fire Underwriters concluded that the arguments in favor of Willis's prediction were more convincing than those against it, especially in light of the evidence adduced by the Coast and Geodetic Survey for a significant buildup of strain.[62]

While they were evaluating Willis's prediction, the Board of Fire Underwriters received the results of a survey of the earthquake-resistance of buildings in California. During late 1925 and 1926, engineers for the board had examined 2,700 office buildings, warehouses, and other commercial buildings in San Francisco, Oakland, Los Angeles, and San Diego. They inspected the strength of the foundation and structural frame and the quality of floor and wall construction in order to estimate how each building would fare in a severe earthquake. The survey suggested that most buildings in

California were simply not designed to resist an earthquake and would suffer a large amount of damage in a strong shock. After considering the building survey and the assessments of Willis's prediction, the Board of Fire Underwriters decided that drastic action had to be taken. In May 1927 the board significantly raised the premiums it charged for earthquake insurance. In most of California, the cost of insurance on steel-frame buildings went to 60¢ per $100 of coverage from 35¢; on reinforced concrete buildings, $1.50 rather than 60 cents; and on masonry buildings (by far the most common), $3.50 instead of $1.25. Coupled with an earlier rate increase in late 1925, this meant that the cost of earthquake insurance had increased five- to tenfold since the Santa Barbara earthquake.[63]

Because many California banks had made earthquake insurance a prerequisite for mortgages on commercial buildings, the steep increase in earthquake insurance rates put a serious dent in the profitability of commercial construction and finally brought the earthquake problem to the attention of California businessmen. Some of these businessmen reacted the way Willis had hoped—that is, they began to demand more earthquake-resistant construction. One group adopting this approach was the California Bankers Association, which formed an insurance committee in 1927 to consider the earthquake insurance problem. After several meetings, the committee concluded "that the best earthquake insurance is proper construction" (an assertion Willis had stressed for years) and thus "that banks and bond underwriters [could] materially assist . . . by eliminating or modifying their requirements for earthquake insurance, and insisting upon proper building construction." Specifically, the committee recommended that each building on which a mortgage was being considered be inspected by a qualified engineer, and that the mortgage be approved only if the engineer certified the building earthquake-resistant. In such cases there would no longer be a need for costly earthquake insurance. This recommendation was adopted by the California Bankers Association as a whole.[64]

Another group that reacted by emphasizing the value of proper construction was the California Development Association, a group of boosters and businessmen that would soon become the State Chamber of Commerce. Several leading members, apparently swayed by the arguments of the bankers' insurance committee, decided to attack the insurance problem by developing a uniform building code for California. If a building code could be developed that required earthquake-resistant construction, and if municipalities across the state could be persuaded to adopt it, then there would no longer be any need for high earthquake insurance rates. Moreover, a uniform building code would be to the advantage of large building contractors, since they would no longer have to adhere to different building regulations in different jurisdictions. Therefore, the California Development Association assembled a committee of eminent California engineers, architects, and contractors to draw up such a building code. This committee in turn retained as its technical editors, who would do the detailed work of putting together the new building code, Henry Dewell and Edwin Bergstrom, the latter a Los Angeles architect who also had strongly supported earthquake-resistant construction after the Santa Barbara earthquake. The association demonstrated its commitment to improved construction practices by raising $20,000 for the committee's work, which began in April 1928 and continued for several years.[65]

Willis's earthquake prediction, while it did prod some bankers and businessmen to call for more earthquake-resistant construction, also drew the ire of numerous southern California real estate promoters and speculators, who saw in it a direct threat to their economic well-being. These boosters regarded the high insurance premiums, collected by companies headquartered in northern California or on the East Coast, as "a tribute taken from the city [of Los Angeles] never to be returned," and they worried that "a reputation for seismicity, once established, may never be overcome; and our whole economic structure is predicated upon a continuance of growth. The enormous drain of our present premiums, and their probable increase, is not only destructive of development, but of our present prosperity."[66] Consequently these promoters focused their considerable political resources on undermining Willis's prediction, in order to remove the threat to southern California's reputation for safety, a reputation that was in part responsible for drawing tens of thousands of new residents to the area each year and thereby creating enormous profits for them.

The first to act against the prediction was Henry M. Robinson, a leading Los Angeles booster who also happened to be a trustee of the California Institute of Technology. In April 1927, he learned that Caltech professors Buwalda and Martel endorsed Willis's prediction and were planning to issue a lengthy public statement of their own setting out the reasons for expecting a destructive earthquake in southern California. In a blunt letter to Caltech's executive director, Robinson endeavored to squash Buwalda and Martel's pronouncement: "I wonder if you have any idea how much damage this loose talk of these two men is doing to the [property] values in Southern California. . . . You can hardly appreciate how serious the situation is here and if we . . . can not stop their talk about the earthquake problem I for one am going to see what I can do about stopping the whole seismological game." The implied threat of cutting off funding for all earthquake research at Caltech had its desired effect: Buwalda and Martel did not go public with their warning. Wood, the Carnegie Institution seismologist who headed the Seismological Laboratory at Caltech, also decided not to issue any more public statements about earthquake hazards in southern California.[67]

While Robinson did his part to silence local seismologists, the Building Owners and Managers Association of Los Angeles concentrated on directly discrediting Willis and his prediction. The main tool they used was Robert T. Hill, best known among his fellow geologists for his work on the geology and petroleum resources of Texas. In the mid-1910s, Hill had conducted some fieldwork in southern California for the U.S. Geological Survey. The survey had not published his report on this work, considering it too rambling and incoherent—a problem he had also encountered with his earlier reports for the survey. In 1920, he had published a short paper on the faults of southern California that generally echoed the dismissive attitude toward earthquake hazards. In 1927, Hill was once again in California, lecturing at the University of California's new campus in Los Angeles and doing some consulting work on the side. He cooperated in some of these consulting ventures with C. A. Copper, the executive secretary of the Building Owners and Managers Association.[68]

At Copper's request, Hill addressed the association on July 13, 1927, to condemn Willis's prediction and the response it had evoked from insurance companies. His strategy was to cast doubt on the prediction by pointing out the many uncertainties of seismological knowledge: southern California's faults had not yet been examined closely enough to distinguish the live ones from the dead; even the location of some faults was still subject to debate among experts; and there was no real reason for supposing that future earthquakes in southern California would be as strong as past ones or as those in northern California. According to a member of the audience, Hill concluded by drawing "attention to the many eminent geologists who are living right in this region, mentioning some who had built their homes right over faults passing through this section and yet they do not fear any catastrophe in the future, immediate or distant." He asked why Bailey Willis's "theoretical fear" should "upset the whole psychology and business structure." Hill's message of reassurance was enthusiastically received by members of the Los Angeles Chamber of Commerce and other businessmen in the audience.[69] Within days the Building Owners and Managers Association asked him to write a formal brief to counteract Willis's prediction among insurance officials; the chronically debt-ridden and underemployed Hill accepted gleefully, not least because he was promised at least $1000 for it. Once he had begun work on the brief, Hill also agreed to write a book-length report on the geology and earthquake potential of the Los Angeles area based on his earlier fieldwork for the Geological Survey.[70]

While laboring on behalf of the Building Owners and Managers Association, Hill discovered a crucial weakness in the scientific underpinnings of Willis's prediction: the Coast and Geodetic Survey was retracting its earlier announcement that it had discovered a significant buildup of strain along the southern San Andreas Fault. The survey had drawn its initial conclusion in 1923, after the second season of fieldwork in its retriangulation of California. At that time, the retriangulated network extended from near Lake Tahoe, where it connected with the national geodetic network, west to the bay area, and then south to Santa Barbara County. Since the newly retriangulated network connected with the national network at only one place, there was no adequate way to double-check the retriangulation for accuracy. Yet when the survey had found evidence of movement of the earth's surface by as much as 24 feet near Santa Barbara, it had gone ahead and published these preliminary data. In his 1924 report setting out these data, William Bowie of the survey did point out that the retriangulation had not yet been completed, but he was also at pains to argue that the possible errors from all conceivable sources were so small that there could be no doubt that the earth's surface had indeed moved by a significant amount in southern California. The survey finally completed the fieldwork for the retriangulation of California toward the end of 1926, when it linked the retriangulated network back to the national network at the California-Arizona border.[71] The final results of the retriangulation were not, however, made known to the public immediately.

In October 1927, Hill telegraphed the survey's headquarters to confirm its earlier announcement of a significant buildup of strain in southern California. R. L. Faris, the survey's acting director, telegraphed back with the news that "readjusting old work . . . shows less movement than preliminary results indicated." In fact, Gaviota Peak, which the preliminary data had shown to have moved northward by 24 feet from the 1880s to 1923, now seemed to have moved only 5 feet, and the movement had been to the southeast. In a subsequent letter, Faris elaborated by pointing out that the original triangulation of California in the nineteenth century and the retriangulation in the 1920s had used different methods for establishing true vertical direction at various key stations. This difference of method had not been taken into account in calculating the preliminary results. When the data were corrected for the difference in method, the retriangulation showed that points on the earth's surface in California had not moved more than seven feet since the mid-nineteenth century—and even this movement might be an artifact that was due to random measurement errors.[72] The explanation advanced by Faris for the survey's reversal was rather subtle and rested on some highly technical matters. Many California seismologists, once they had heard of the survey's retraction of its earlier announcement, had to admit that they could not really follow Faris's explanation. Yet the seismologists, who for so long had been extolling the precision and carefulness of the survey's work, had no choice but to accept the survey's assertion that it had not in fact discovered a severe buildup of strain in southern California.[73]

Hill lost little time in trumpeting his discovery of the survey's reversal. On December 1, 1927, he addressed "a group of Los Angeles financial and scientific authorities" assembled by the Building Owners and Managers Association. In his speech, Hill quoted triumphantly from the communications he had received from the survey. After demolishing the most quantitative piece of evidence cited by Willis, Hill went after the more qualitative arguments used to support his prediction. He asserted, for example, that the earthquakes of 1812 and 1857, noted by Willis and others as having been catastrophic, had damaged only the most fragile adobe buildings in southern California; thus, according to Hill, southern California had never really suffered a severe earthquake in the past. Hill concluded, "In no portion of the United States or any of its territories over which the flag flies is there a place so free from natural disasters as Southern California," and he challenged Willis to "withdraw his prophecies" which had "deeply hurt" the "good name of Los Angeles." Officials of the association made sure to bolster Hill's conclusions by having fellow geologists attest to his high scientific reputation in an introduction to his speech. Several Los Angeles newspapers obliged by calling Hill a "world authority" and a "geologist of international repute" in reporting on his address. The president and the executive secretary of the Building Owners and Managers Association also were careful to maximize the impact of his speech by widely disseminating a mimeographed transcript.[74]

The meeting of the Building Owners and Managers Association on December 1 was not the only occasion on which Hill showcased his arguments against Willis's prediction. Late in that month Hill, at the association's expense, traveled to Cleveland, where he presented his view of "earthquake conditions in southern California" at the annual meeting of the Geological Society of America. Willis, the Geological Society's incoming president, was in Hill's audience, but apparently remained unswayed by his arguments. Subsequently, Hill went to Washington, D.C., where he conferred with officials of the Geological Survey and the Coast and Geodetic Survey on earthquake matters without, however, receiving any official endorsement of his views. Meanwhile Copper, the association's executive secretary, kept up his public relations attack on Willis's prediction. In mid-January 1928, he planted a lengthy article in the *Los Angeles Times* reiterating Hill's arguments and pointing out that they "completely refuted" Willis's prediction. Several weeks later Copper went so far as to claim in another newspaper article that the Geological Survey and the Coast and Geodetic Survey had both endorsed Hill's conclusions and that Willis had abandoned his position, thus leaving no rational basis for the "exorbitant" earthquake insurance rates. Copper's exaggerated claims had their intended effect of attracting public notice. In late February 1928, for example, *Time* magazine, which had earlier publicized Willis's prediction, printed an eight-paragraph story setting out Hill's arguments, as condensed by Copper, and implying that Willis had been motivated by northern Californians' jealousy of southern Californians' success.[75]

The campaign by the Building Owners and Managers Association culminated on April 16, 1928, when Hill's book *Southern California Geology and Los Angeles Earthquakes* appeared in local bookstores. On its title page the book carried an indication that it had been published by the Southern California Academy of Sciences—an arrangement that had been made with the academy to give the book greater scientific credibility. In fact, it was published and distributed by Copper, who also retained copyright in the book. The bulk of the 230-page book was given over to a detailed description of southern California's various faults, their history, and their effect in producing the region's landscape of rugged mountains and broad valleys. Hill admitted that faults had been quite active in the area in the past, but he attributed this activity to the geological stresses produced by heavy ice loads during the glacial age and asserted that, ever since the long-ago end of that age, southern California had been "passing through a period of decline in seismic activity." Hill also argued that southern California was quite different geologically from northern California, so that there was no logical reason for presupposing that it could experience an earthquake as devastating as the 1906 San Francisco shock. Prefacing this lengthy scientific treatise were two chapters setting out Willis's prediction (under the heading "Southern California Attacked") and Hill's by then familiar arguments against it.[76]

The Building Owners and Managers Association, in order to drive home the point that it wanted readers to carry away, distributed Hill's book with a red dust cover prominently bearing the statement: "This book completely refutes the predictions of Professor Bailey Willis that Los Angeles is about to be destroyed by earthquakes. It proves that this area is not only free from a probability of severe seismic disturbances but has the least to fear from 'Acts of God' of any city under the American flag."[77] Copper also made sure to generate significant positive publicity about Hill's book in Los Angeles newspapers on the day of its release. This public relations campaign soon brought about exactly the desired result: in mid-May 1928, the Board of Fire Underwriters of the Pacific reduced premiums on earthquake insurance to 1925 levels, a move the Los Angeles press attributed to the salutary influence of Hill's book.[78]

While the Building Owners and Managers Association was busily hammering away at his earthquake prediction, Willis in 1928 was no longer in a position to defend it publicly. His overenthusiasm in naming a specific time frame during which southern California would experience a seismic catastrophe had opened him up to attack on scientific grounds, especially after his most persuasive piece of quantitative evidence, the Coast and Geodetic Survey's measurement of strain buildup, turned out not to be valid. But Willis had also lost his institutional standing as president of the Seismological Society of America, after becoming embroiled in a bitter conflict with other officers of the society. His main opponent within the society was S. D. Townley, the Stanford professor of astronomy who had served as the society's secretary and treasurer as well as editor of its *Bulletin* since 1911. During the mid-1920s, Townley became increasingly upset at Willis's emphasis on public education. In 1926, for example, he bitterly opposed Willis's plan to publish a lengthy article by a consulting engineer celebrating the future growth of the San Francisco metropolitan area and incidentally noting the provisions that should be taken to guard bridges, pipelines, and other infrastructure against earthquakes. While Willis argued that this article was necessary for swaying the minds of the many influential bay area businessmen who had joined the society, Townley believed that it had no place in a scientific journal.[79] Townley also from the beginning regarded Willis's earthquake prediction as scientifically unsound, and he therefore came to judge Willis's knowledge of seismology as "quite superficial."[80] Townley's opposition came to a head in the summer of 1926, when he refused to nominate Willis as the Seismological Society's official representative to a scientific congress in Japan. In the end Willis did attend the congress, but Townley's public opposition led him to resign from the society's

presidency.[81] Consequently, when the Building Owners and Managers Association attacked his prediction in 1928, Willis no longer could count on
the support of the Seismological Society.

Apparently chastened by the reception he was receiving both from
California's boosters and from fellow officers of the Seismological Society,
Willis gradually withdrew from seismology in the late 1920s. By 1929, he
was no longer even a member of the society's board of directors, and a fellow seismologist noted that his membership in the society was now only
nominal. Except for a few articles on the earthquake engineering laboratory at Stanford that he had helped to establish, Willis also ceased to write
or lecture on seismological topics and instead turned his attention more
fully to abstruse topics of geological theory, such as the possibility of continental drift.[82] With the discredited Willis no longer enthusiastically advocating greater earthquake preparedness, southern California's boosters by
the end of the 1920s had succeeded in removing his challenge to their assertions of seismic immunity.

IN THE END, WILLIS'S DECISION to push the campaign for greater seismic safety by predicting a catastrophic earthquake in southern California
within a decade resulted in a serious setback for the campaign. This turn
of events should not, however, obscure Willis's achievements. Not only did
he recruit new members into the Seismological Society and put its finances
back on a solid footing; he also refined the rhetorical stance on earthquake
hazards first developed by G. K. Gilbert and John Branner and for the first
time brought together seismologists, engineers, architects, and public officials to promote earthquake preparedness. Willis also organized new scientific resources for the society, including the fault map and, somewhat
later, a seismograph network in the bay area and an earthquake engineering laboratory at Stanford. When deployed to interpret a recent earthquake
and draw moral lessons from it, the scientific and organizational resources
developed by Willis allowed him to make some headway against the dismissive pronouncements of California's boosters. But his failure to scare
the California public into greater earthquake consciousness by prophesying a seismic catastrophe revealed the limits of a seismic safety campaign
directed at the general public. Boosters and real estate speculators could
draw upon far more powerful organizational and financial resources than
seismologists could to present their view of California's earthquake hazards. Moreover, the kind of stark warning necessary to grab the public's attention opened scientists up to the risk of losing one of their most valuable

resources, their scientific credibility. Earthquake researchers took the lessons of Willis's failure to heart. After 1930, they focused on working quietly behind the scenes to gain support among engineers and bureaucrats who might impose earthquake-resistant construction standards rather than trying to sway the general public to be more earthquake-aware. Only in the immediate aftermath of an earthquake, when the carnage and devastation was still fresh in the public's mind, would these careful Progressives seek the support of public opinion.

CHAPTER 4

Engineering a Regulatory-State Apparatus

Seismic Safety in the 1930s

With the failure of Bailey Willis's public relations campaign for greater seismic safety, earthquake researchers in California changed their tactics. Rather than trying to scare the general public, they would now cultivate alliances among professional groups that had an interest in greater building safety. In the early 1930s, seismologists gained valuable assistance from structural engineers, a new subdiscipline of engineering that saw earthquake-resistant building design as a central part of its mission. In 1933, these structural engineers and seismologists saw an opportunity to advance the cause of seismic safety when an earthquake caused widespread destruction in Long Beach and adjoining areas of Los Angeles County. Seismic safety advocates embarked on a vigorous publicity campaign to paint that earthquake as a manifestation of a general hazard rather than an isolated occurrence. Their main objective was to inspire emplacement of building regulations that would ensure earthquake-resistant construction even after the memory of the Long Beach earthquake had faded. In this they succeeded: not only did several southern California municipalities pass building codes requiring earthquake-resistant construction, but the state legislature enacted a law requiring schools to be earthquake-resistant and, even more importantly, gave far-reaching powers to a state agency to enforce this law. In this manner, seismologists and structural engineers enrolled the power of the state for their cause and created the first regulatory-state apparatus for earthquake hazard mitigation.

Applied Seismology and Earthquake Engineering circa 1930

The campaign for greater seismic safety in California continued after Willis's withdrawal from seismology in the late 1920s, albeit without the scare tactics he employed in his prediction. In southern California, the regional seismograph network put together by Harry Wood finally began to record earthquakes in 1927, and Wood used these records to delineate the region's zones of seismic activity. Meanwhile, a number of engineers throughout California started to involve themselves more deeply in making buildings earthquake-resistant. In the early 1930s, they devised both improved methods for studying the effects of earthquakes on buildings as well as their own organizations for pushing greater earthquake preparedness. In this manner, the engineering profession finally became an effective partner in the campaign for greater seismic safety.

Throughout the late 1920s and early 1930s, Wood continued to devote himself to studying local earthquakes and the faults on which they originated. In 1927, the Carnegie Institution's southern California regional seismograph network, which Wood had labored for years to establish, finally went into operation, with local earthquakes being recorded on Wood-Anderson seismographs at stations in Pasadena, Riverside, Santa Barbara, La Jolla, and atop Mount Wilson. Several years later, two more stations were added, in the Owens Valley north of Los Angeles, near the main aqueduct bringing water to the metropolis. Wood further expanded the capabilities of his network by hiring Charles F. Richter, a Caltech graduate student in physics, to assist him in processing the vast amount of data produced by the seismographs. The routine registration of local earthquakes, many of them too weak to be felt by humans, soon brought significant results. In 1927, the location of earthquake epicenters showed that the Whittier Fault in the Los Angeles Basin was indeed seismically active. A year later, Richter delineated an active submarine fault zone off the coast of Ventura County, just west of Los Angeles. In the summer of 1929, analysis of records accumulated during the first two years of the network's operation showed that as many as 35 fault segments were seismically active in southern California; in subsequent years, Wood and Richter expanded this list of active faults even further.[1]

In January 1931, Wood and Richter began to issue a monthly bulletin of the stronger shocks, both local and distant, registered by their southern California network. Two years later, seismologist Perry Byerly at the University of California started a bulletin of local earthquakes registered by the northern California regional seismograph network. This network, which had had its inception in the fundraising efforts of Bailey Willis immediately after the Santa Barbara earthquake, now included stations in Berkeley, San Francisco, Palo Alto, and atop Mount Hamilton. Besides collecting instrumental records of earthquakes, Wood and Byerly in the late 1920s also began cooperating with the U.S. Coast and Geodetic Survey to collect noninstrumental observations of earthquakes strong enough to be felt, a job that in the 1910s had been undertaken by the Seismological Society and the U.S. Weather Bureau.[2]

In addition to recording earthquakes, Wood and Richter contributed improved methods for grading earthquake strength. During the first three decades of the twentieth century, American seismologists had assessed the intensity of earthquakes using the Rossi-Forel scale. This scale consisted of ten grades of intensity, ranging from barely perceptible to catastrophic. Each grade of intensity was marked by observable earthquake effects characteristic of that intensity. Creaking of floors during a quake, for example, marked grade IV, while ringing of church bells marked grade VII. The scale had been designed by Swiss and Italian seismologists around 1880. But with fifty years worth of technological innovations and rather different standards of building construction in the United States, so Wood pointed out, earthquakes produced effects in California in the 1930s that did not match those of the Rossi-Forel scale. Consequently Wood, in cooperation with a Coast and Geodetic Survey seismologist, produced an updated intensity scale appropriate for contemporary conditions in California. This new scale enabled Wood to assess more precisely the strength of earthquakes occurring in California.[3] Meanwhile, Richter in the early 1930s devised a numerical scale for grading instrumentally recorded earthquakes. This scale, the Richter magnitude scale, relied on the amplitude of the record produced by an earthquake on a standard Wood-Anderson seismograph. It provided a far more objective measure of the strength of earthquakes than the earlier intensity scales, and since the mid-1930s it has become a universally recognized means for rating temblors.[4]

Wood readily provided his data to engineers, insurance companies, and others, and thereby contributed to the campaign for greater seismic safety. In 1929, for example, he assisted an insurance executive interested in earthquake insurance by commenting on the seismic potential of various parts of the United States. A year later, he provided a Los Angeles city engineer with detailed data on the active faults underlying that city. In 1933, Wood supplied considerable information for a listing of strong earthquakes that had occurred in California since 1769. This list, drawn up by the Coast and Geodetic Survey at the request of the American Red Cross, was intended for the use of engineers, rescue officials, and others who needed to know where destructive earthquakes might strike.[5] Wood also contributed more directly to the cause of earthquake preparedness when, in a letter published in 1928 by the American Society of Civil Engineers, he pointed out that dams would have to resist not only direct shaking by earthquakes, but also battering from adjacent bodies of water set into motion by the tremor. This observation eventually led some California engineers to alter the way they designed dams.[6]

Thus, throughout the late 1920s and early 1930s Harry Wood and fellow seismologists made sure that knowledge about earthquakes and faults in California continued to increase. The campaign for greater seismic safety, however, was no longer a concern merely of scientists. Engineers now also took a prominent role. Foremost among them, at least until he suffered a nervous breakdown in 1932, was Henry Dewell, whom Willis had drawn into the campaign. After the 1925 Santa Barbara earthquake, Dewell repeatedly urged his fellow engineers to incorporate earthquake provisions in their designs. He echoed Willis in insisting that strong earthquakes would recur in California and in railing against "certain types of speculative owners, whose sole object is to build something as cheaply as possible, which will sell quickly." Like Willis, he also pointed out that "sound materials and honest skillful workmanship" were not enough to provide protection against earthquakes; instead, engineers had to consciously design their buildings to resist earth tremors. As a basic principle of design, Dewell recommended that structures be constructed so as to withstand a horizontal acceleration of 0.10g. Dewell relied heavily on the pronouncements of Willis, but also on the work of Japanese engineers who had investigated the catastrophic destruction in Tokyo caused by an earthquake in 1923. Dewell especially promoted the design method developed by Tachu Naito for calculating the proper rigidity and strength necessary for walls and floors to withstand a given horizontal force.[7]

Other California engineers also became strong advocates of greater earthquake preparedness in the late 1920s. Among them was Henry M. Engle, an engineer with the Board of Fire Underwriters of the Pacific. Like Dewell, Engle urged that buildings be constructed to withstand a horizontal acceleration of 0.10g, and he also promoted Naito's method for properly designing such buildings.[8] R. R. Martel, the professor of civil engineering at the California Institute of Technology who had supported Willis's prediction but was discouraged from going on the record, likewise joined the campaign. In 1926 and again in 1929, he traveled to Japan as a delegate to scientific or engineering conferences; on these occasions, he learned much about Japanese methods for building against earthquakes. Back in California, he worked to have earthquake provisions included in building codes. He also put his engineering knowledge into practice by designing a bank building in Pasadena to withstand earthquakes and by developing a valve that would automatically shut off gas lines during a strong earthquake in order to prevent the outbreak of fire. A few other engineers joined Martel in designing bridges, commercial buildings, and other structures against earthquakes, even though local building codes did not require them to do so.[9]

While Dewell, Engle, and Martel campaigned in public for greater earthquake preparedness, others worked to increase the technical engineering knowledge on how to design against earthquakes. At Stanford University, by then professor Lydik Jacobsen was simulating earthquakes with the shaking table—essentially a large platform on wheels that could be vibrated in several different ways—built in 1926 with funds raised by Willis. Using this experimental apparatus, Jacobsen determined that timber walls that had diagonal sheathing were better able to resist ground shaking than those with horizontal sheathing. At the request of Dewell and other engineers, Jacobsen also confirmed experimentally that earthquake waves did become amplified in loose, water-logged soil, as numerous seismologists had previously asserted on the basis of observations of earthquake damage. In the early 1930s, Jacobsen built a model of an unconventional 26-story building that several engineers had designed for a San Francisco club, and he tested it for its resistance to horizontal accelerations on the shaking table. In the process he gained much insight into how earthquake resistance might be increased by varying the stiffness of different stories in the building. Jacobsen also tested the valve that Martel had designed to shut off gas in the event of a strong earthquake, and he conducted experiments on the battering of water against a dam during an earthquake.[10]

In their efforts to improve earthquake engineering in their state, California's engineers received significant assistance from engineers living elsewhere. Especially important was John R. Freeman, a nationally prominent hydraulic engineer and fire insurance executive in Providence, Rhode Island, who had become interested in earthquake preparedness after a moderate tremor shook New England in 1925. In the early 1930s, Freeman arranged a national lecture tour by Kyoji Suyehiro, the eminent engineer who headed the Japanese Earthquake Research Institute. The American Society of Civil Engineers (of which Freeman was a former president) published Suyehiro's lecture notes, thereby providing American engineers with an up-to-date treatise on earthquake engineering. In 1932, Freeman published a massive book of his own, which discussed the historical record of earthquakes in the United States; the extent and kinds of damage produced in California, Japan, and elsewhere; and the measures taken by engineers, insurance companies, and writers of building codes around the world to reduce damage.[11] This information greatly added to what engineers in California had already learned from their own research on earthquakes.

Freeman's greatest contribution, though, lay in the inauguration of strong-motion seismology. As early as 1927, he had complained that the seismographs operated by geophysicists did not provide data of use to structural engineers. These seismographs were delicate enough to catch the tiny motion from the seismic waves of small earthquakes, but the strong motion from the violent waves of large nearby earthquakes threw them out of order. As a result, engineers did not possess an accurate record of the strong motion which a building would need to resist. The assumption that earthquakes produced a horizontal acceleration of at most 0.10g was based on estimates and rough calculations rather than hard data. After seeing at first hand in 1929 the efforts of Japanese engineers to obtain such data, Freeman redoubled his efforts to introduce strong-motion seismology in the United States. At his urging, the U.S. Coast and Geodetic Survey agreed to take on the project and obtained an annual appropriation of about $25,000 from Congress. By 1932, laboratories at Massachusetts Institute of Technology and the National Bureau of Standards had developed for the survey several rugged instruments capable of recording even the most intense earthquake motion. Over the next three years, the survey, advised by California seismologists and engineers, placed strong-motion seismographs at 45 sites across the state. Several earthquakes recorded in these years showed spikes of motion in which accelerations exceeded 0.10g, clearly demonstrating that earthquake-resistant buildings would have to be designed to resist at least this level of force.[12]

During the 1930s, California engineers also strengthened existing organizations for promoting greater earthquake preparedness and created new ones. The Seismological Society of America, with its quarterly *Bulletin* reaching hundreds of members interested in seismology, continued to serve as an important forum for disseminating the latest ideas for increasing seismic safety. Some scientifically inclined members wanted to turn the *Bulletin* into a more purely technical magazine devoted exclusively to advanced scientific discussions, but the society continued to cater to the engineers, architects, businessmen, and others among its membership interested in practical matters of earthquake protection by publishing articles on earthquake-resistant construction, earthquake insurance, and similar subjects. California engineers also made sure that meetings of the Seismological Society continued to include sessions on earthquake engineering as well as theoretical seismology.[13]

Around 1930, engineers interested in earthquake-resistant construction gained an additional platform with the formation of structural engineers' associations in northern and southern California. According to a contemporary engineer, a structural engineer was someone "whose business it is to calculate the forces that act upon a [building] structure, and to design the frame or structural part of the building to adequately resist these forces."[14] Earlier in the twentieth century, structural engineering had been part of the larger profession of civil engineering. By the late 1920s, structural engineers, who concerned themselves specifically with buildings, began to develop a professional identity separate from that of civil engineers who focused on dams, bridges, pipelines, and similar structures. In 1929, about a dozen southern California structural engineers formed the Structural Engineers Association for their region; a year later, their northern California colleagues followed suit. By 1933, these two associations together had about 135 members, including Dewell, Engle, and Martel. Both associations were founded mainly to secure professional standing and recognition for structural engineers. They worked to obtain state recognition through licensing of structural engineers, and they promoted the use of structural engineers in addition to or instead of architects in building design.[15] Yet they also provided a ready forum for discussing the design of buildings against earthquakes. After all, as a number of structural engineers came to argue, seismic forces were among those that acted on a building structure. Throughout the early 1930s, Jacobsen repeatedly presented his shaking-table research at meetings of the northern California association. These talks were sometimes accompanied by film demonstrations of exactly how models of buildings reacted to horizontal shaking. Other association members expounded on their theories of earthquake-resistant design and debated the kinds of earthquake provisions that should be included in building codes.[16] By 1933, members of both associations were in agreement on the need for greater earthquake preparedness, and consequently both groups were ready to engage in public campaigning for seismic safety.

Yet another channel for spreading the idea of earthquake-resistant construction appeared with the initiation of work on the Uniform Building Code for California. In 1927, the California Development Association responded to the substantial hike in earthquake insurance rates occasioned by Willis's earthquake prediction by appointing a committee of engineers, architects, and contractors to develop a model building code for California. This so-called Uniform Building Code was intended to include provisions for earthquake-resistant design and thereby obviate the need for earthquake insurance. The committee in turn retained Dewell, the engineer who was most outspoken on the need for earthquake preparedness, as one of the two technical editors who would actually compile the code. Work on the Uniform Building Code dragged on for years. After all, it was supposed to standardize all aspects of construction practice in California, not just earthquake-resistance. Moreover, after earthquake insurance rates dropped again when Willis's prediction was undermined, the Uniform Building Code appeared to be a less urgent matter to many of its original backers. Nevertheless, by 1933 Dewell and his colleagues on the committee had drawn up a code, and it incorporated a number of provisions for earthquake-resistant construction. It specified how columns, beams, and other components of a structural frame should be tied together to provide a unified whole capable of resisting horizontal shaking; it made provisions for strengthening parapet walls, ornamentation, and other protrusions that might fall off during an earthquake; and it included a general requirement that buildings be designed to withstand not only vertical forces but also a horizontal acceleration of 0.10g.[17] These provisions drew upon and codified the knowledge of earthquake-resistant design accumulated by California seismologists and engineers over the previous three decades.

Thus, in 1933 California seismologists and engineers were far better prepared for a destructive earthquake than they had been in 1906. At the time of the San Francisco earthquake and fire, scientists had given very little thought to the possibility of seismic activity in California, and engineers had been preoccupied with such campaigns as those for better fire protection and for the use of reinforced concrete as a building material. As a consequence, the lone engineer who had argued for increased earthquake preparedness after the destruction of San Francisco, Charles Derleth, had not been able to make himself heard among his fellow engineers. By 1933, however, a number of engineers as well as scientists had become convinced of the need for greater earthquake preparedness in California, and they had a wide variety of resources to draw upon that had been unavailable in 1906: the detailed lists of historical earthquakes and of dangerous faults drawn up by seismologists; the technical information on earthquake-resistant construction provided by American and Japanese researchers; the organizational support of the Seismological Society of America and the California structural engineers associations; and the Uniform Building Code's earthquake provisions as an explicit guide to how buildings should be constructed to resist tremors.

The Long Beach Earthquake and Its Aftermath

At 5:54 P.M. on March 10, 1933, a moderately strong earthquake (of magnitude 6.2 on Charles Richter's new earthquake scale) struck just off the southern coast of Los Angeles County. This earthquake was of about the same intensity as the 1925 Santa Barbara earthquake, but, because it struck in a more densely populated region, it produced considerably more damage. The 1933 earthquake caused 120 deaths and property damage amounting to at least $40 million (about $400 million in 2001 dollars), primarily in the port city of Long Beach and in adjoining suburbs south of Los Angeles. The area of damage reached northward into Los Angeles proper and eastward to Orange County.[18]

In a sense, this earthquake fulfilled Willis's prediction of a devastating shock in the Los Angeles area within ten years of 1926. Yet, as a few seismologists were quick to point out, the Long Beach earthquake was not the catastrophe Willis had foreseen. The 1933 shock was certainly not as intense as the 1906 San Francisco earthquake or even the 1857 southern California earthquake had been. Moreover, as Wood concluded from his seismographic recording of the earthquake, it originated not along the San Andreas Fault but along the Inglewood fault zone, the subordinate zone that had also been responsible for the moderate Inglewood and Los Angeles earthquakes of 1920. To Wood and some of his colleagues, there still remained the threat of a truly devastating earthquake in southern California at some undetermined time in the near future.[19] Nevertheless, the Long Beach earthquake was sufficiently destructive to provide engineers and seismologists with the opportunity they had been awaiting to convince Californians of the need for earthquake preparedness.

After the Long Beach earthquake, many southern California boosters reacted, just as they had after previous earthquakes, by dismissing the damage and proclaiming the overall safety of their region. Three days after the earthquake, the *Los Angeles Times* published a lengthy article that cited a variety of statistics, to demonstrate that hurricanes and other natural disasters elsewhere in the United States caused far more deaths, both in absolute numbers and as a percentage of the total population, than did earthquakes in southern California. The newspaper reinforced this point several days later with a front-page editorial cartoon depicting a puny and scared-looking "Wild Bill" Earthquake facing a "bunch of hard-boiled Eastern Gangsters," including "Killer" Hurricane, "Twister" Tornado, Flood "the Blood," "Legs" Lightning, "Old Kid" Blizzard, and "Gorilla" Heat. Twister Tornado, looking much more menacing than Wild Bill Earthquake, growled in disbelief: "So *you're* that desperate character we've been hearing so much about!" The paper also urged its readers to tell their acquaintances across the nation "that no place on earth offers greater security to life, and greater freedom from the dangers of natural elements, than Southern California." Several other southern California newspapers similarly appealed to their readers to reassure their friends elsewhere that damage from the earthquake had been quite limited.[20] Other boosters used a large aerial photo of Long Beach's skyline to proclaim that the earthquake had not toppled the city's large buildings; the cracks that rendered many of these buildings structurally unsound could not, of course, be seen in this photo. Armed with this and similar photos, a representative of the Los Angeles Chamber of Commerce went on a "nation-wide barnstorming tour," lecturing to audiences in 39 cities and giving talks on 76 radio stations to convince the public that "earthquakes on the American continent have, in only one or two instances, proven as destructive as severe storms experienced in other parts of the country."[21]

Compared with the all-out public relations campaign orchestrated after the Santa Barbara earthquake, however, the efforts of southern California boosters in 1933 were rather anemic. A number of economic factors contributed to this lack of vigor. The Long Beach earthquake occurred at one of the bleakest times during the Great Depression. The massive real estate boom that had fueled the boosters' efforts in the mid-1920s was over; the building permits issued in Los Angeles in March 1933 had a total value of less than $1 million, far less than the $15 million value of building permits issued in June 1925.[22] Moreover, when the Long Beach earthquake struck on March 10, 1933, the nation's banks were in the midst of a forced holiday, imposed by President Roosevelt to stop panic-driven withdrawals and thereby save the banking system from complete collapse. At this time, when confidence in the nation's economy was at its nadir, most southern California businessmen simply had too many other things on their mind to worry much about the damage the earthquake might cause to the region's reputation. Furthermore, because of the depressed state of the economy, southern California boosters could not attempt, as they had done in 1925, to limit national advertisement of earthquake damage by soliciting donations for reconstruction only from Californians. In 1933, only the federal government had the resources necessary to cope with the earthquake-induced devastation. Consequently, congressmen from California promptly sought $5 million in federal disaster aid. In order to secure this aid, though, they had to admit to the nation that the earthquake had indeed produced significant damage.[23]

With the public relations efforts of southern California boosters weakened, those arguing for greater earthquake preparedness were able to mount a stronger campaign. In addition to being better prepared for such a campaign now than in 1925, these seismologists and engineers benefited from certain specific features of the Long Beach earthquake. It struck a large metropolitan area rather than an outlying resort city; consequently, a far greater number of Californians were able to witness for themselves the strength of this earthquake and the destruction it caused. Also, the Long Beach earthquake did a disproportionate amount of damage to school buildings, so the impression made on the minds of worried parents was magnified. In the city of Long Beach, 15 of the 35 schools were completely ruined by the quake, and the overall damage amounted to about two-thirds of the total value of the school buildings. The quake also damaged a number of schools in the City of Los Angeles. All schools in Los Angeles were closed for a week to allow structural inspection of the buildings; at the end of this period, 41 schools were deemed unsafe for occupancy and remained closed. Despite the widespread damage, there were few fatalities among school children, mainly because the earthquake occurred after classes were over for the day. But, as a number of commentators emphasized, if the earthquake had struck several hours earlier, tens of thousands of children might have been killed or injured.[24] The Hearst-owned *Los Angeles Examiner* fanned the anxiety and outrage of southern California parents by repeatedly printing dramatic pictures of destroyed school buildings. In the first few weeks after the earthquake, angry parents and public officials, backed by the *Examiner* and other newspapers, demanded a criminal prosecution of contractors for building schools that collapsed so readily.[25]

Seismologists and engineers took advantage of this outrage but redirected its aim away from retribution and toward enacting legislation that would ensure earthquake-resistant construction. On March 14, four days after the quake, Harry Wood told a reporter that "the practical lesson of the recent temblor, as of all others, . . . is build well and choose or prepare strong foundations. Design for strength and [construct] conscientiously using good materials. Avoid what experience has shown to be faulty." Wood exhorted the public to "frame and enforce regulations to insure good building." The *Long Beach Press-Telegram* promptly endorsed this appeal in an editorial.[26] A day later, John Buwalda, still a professor of geology at Caltech and a close colleague of Wood, informed the readers of the *Los Angeles Examiner* that "the lesson of the earthquake with reference to construction is . . . that buildings should be properly designed and built of earthquake-resistant materials. The loss of life is related to weak walls tumbling down, loose cornices and loose decorations falling from the sides of buildings and the collapse of some weak structures."[27]

On March 18, the *Examiner* printed articles by 11 southern California engineers and architects explaining why schools had received so much damage in the earthquake. Most of the authors took this opportunity to emphasize the need for solid construction using steel framing or reinforced concrete rather than brick. They also stressed the importance of good foundations and good mortar as well as the folly of including cornices and other ornamentation that might easily fall down. Caltech engineer R. R. Martel further urged Californians to enshrine earthquake-resistant construction in revised building codes. The *Examiner* summarized these recommendations in a front-page article and followed up with a strong editorial several days later insisting, "Building codes must be revised to bring the minimum requirements to accepted standards of safety."[28]

Seismic safety advocates found yet another forum in a coroner's inquest. Responding to the general outrage over the ready collapse of school buildings, Los Angeles County Coroner Frank Nance decided to hold a formal inquest that would go "far beyond the actual circumstances of death" to include "testimony concerning safe building construction, especially in public schools," as well as the possible criminal liability of contractors for negligence in the deaths. To conduct this wide-ranging inquiry, Nance appointed a jury of nine prominent architects, engineers, and building contractors. He also enlisted John C. Austin, a leading architect and former president of the Los Angeles Chamber of Commerce. During the two days that the coroner's jury heard testimony, Austin, who had advocated greater earthquake preparedness in an article for the *Los Angeles Examiner,* directed most of the questioning of witnesses.[29]

The testimony heard by the coroner's jury included a considerable amount of information on seismic hazards and the principles of earthquake-resistant construction in southern California. A local geologist described the various fault zones affecting Los Angeles County, while Wood summarized data on the origin of the Long Beach earthquake gathered from his seismograph network. Wood took this opportunity to urge once again that buildings be designed against horizontal as well as vertical stresses. Martel provided more detailed guidelines on earthquake-resistant construction, including diagonal bracing of structural frames and the use of metal laths with plaster. He also insisted that, in general, buildings be designed to withstand a horizontal acceleration of 0.10g. Another local engineer, in explaining why so many buildings had failed, criticized "unsupported firewalls, steel girders resting on brick columns that crumbled, and sandy mortar in the chimneys of residences" as well as "ornamentation apparently stuck on with chewing gum." A number of architects, engineers, and building inspectors joined the chorus calling for more earthquake-resistant construction. The coroner's jury, after deliberating for a week, decided not to point a finger at anyone for criminally careless or negligent construction. Rather, in its verdict it repeated and endorsed many of the recommendations for earthquake-resistant construction. Both the testimony and the verdict were widely reported in southern California newspapers, often on the front page, and in the trade journals of the construction industry.[30]

Those interested in greater seismic safety did not rest content; they also wrote several lengthy reports to reinforce their public relations efforts on behalf of greater seismic safety. In mid-May, Wood assembled a report on the earthquake for the Seismological Society's *Bulletin*. In it, he reviewed the seismographic records produced by the shock, described the damage in various portions of southern California, and concluded that it had been "in large measure avoidable." The article was accompanied by 22 photos that not only showed dramatically the kinds of devastation wrought by the earthquake but also demonstrated that the few structures built according to the principles of earthquake-resistant construction had survived the earthquake unscathed. At about the same time, several southern California structural engineers working for the National Board of Fire Underwriters assembled a lengthy report on the Long Beach earthquake that was issued in pamphlet form by their employer. The engineers analyzed in detail the damage suffered by brick, steel, and reinforced concrete buildings as well as by water and power systems, gas pipelines, and telephone lines. Like so many other seismologists and engineers, they urged the adoption of earthquake provisions in local building codes.[31]

The most influential report on the earthquake came from a committee chaired by Robert Millikan, the Nobel Prize–winning physicist who headed Caltech. In the weeks immediately after the earthquake, separate groups of engineers, architects, geologists, and contractors had been commissioned by the Los Angeles Board of Education to investigate the damage to the city's schools. Soon, leaders of these various groups came to realize that they would gain by coordinating their surveys. They approached Millikan, who had acquired a reputation as a sagacious leader on matters of public interest, to head a coordinating committee. In early April, Millikan accepted their invitation and formed the Joint Technical Committee, consisting of representatives from the various investigating groups, from the Board of Education, and from other civic groups. This committee included a number of outspoken advocates of greater seismic safety, such as engineer Martel (made vice-chairman of the committee), architect Austin, and geologist Buwalda. Millikan even managed to secure the cooperation of the Los Angeles Chamber of Commerce, which defrayed the committee's secretarial expenses.[32]

On June 6, Millikan's committee issued its report, which was a masterful brief on the need for greater seismic safety. The report turned first to a discussion of the evidence for continuing seismicity in southern California. It stated:

> Certain fundamental conclusions are evidenced by years of geological and seismological investigations in California and other parts of the Pacific Southwest.
>
> (a) Earthquakes of damaging or destructive intensity will continue to occur in California from time to time in the future.
>
> (b) An earthquake is apt to occur in this region comparable in intensity and duration with the San Francisco earthquake of 1906. . . .
>
> (d) The degree of risk is such that earthquake resistant construction is absolutely necessary in this region in order to avoid great loss of life and heavy damage to property.

The report then reviewed the kinds of damage caused by the Long Beach earthquake and asserted that "damage to buildings and other structures in San Francisco, in Santa Barbara, and in Long Beach, and similar destructive effects during earthquakes which have occurred in other parts of this country and in foreign countries, is ample evidence that designs which do not include allowance against forces produced by earthquakes are not satisfactory." The report summarized the principles of earthquake-resistant construction that had been developed by Japanese, Italian, and American seismologists and engineers, and it urged that these principles (including a requirement that buildings be designed for a horizontal acceleration of 0.10g) be embodied in building codes. But the report went even further: "Any building code adopted as part of the law of any community is applicable to all new construction, but perhaps such laws cannot be made retroactive. This situation is of tremendous importance for the number of buildings already existing is many times the number to be built during the next ten or twenty years. . . . Many such buildings would be unable to with-stand an earthquake of even the intensity of that which occurred in Long Beach." The report argued strongly that existing buildings, particularly schools, should be redesigned and strengthened so that they, too, would be able to resist earthquakes. The report concluded by warning that "at some unknown time in the future an earthquake of major intensity will occur in this region, and unless existing evils are corrected by adequate protection against earthquakes, disaster must follow."[33]

The Millikan Report soon drew a number of influential endorsements, even from those more commonly engaged in boasting about the safety of southern California. The Los Angeles Chamber of Commerce, in a cover letter accompanying the report, strongly supported its findings and recom-mendations. The chamber's president stated, "Southern California is no different from other sections of the United States or the world, with respect to conditions of hazard, for other areas have their floods, tornadoes, hurri-canes, storms and even earthquakes, and they can be justly criticised when they, like ourselves, sit idly by and fail to recognize that unless known pro-tective measures are taken, such natural hazards may cause certain dam-age which otherwise might reasonably be avoided."[34] While the chamber in this statement still insisted on pointing out that other sections of the country also faced hazards, it for the first time asserted unequivocally that southern Californians had to take cognizance of their own earthquake haz-ard. The Millikan Report's recommendations were also endorsed by the *Los Angeles Times,* the newspaper that had most strongly voiced the boost-ers' views immediately after the Long Beach earthquake.[35]

While seismologists and engineers in the greater Los Angeles area were particularly energetic after the Long Beach earthquake, seismic safety ad-vocates elsewhere in California also took the opportunity to urge greater earthquake preparedness. In the first few weeks after the earthquake, for example, a number of San Diego architects who had visited the Long Beach area bemoaned the failure of school buildings there and argued that earth-quake-resistant construction should also be required in San Diego, even though it had not yet experienced a similar earthquake.[36] Meanwhile, in San Francisco, the city's chief building inspector, a prominent structural engineer who had been heavily involved in formulating the Uniform Build-ing Code, asserted that "earthquakes in California never should have caused loss of life nor ever should again." The *San Francisco Chronicle* took up this call in an editorial urging that it was "time to adopt strong building codes" that would require earthquake-resistant construction. This message was once again reiterated by Bailey Willis and others at a "Face the Facts" luncheon sponsored by several San Francisco civic groups on April 18, the anniversary of the 1906 San Francisco earthquake.[37]

The campaigners for greater seismic safety were finally suceeding in shaping public opinion, as well as prodding governmental bodies into action. In Long Beach, the move to require earthquake-resistant construction was led by C. D. Wailes, Jr., the city's chief building inspector, who had been a member of the Seismological Society of America since 1930. After the earthquake, Wailes was impressed by how poorly brick buildings had fared compared to buildings of reinforced concrete or wood. Consequently, within ten days he promulgated new rules requiring that mortar contain cement and that all brickwork be reinforced with horizontal and vertical steel bars. These rules were particularly intended to strengthen brick chimneys and parapet walls, many of which had collapsed during the earthquake. Just several days later, the Long Beach City Council incorporated the new rules into the city's building code. With the help of an insurance engineer who was a member of the Structural Engineers Association of Southern California, Wailes quickly distributed copies of his rules to building departments in neighboring communities. A number of these towns soon also required that masonry be reinforced.[38] On March 20, the Los Angeles County Board of Supervisors adopted the Santa Barbara building code, which required that buildings be designed against horizontal accelerations, to regulate construction in all unincorporated areas in the county. The Board of Supervisors also provided for the employment of several dozen building inspectors to enforce the new code. Several weeks later, the board, following the recommendation of the coroner's jury, replaced the Santa Barbara code with the Uniform Building Code, which contained more elaborate and up-to-date requirements for earthquake-resistant design.[39]

Although it moved more slowly than neighboring jurisdictions, the City of Los Angeles soon also included earthquake-resistant construction in its building code. At a meeting of the city's Building and Safety Commission on March 29, commissioner Blaine Noice, a member of the Structural Engineers Association of Southern California, introduced a resolution urging that design against horizontal acceleration be required. In response, the commission set up a committee to study possible amendments to the city's building code; this committee included a number of leading structural engineers. The city's mayor appointed a separate committee to consider changes to the building code; it was headed by John Austin, the architect who had led the questioning of witnesses during the coroner's inquest. In June 1933, these two committees forwarded their recommendations to the city council. As adopted by the city council, the additions to the building code required that buildings be designed against horizontal accelerations with a seismic factor of 0.08g (0.10g for schools) and that mortar contain cement. Reinforcing of brickwork, while not required, was strongly encouraged. Similar requirements were also adopted by the northern Los Angeles suburbs of Santa Monica, Beverly Hills, and Pasadena.[40]

Meanwhile, structural engineers helped push through two state laws requiring earthquake-resistant construction for all of California. The first of these acts, submitted to the state legislature by assemblyman C. Don Field, of northern Los Angeles County, on March 22, concerned itself with ensuring earthquake safety in schools. The impetus for this bill came from parents outraged over the widespread collapse of school buildings during the Long Beach earthquake. In drafting his bill, Field relied heavily on the advice of structural engineers employed by the state government. As originally worded, Field's bill would have required that all plans for new schools or major repairs be prepared by a certified architect or structural engineer and that these plans be approved by the State Division of Architecture. It was implicitly understood that structural engineers within the Division of Architecture would approve only those plans that included sufficiently strong measures for earthquake resistance. After hearing portions of the verdict from the coroner's inquest into the Long Beach earthquake deaths, a state Senate committee amended Field's bill to give the Division of Architecture the power not only to approve plans but also to inspect the actual construction of schools, to ensure that earthquake-resistant methods of construction were indeed used. Under the amended bill, the division was also authorized to inspect existing school buildings at the request of local school districts, to see whether they met current safety standards.[41] Many of the state's leading newspapers strongly supported the Field bill in editorials that often repeated the appeals of seismologists and engineers for greater seismic safety. The state legislature, taking into account the bill's endorsement by teachers' organizations, women's clubs, and business groups as well as architects and engineers, quickly passed the bill, and the governor signed it into law on April 10.[42] With this measure, the state government of California for the first time officially admitted the need for earthquake hazard mitigation.

The Division of Architecture entrusted enforcement of the Field Act to its principal structural engineer, Clarence H. Kromer, who as early as 1928 had been involved in framing the Uniform Building Code. He rapidly adopted the earthquake provisions of that code as the standard by which the Division of Architecture would judge the earthquake safety of school building plans submitted for approval. Under his guidance, the division also hired several dozen structural engineers to assist in checking plans for earthquake safety; by this means, the state government became the major employer of structural engineers in California during the Depression years. In addition, the Division of Architecture drew on an advisory board of architects and engineers appointed in the summer of 1933 for further guidance on the details of earthquake-resistant construction.[43] In this manner, the division gained a vested interest in the Field Act and became a major government bulwark for the establishment of seismic safety in California.

Yet seismologists and engineers were not content with requiring earthquake-resistant construction only for school buildings. Over the next two months, they agitated for a statewide law requiring earthquake resistance in all buildings. Such a law had been drafted as early as March 29 by a committee of the Structural Engineers Association of Northern California. On April 25, state assemblyman Riley of Long Beach submitted a modified version of this bill, which would require all buildings in the state (with a few exceptions for rural areas) to be designed to withstand horizontal accelerations of 0.02g. Over the next month, the structural engineers' and architects' organizations of California, the State Chamber of Commerce, and the chambers of commerce of San Francisco and Los Angeles all endorsed this bill. Given such organized support, the state legislature approved the bill, and the governor signed it into law on May 27. As some engineers pointed out, the law, soon called the Riley Act, had a number of weaknesses. The seismic factor of 0.02g it required was rather low, since strong motion records suggested that horizontal accelerations in the Long Beach earthquake had reached as high as 0.23g. Moreover, enforcement of the act was left to local building inspection departments, which were generally understaffed, rather than to a state agency. Indeed, as it turned out, many building departments, especially in the inland portions of California that were far away from active faults, did not enforce the Riley Act. Still, as the *San Francisco Chronicle* pointed out, this law, despite its many weaknesses, stood as an official acknowledgment that earthquakes do happen in California.[44] Over and over again, seismic safety advocates had hammered home the need for earthquake-resistant construction. With public sentiment finally mobilized, the new building codes at last put government in charge of ensuring seismic safety in the future.

Earthquake-Resistant Construction in Post-1933 California

The Field Act and the numerous local building codes that now included earthquake provisions did serve to increase the earthquake resistance of much of California's building stock, especially its schools. Despite the protests of some local school officials that its requirements were too stringent, the State Division of Architecture continued to insist that plans for new school construction submitted for approval satisfy the earthquake provisions of the Uniform Building Code. During its first two years of enforcing the Field Act, the division certified slightly more than 1,000 school construction plans as meeting requirements for earthquake safety. In order to satisfy these requirements, California architects and engineers devised a number of innovative designs for school buildings. Rather than the traditional two-or three-story brick buildings with elaborate cornices and other ornamentation, they now preferred one-story reinforced concrete or reinforced masonry buildings with unadorned facades. These design elements meshed well with the tenets of architectural modernism that were gaining favor among California architects. Modernist architects in general relished the use of reinforced concrete and other relatively new structural materials and rejected as outmoded the use of elaborate ornamentation. Many California architects enthusiastically incorporated into their designs modernistic features that happened also to be earthquake-resistant.[45]

The strengthening of existing structures proved a large task. In the summer of 1933, the Los Angeles Board of Education moved to implement the recommendations of the Millikan Report, including its suggestion that existing school buildings be brought up to the safety standards adopted by the State Division of Architecture. The board hired architect John Austin, a geologist, and a structural engineer to draw up detailed plans for such reconstruction. This committee concluded that more than 90 percent of the city's school buildings needed strengthening, while the rest were so unsafe that they had to be torn down. They estimated that the entire reconstruction project would cost more than $30 million. Over the next several years, the Board of Education managed to raise this money with the help of geologists, structural engineers, and architects who emphasized the sheer necessity of reconstructing existing school buildings. A large part of the money came from the federal government; the remainder, in the form of several bond issues, was provided by sometimes reluctant local voters.[46] With this money, the board went ahead with the strengthening of the city's school buildings. In a number of brick structures, steel or reinforced concrete columns and beams were introduced to strengthen the structural frame. In other brick buildings, the outermost layer of bricks was removed from all walls and replaced with a layer of reinforced concrete that would serve much better to hold the building together in case of an earthquake. A few masonry buildings were completely replaced with one-story wooden structures covered with fire-resistant plaster. Finally, in many cases existing foundations and floors were strengthened through the addition of reinforced concrete, and ornaments and other protusions were removed. All of these measures helped bring Los Angeles's schools up to the standards of seismic safety required by the State Division of Architecture.[47]

Other southern California cities also followed the recommendation of the Millikan Report and upgraded the earthquake resistance of their existing school buildings. Pasadena, home of numerous Caltech geologists and engineers who had long urged greater earthquake preparedness, spent more than $3 million to strengthen its schools. Glendale, Alhambra, and other cities in Los Angeles County also moved forward to increase the safety of their schools. While less thorough in its efforts, San Diego nevertheless expended more than $100,000 to remove ornamentation from and add bracing to its school buildings. Even in northern California, far removed from the destruction of the Long Beach earthquake, a number of school districts reconstructed some of their school buildings for greater seismic safety.[48] In many cases, school officials were prompted by a ruling of the state's attorney general, issued in November 1933, which stated that members of local school boards would be held personally liable for injuries or deaths caused by earthquake damage to school buildings. Frightened by the prospect of such an enormous liability, many school board members asked the State Division of Architecture to inspect their existing schools for safety and sought ways to raise the money necessary to bring them up to current safety standards.[49]

While the Field Act was prodding Californians to render their school buildings more earthquake-resistant, other kinds of buildings also began to be modified for greater seismic safety, especially those in southern California. In repairing earthquake-damaged commercial and public buildings in Long Beach, for example, contractors often replaced broken masonry walls with reinforced concrete in such a way as to increase their earthquake resistance. Even some buildings in Los Angeles County that had received only minor damage were extensively remodeled and strengthened against future earthquakes. Elevated tanks for holding water, which had sustained considerable damage during the Long Beach earthquake, were generally strengthened throughout California, as were power plants. As an editor of the *Engineering News-Record* remarked in 1938, five years after the Long

Beach earthquake, recently constructed buildings in southern California "reflect[ed] earthquake design quite plainly."[50]

Of course, the widespread use of earthquake-resistant construction after the Long Beach earthquake did not mean that engineers and seismologists had won a complete or permanent victory in their campaign for greater seismic safety. As early as 1935, a number of elected officials, businessmen, and others, arguing that earthquake hazards in California had been vastly exaggerated, sought to weaken or eliminate the Field and Riley Acts. In particular, school board officials objected to being exposed to liability in case a building collapsed during an earthquake and killed or injured someone. Structural engineers, seismologists, and officials of the State Division of Architecture had to rally their organizational resources and lobby the state legislature to preserve the two laws. In 1939, the Garrison Act did remove personal liability for school board members and thereby eliminated the major incentive driving the strengthening of existing schools. But seismic safety advocates, with their ever-vigilant allies within the Division of Architecture alerting them to every threat to the new regulatory-state apparatus, did succeed in maintaining the chief provisions of the Field and Riley acts.[51]

Thus a regulatory approach to earthquake hazard mitigation took root in California, marking a substantial change in attitudes toward seismic hazards. As late as 1906, California scientists and engineers had joined the state's boosters in belittling seismic hazards, and as a result little was done in the rebuilding of San Francisco to provide against future earthquakes. But by the mid-1930s many southern California towns required earthquake resistance in their building codes, and a state agency ensured that schools would be able to withstand future shocks. These seismic safety regulations were supported not only by the building inspectors and state bureaucrats charged with their enforcement but also by organized groups of seismologists and earthquake engineers, who stood ready to defend the need for earthquake preparedness with both scientific data and lobbying muscle.

This regulatory-state apparatus had its roots in the work of the Progressive scientists who had been awakened to California's seismic hazards by the 1906 San Francisco earthquake. G. K. Gilbert in early 1909 first issued a general plea to the California public for seismic safety, arguing that it was the duty of scientists and engineers to ensure that buildings could resist the earthquakes that were bound to strike at any time. John C. Branner and Harry Wood, taking up Gilbert's call, created a scientific infrastructure capable of gathering data on the areas particularly susceptible to earthquakes: Branner by revitalizing the Seismological Society and its collection and dissemination of geological observations, Wood by securing the Carnegie Institution's support for a regional seismograph network in southern California. Bailey Willis added another significant component to the campaign for seismic safety by embarking on a concerted public relations effort and by involving engineers and architects more actively in the search for how to make buildings earthquake-resistant. Other seismologists continued Willis's attempts to foster alliances with engineers and architects. By the early 1930s, structural engineers had become full-fledged members of the campaign for earthquake preparedness and were poised to become critical players in its implementation. Scientists and engineers were prepared in 1933 to shape public opinion after the Long Beach earthquake and use it to create an effective regulatory apparatus for ensuring earthquake-resistant construction.

The scientists and engineers who created this apparatus remained deeply committed to the values of the Progressive movement. They firmly believed that the application of science and technology were the key to human progress and that it was the duty of scientists and engineers to use their expertise for human betterment. As a result, they energetically sought ways to increase scientific knowledge about earthquake hazards and mitigation strategies. They also were appalled at the inefficiency and waste produced by short-sighted developers and greedy building contractors. Rickety buildings that readily collapsed in an earthquake were just as bad as forests clearcut without replanting or unmanaged water resources that were being frittered away—they posed a threat to the stability of a growing industrial economy. Consequently, like their colleagues in the conservationist environmental groups, seismologists and structural engineers sought to introduce construction methods that were more efficient and sustainable in the long run. In doing so, they adhered to the Progressive belief that the application of technical expertise was the best way to bring about such efficiency. And after the collapse of Willis's more broad-based public relations campaign in the late 1920s, seismic safety advocates grew to believe that quietly building consensus among technical experts and fostering alliances with government bureaucrats, who could bring the power of the state to bear, was the best strategy for achieving their goals. Thus, the regulatory-state apparatus created in the 1930s crystallized around a set of Progressive values—values that would persist through the remainder of the century and bring the regulatory apparatus, with its marriage of science and the state, to full bloom.

CHAPTER 5

Earthquake Experts and the Cold War State

World War II marked a major hiatus in earthquake research. As scientists and technicians left for war work, the routine determination of earthquake epicenters from seismographic records essentially ceased. Earthquake engineers also departed for the front, and the U.S. Coast and Geodetic Survey's work in California came to a halt. Even those who remained at their academic posts, such as Perry Byerly at the University of California and Charles Richter at the California Institute of Technology, turned their attention from earthquake investigation to teaching physics to servicemen or conducting defense-related research. The Seismological Society of America suspended annual meetings in 1943, and its *Bulletin* shrank to half its prewar size.[1] Once the war ended in 1945, scientists and engineers could again focus on earthquakes, and they sought to reestablish and expand the programs of research and advocacy they had conducted before the war. But they did so in a radically altered context, one shaped by a new relationship between science and the state.

During the war, scientific research and engineering ingenuity had produced many of the tools—radar, the proximity fuse, penicillin, and most prominently the atomic bomb—that had brought about Allied victory. Scientific experts thus had demonstrated their value to the state. As the continuing Soviet military threat after 1945 sparked a permanent "cold war" mobilization in the U.S., the federal government sought to maintain its access to scientific expertise by funding research through such agencies as the Office of Naval Research, the Atomic Energy Commission, and eventually the National Science Foundation. These agencies made an ever-increasing supply of highly trained experts available and facilitated continuation of research that served American military and economic interests. At the same time, government agencies came to rely increasingly on scientific experts in formulating policy in arenas as diverse as national security, the economy, and social welfare. The growing alliance between science and the state clearly benefited both sides.[2]

The increasing reliance of the state on science and technology had a particularly marked effect on California's economy. During World War II, the federal government had spent $35 billion in California, including $12 billion for aircraft production and another $3 billion for shipbuilding. As a result, California became a center for defense contractors, with such companies as Lockheed, Northrup, and Hughes being headquartered there. After 1945, the flow of federal dollars continued; the Department of Defense alone spent more than $5 billion annually on research and development of new technologies. By 1970, defense and aerospace companies in California employed more than 600,000 people, accounting for more than a third of the state's manufacturing labor force. The growth of this military-industrial establishment went hand-in-hand with a boom in California's academic sector. Stanford University, for example, used abundant federal funding in the Cold War years to make itself into both a nationally known research university and the entrepreneurial center of "Silicon Valley," where technological ingenuity combined with defense contracting to produce a flourishing electronics industry. The economic expansion in California was accompanied by a demographic explosion; from 1940 to 1970, the state's population nearly tripled, from 7 million to 20 million, making it the nation's most populous. California's growth brought with it increasing political power; Eisenhower's vice president, for example, was native Californian Richard Nixon. Perhaps emblematic of the new postwar power relationships between California scientists and engineers and the federal government was the appointment in 1961 of Glenn Seaborg, a physicist who had been president of the University of California, as chairman of the Atomic Energy Commission.[3]

While California was at the forefront of developing the new military-industrial-academic complex, it also provided the seedbed for the protests that arose against this establishment in the 1960s. In the 1950s, the close intertwining of science and the state had produced a technocratic mode of politics, in which complex policy decisions were made in insulated policy arenas dominated by technical experts, with the public deferring to both scientific expertise and state authority. In the 1960s, though, first civil rights activists and then anti–Vietnam War protesters came to challenge both the legitimacy of state authority and the wisdom of scientific expertise. To them, the delegation of important decisions to closed, expert-driven policy-making circles was elitist, antidemocratic, and detrimental to the public welfare. One of the first challenges from the grass roots occurred at the University of California at Berkeley in the fall of 1964, when the Free Speech Movement erupted to protest attempts by the university administration to limit discussion of civil rights and other politically charged topics. As the war in Vietnam intensified, student protests at Berkeley and elsewhere also escalated; activists staged sit-ins, teach-ins, and draft card–burning ceremonies to challenge the military-industrial-academic alliance that they blamed for bringing about the war in Southeast Asia.

Environmentalists, too, convinced of the moral necessity of preserving undisturbed nature and concerned that elite policymakers and experts gave insufficient attention to such moral imperatives, joined in the attack on expert authority and became more confrontational in their efforts to bring about greater environmental protection. Here also a Californian led the attack. David Brower, the Sierra Club's outspoken executive director, mobilized effective campaigns to force the protection of wilderness areas and stop new development projects across the state and the nation.[4]

These developments in the broader relationship between science and the state affected earthquake research in California. In the years after 1945, seismologists and earthquake experts mostly sought to just maintain and expand the programs of research and advocacy they had conducted before the war. During these years, however, first earthquake engineering and then seismology came to depend more and more on funding from military agencies and the National Science Foundation, as federal support for earthquake research increased a hundredfold. Seismologists also turned to the new civil defense apparatus, set up in the 1950s to guard the country against the effects of a Soviet nuclear attack, to spread their message about earthquake safety. In these ways, earthquake researchers came to feel ever more comfortable as experts in the service of the state.

The depth of their commitment to this relationship became evident in the early 1960s, when environmental activists started challenging both the earthquake safety of nuclear power plants in California and the cozy relations within the military-industrial-academic complex that allowed such plants to be built with little interference by regulators. Seismologists and engineers generally recoiled from the tactics employed by grassroots activists and instead emphasized that decisionmaking needed to remain within the insulated policy arena they had established with their allies from the state bureaucracy.

Reestablishing Earthquake Research after the War

In many ways, earthquake research after World War II looked much like it had before the war, using many of the same techniques to pursue the same goals and seeing change in only small, incremental steps. At both Caltech and the University of California at Berkeley, for example, seismologists in the late 1940s reestablished routine operation of their seismograph networks for recording local earthquakes and determining their epicenters. Over the next decade, these networks grew slowly. The southern California network, which Caltech had taken over from the Carnegie Institution in 1937, expanded to 16 stations by the late 1950s, while Berkeley's northern California network reached 11 stations. The main purpose for operating these networks also remained what it had originally been in the 1920s: to locate active faults by noting where barely perceptible earthquakes clustered. At Caltech, Richter by 1948 had concluded that areas of frequent small earthquakes did not necessarily correlate with areas where stronger earthquakes might strike. But at Berkeley, Byerly persisted in trying to locate active faults by this means, and even the Caltech group continued to accumulate data on local seismicity in order to detect spatial patterns in the release of stress by earthquakes.[5]

The U.S. Coast and Geodetic Survey also returned to earthquake work in California. In the mid-1940s, several California seismologists, led by Byerly, urged the survey to repeat its triangulation of coastal California. They wanted once again to test the elastic rebound theory, first proposed by Harry Fielding Reid after the 1906 earthquake, which asserted that strain built up gradually in the earth's crust before being released in an earthquake. Previous geodetic surveys in the San Francisco Bay area had failed to show beyond the errors of measurement that strain was building up.[6] In 1947, the survey once again triangulated the bay area. By this time, the slow accumulation of strain over time had finally produced movements that were clearly discernible beyond the errors of measurement. A comparison of the new results with those from 1882, 1906, and 1922 showed unequivocally that points several miles west of the San Andreas Fault had consistently moved northward relative to those east of the fault at a rate of about 2 inches a year.[7] Since no significant earthquake had occurred in the bay area since 1906, this meant that about 7 feet worth of unrelieved strain had now built up along the fault in that area. Moreover, since movement along the fault in 1906 had amounted to about 16 feet (192 inches), an annual strain rate of 2 inches meant that an earthquake of the size of the 1906 event should happen about once every hundred years. The hope of Willis, Lawson, and others in the 1920s that geodetic surveys would produce convincing evidence for quantifying the recurrence interval of great earthquakes in California had finally been fulfilled.

The Seismological Society of America resumed its prewar activities and began to rebuild its membership, which grew from about 600 in 1945 to slightly more than 800 in 1961.[8] Despite encountering an acute financial crisis due to rising publishing costs in the inflationary period of the late 1940s, the society returned to publishing a full-size *Bulletin* averaging about 400 pages annually. As before the war, many of the articles appearing in the *Bulletin* were technical disquisitions on theoretical seismology with such esoteric titles as "Dependence on Azimuth of the Amplitudes of P and PP" or "On the Layer of Relatively Low Wave Velocity at a Depth of About 80 Kilometers." Such articles drew protests from the many engineers, insurance people, and other nonseismologists in the society's membership, who found them incomprehensible.[9]

The society's editors, however, also continued to include accessibly written articles of general interest. Like the pieces that characterized the *Bulletin* during its early existence in the 1910s, these articles frequently focused on the damaging effects of earthquakes and railed at poor design and construction methods. In the 1950s such articles issued primarily from the pen of Karl Steinbrugge, a structural engineer who had worked for the State Division of Architecture before joining the Pacific Fire Rating Bureau, a center of the Pacific Coast fire insurance industry. In 1954, he and a colleague from the bureau contributed a detailed and lavishly illustrated account of the damage produced in 1952 by several strong earthquakes near Bakersfield, in California's Central Valley. After cataloguing the damaged structures and their weaknesses, Steinbrugge concluded that when buildings were "specifically designed to resist a high degree of shock, little loss should be expected," a sentiment that had been voiced repeatedly in the *Bulletin*. Over the next several years, Steinbrugge also described earthquake damage in western Nevada and northern California, pointing out in each case the shortcomings of structures that had failed.[10] In later years, he would become the leading advocate for greater seismic safety in California.

California earthquake engineers, meanwhile, established a new organization to promote research in their discipline. In the spring of 1947, the Coast and Geodetic Survey consulted earthquake engineers from both northern and southern California on the future direction of its work in strong-motion seismology. These engineers in September of 1947 constituted themselves as the Advisory Committee on Engineering Seismology. With their support, the survey expanded its program; by the late 1950s, it operated strong-motion seismographs at 49 sites in California and 22 elsewhere.[11] But the Advisory Committee at its first meeting also decided to push for the establishment of an independent Earthquake Engineering Research Institute to supplement the survey's research efforts. As originally envisioned, the institute's objective was "the thorough investigation of all matters which will lead to the most economical yet reliable design and construction methods and locations for various structures to resist earthquake damage." At first, the earthquake engineers hoped that the institute (which was formally inaugurated in 1949 with 15 members) would establish its own laboratory to conduct engineering research. Failing to attract either funding or a laboratory director to carry out the research, institute members soon abandoned these plans. Instead, the institute turned to hosting earthquake engineering conferences (including the first World Conference on Earthquake Engineering, held at Berkeley in 1956) and endorsing engineering research at already existing academic laboratories. In this manner, it joined the structural engineering associations of California as an institutional voice for earthquake engineering.[12]

California earthquake researchers continued their prewar outreach work also. Seismologists at both Caltech and the University of California, for example, responded to inquiries from insurance companies about relative earthquake risks in the state. Richter in particular also aided school board members who needed information on the seriousness of earthquake threats, to use in campaigns for support of local bond issues to strengthen schools.[13] Moreover, seismologists and engineers consulted on various large construction projects. When the state's Department of Water Resources announced plans in 1958 for a massive State Water Project that would include several large dams in northern California and an aqueduct to transfer the water to the Los Angeles region, Caltech seismologist Hugo Benioff convinced the department to include earthquake considerations in its designs. By 1962, a consulting board comprising some of the state's most prominent seismologists and engineers was advising the department, which had decided to use more costly surface aqueducts rather than tunnels in several places so that earthquake-damaged water conduits could be repaired more easily.[14]

Seismologists also continued their involvement in legislative matters. For example, they championed an ordinance passed by the City of Los Angeles in the late 1940s requiring parapets, cornices, and other overhanging ornamentations that might come crashing down during an earthquake to be either strengthened or eventually removed.[15] Yet most lobbying by seismologists was geared not toward new initiatives but toward protecting the achievements of the 1930s, especially the Field Act. As school construction boomed in rapidly growing California in the late 1940s and early '50s, school officials and contractors became increasingly critical of the act's earthquake design provisions, claiming that they significantly increased the cost of new schools and produced unnecessary delays.[16] In 1951, a bill was introduced in the legislature to eliminate them. At a hearing before a special State Assembly committee, earthquake researchers defended the Field Act. Byerly explained at length the history and geography of earthquakes in California, pointing out that earthquake damage could occur anywhere in the state and ending his presentation with a dramatic picture of a student killed in 1949 in a Washington State earthquake. A representative of the California Council of Architects then explained that Field Act provisions added less than 5 percent to the cost of schools; the rapidly rising cost of construction was to be blamed on increasing educational requirements for school buildings rather than on earthquake safety. These arguments swayed the Assembly to preserve the Field Act, at least temporarily.[17]

Attacks on hazard mitigation measures came from inside the engineering community as well. In particular, some northern California earthquake engineers, concerned primarily with the economics of construction, considered the earthquake provisions then commonly found in building codes to be excessive. San Francisco had not been among the cities that adopted seismic design requirements in their building codes following the Long Beach earthquake. In 1947, the city's director of public works, Harry Vensano, finally succeeded in having a code enacted that required buildings to be designed for lateral forces ranging from 0.037g to 0.08g. Although these provisions were less than the 0.10g commonly required in southern California building codes, San Francisco engineers protested strenuously, asserting that because of these "uneconomical and impractical" earthquake provisions, "the cost of skyscrapers would be jumped so sharply that no more would be built."[18]

In early 1948, the San Francisco Section of the American Society of Civil Engineers and the Structural Engineers Association of Northern California appointed a joint committee to devise what they considered more rational seismic design provisions. They eventually settled on requiring buildings to be designed for lateral forces ranging from 0.02g to 0.06g. These low values were based essentially on the bald assertion that numerous structures not designed to resist lateral forces of more than 0.02g had survived even the strong shaking of the 1906 earthquake. Caltech earthquake engineers, led by R. R. Martel, attacked these provisions as inadequate, but the northern California engineers dismissed the criticisms as the harping of academic researchers who did not have to face the practical task of designing buildable structures.[19] Throughout the 1950s, structural engineers continued to debate the merit of the building code provisions passed following the Long Beach earthquake. Little thought was given to the possibility that these provisions might be too lax rather than too stringent, even though some strong-motion records showed that horizontal accelerations might reach as high as 0.33g during an earthquake.[20]

The defensive stance of earthquake researchers also affected the way they reacted to several noteworthy earthquakes that struck California in the 1950s. Rather than using them as occasions to push for new hazard mitigation measures, seismic safety advocates used them to stress the value of the measures that had been enacted in the 1930s, thereby contributing further to a sense of complacency concerning seismic hazards. On July 21, 1952, an earthquake of magnitude 7.5—the strongest in California since 1906—hit Kern County, at the southern end of the Central Valley. This temblor and a magnitude 5.8 aftershock near the city of Bakersfield several weeks later together killed 14 people and, by contemporary accounting, caused at least $38 million in property damage, $10 million of that in Los Angeles County to the south.[21] Yet engineers quoted in the *Los Angeles Times* immediately after the quake focused not on the damage that had occurred but on the fact that several Field Act schools near the epicenter had sustained no structural damage and that buildings designed under modern building codes had remained intact. In essence, they proclaimed that engineering had already solved the earthquake problem and that no new mitigation measures were needed, as only old, premodern buildings had sustained any damage.[22] A similar response greeted the magnitude 5.3 earthquake that struck Daly City, a suburb just south of San Francisco, on March 22, 1957. The strongest earthquake to hit the bay area since 1906, it caused $1 million in damage, mostly superficial plaster cracks. Once again, local engineers and public officials concluded that modern buildings had performed admirably, that no upgrading of the weak San Francisco building code would be necessary. Charles Richter and some other earthquake researchers insisted to San Francisco newspapers that the Daly City earthquake was rather minor, in no way comparable to the 1906 earthquake, and thus that the small amount of damage was no indication of the toll the coming big earthquake would exact. These warnings, however, fell on deaf ears; San Franciscans, echoing the views of local earthquake engineers, now smugly asserted that there was "no cause to fear a quake like '06."[23]

Federal Support for Earthquake Experts

Earthquake experts in the postwar years sought mainly to reestablish the research programs they had carried on before the war and to defend the gains in seismic safety they had achieved in the 1930s. Yet their community did undergo one significant and far-reaching change: a vast expansion in the role of the federal government in their work. This expansion was stimulated in particular by the many ways in which earthquake work became tied to that central technological artifact of the Cold War years, the nuclear bomb. Earthquake experts obtained substantial federal funding because they could aid in designing nuclear blast–resistant shelters and could help monitor underground explosions for violations of nuclear test bans. Even advocacy for greater seismic safety obtained a new outlet in the Cold War years, as a few earthquake experts discovered that the governmental civil defense apparatus established to guard against a Soviet nuclear attack could also be used to prepare for earthquakes. Through all of these links, earthquake experts became ever more enmeshed in the Cold War state.

Among the areas involved in earthquake research, earthquake engineering was the first to receive the patronage of the federal government. It attracted the attention of military agencies because its research could be applied to designing buildings to withstand the effects of nuclear blasts. As one San Francisco engineer noted in 1952, "most of the major and many of the minor features of aseismic and blast resistant design are similar." In both cases, the main goal was to design a structure that would resist the effect of lateral wave forces, transmitted through the earth in the case of earthquakes and through the air in the case of nuclear blasts. Thus, the methods developed by earthquake engineers to provide for the horizontal rigidity of structural elements and the strength of joints and other connections could easily be applied to designing nuclear blast shelters.[24] Blast-resistant design became important to U.S. military agencies in 1949, when the Soviet Union exploded its first atomic bomb, thereby ending America's nuclear monopoly. In succeeding years, the American military establishment, while pursuing a program to develop ever more powerful nuclear weapons, sought to ensure the safety of both military installations and civilian populations in the case of a Soviet nuclear attack. In this context, government funding for earthquake engineering took off.

Earthquake engineers did their part in pointing out the relevance of their discipline to nuclear blast design. In 1952, when the Earthquake Engineering Research Institute held its first conference, it brought together earthquake engineers and structural engineers working on blast effects to exchange information on methods common to both fields. Conference organizers adopted this approach in order to attract greater attention for their field and in hopes of securing financial support for the institute. The conference was a great success, with about 200 engineers in attendance, and the published conference proceedings became a standard reference source for structural engineering for both earthquake and nuclear blast effects.[25]

Meanwhile, such government agencies as the Office of Naval Research, the U.S. Army Corps of Engineers, and the Air Force, as well as the Sandia Corporation, a private contractor operating a nuclear weapons laboratory in New Mexico for the Atomic Energy Commission, began funding work by structural engineers, in California and elsewhere, on the problem of blast-resistant design. At Caltech the Office of Naval Research sponsored basic earthquake engineering research that might strengthen the theoretical underpinnings of structural engineering.[26] In at least one case, that of Lydik Jacobsen at Stanford, government funding led an engineer to abandon research on earthquake problems and devote himself completely to blast-resistant design.[27] But government funding also drew people to earthquake engineering. At the University of Illinois, for example, Nathan Newmark had concentrated in the 1940s on general problems of designing reinforced concrete structures. In the 1950s, with funding from the Air Force, the Corps of Engineers, and other agencies, he acquired an interest in designing such structures against lateral forces from both nuclear blasts and earthquakes and developed methods for calculating the response of structures to lateral forces using high-speed computers.[28] In this manner, government funding in the nuclear age greatly strengthened the research foundation for earthquake engineering.

In the 1960s, earthquake engineers became even more closely tied to the federal government through another link between nuclear weapons and earthquake engineering. In the early years of nuclear weapons development, testing of nuclear explosives had generally been conducted in the atmosphere. In response to growing protests over the effects of radioactive fallout from atmospheric tests, however, the Atomic Energy Commission, which managed the American nuclear testing program, in 1957 initiated underground tests. Buried in tunnels or shafts at depths of a few hundred to a few thousand feet, the nuclear bombs exploded in these tests imparted considerable energy to the surrounding rock. An underground explosion of a typical nuclear bomb, with a yield somewhat more than 20 kilotons of TNT equivalent (roughly the size of the bombs that destroyed Hiroshima and Nagasaki) produced earth shaking equivalent to a magnitude 5 earthquake. During the 1960s, the size of bombs exploded at the Nevada Test Site gradually increased, and by 1963 large tests were causing buildings in Las Vegas, about 90 miles south of the test site, to sway. At this point, the Atomic Energy Commission contracted with John Blume, a prominent San Francisco earthquake engineer, to ensure public safety during future underground tests. Blume prodded nuclear engineers into altering a few test shots to minimize the ground shaking they produced, and he induced Las Vegas officials to upgrade their building code by including more stringent seismic design requirements. For the most part, though, Blume served as a public relations agent for the commission, reassuring local residents that no serious damage would result from the nuclear explosions. In return for this service, Blume was able to set up instruments in various Las Vegas buildings and gather valuable information on how structures respond to seismic waves. For such purposes, nuclear explosions of course had the advantage over earthquakes that their occurrence could be known beforehand.[29]

Around 1960, underground nuclear testing also provided the occasion for a vast increase in government funding for seismology. Responding to increasing calls for an end to all nuclear tests, Western and Soviet experts met in 1958 to discuss the means by which a test ban might be monitored. They quickly agreed that detecting above-ground nuclear explosions posed few problems, but monitoring a ban on underground nuclear tests proved far more difficult. Since radioactive products were generally contained underground in such tests, the only way to determine from afar that a test had occurred was from the seismic waves it gave off. Soviet experts asserted that it would be relatively easy to distinguish the seismic waves produced by underground explosions from those emitted by earthquakes, as their spatial distribution and shape would be quite different. American experts, however, were less sanguine, especially after tests in late 1958 suggested that small explosions could be hidden among the noise produced by the many small earthquakes occurring daily. In the end, the two superpowers could agree only on a limited test ban treaty, signed in 1963, that outlawed nuclear tests in the atmosphere, under the oceans, and on the earth's surface, but not underground.[30] The problem of discriminating nuclear explosions from earthquakes did, however, trigger government interest in seismological research. In late 1959, the U.S. Department of Defense assigned the task of increasing American capability for detecting underground nuclear explosions to the Advanced Research Projects Agency, a newly established military agency that sponsored and coordinated research. Over the next 12 years, the agency, in a program code-named Vela Uniform, spent nearly $250 million on seismological research.[31]

Vela Uniform was of course not the first source of federal support for seismological research at universities. By the early 1950s, seismologists at Columbia University's Lamont Geophysical Observatory near New York City were receiving research funding from the Office of Naval Research and the Air Force. And Lamont graduate Frank Press, who headed Caltech's Seismological Laboratory from 1957 to 1965, shifted that institution toward a far greater reliance on federal funding, encouraging even veteran local seismologist Charles Richter to seek a grant from the National Science Foundation.[32] Still, the underground-test detection problem brought about a significant increase in federal funding. In only three years, federal support for seismology jumped sixtyfold, from $500,000 in 1958 to $30,000,000 in 1961.[33]

This sudden influx of money dramatically affected the seismological community. As new graduate students flocked to the field, the membership of the Seismological Society of America swelled from about 800 in 1961 to nearly 1,100 in 1964. The increasing amount of seismological research being published also led the society to expand its *Bulletin* from four to six issues per year. The Air Force funded this expansion as part of the Vela Uniform effort, through a $15,000 publication grant to the society.[34] Naturally, most of the research reported in the *Bulletin* was also funded by the Advanced Research Projects Agency; of the 51 research articles published in the *Bulletin*'s regular issues for 1963, as many as 31 acknowledged support from Vela Uniform.[35] Moreover, government interest in the nuclear detection problem led to a shifting in the seismological community's main research interests. Many of the papers appearing in the *Bulletin* in the early 1960s dealt with the shapes and relative amplitudes of waves given off by distant earthquakes and nuclear explosions and the various factors (such as geological characteristics at the source) that might affect them. Richter complained that Caltech's Seismological Laboratory was spending less and less time on its historic mission of recording and analyzing local earthquakes; this was also true of the research community as a whole.[36]

Even though federal support from the Vela Uniform program moved the seismological community toward a greater focus on the measurement and analysis of waves from distant earthquakes, the Cold War state also provided those interested in local earthquakes with a new avenue through which to promote seismic safety. This avenue arose through the establishment of a civil defense apparatus in the early 1950s, after the Soviet achievement of nuclear capability had raised the specter of an attack. In 1950,

President Truman established the Federal Civil Defense Administration. It underwent a number of bureaucratic metamorphoses over the next several decades and eventually merged into the Federal Emergency Management Agency, which had been created in 1979. But in the 1950s and 1960s the Civil Defense Administration's mission was to prepare the country for an enemy military attack by encouraging the construction and designation of fall-out shelters, promoting "duck-and-cover" exercises and other emergency drills, stockpiling critical materials, and laying out plans for how to restore governmental functions and economic activity after an attack.[37] State-level activities by such groups as the California State Office of Civil Defense (later renamed the California Disaster Office), established in 1950 to help prepare the Golden State for enemy attack, augmented the work of the federal civil defense apparatus. At the local level, too, government officials and interested citizens established civil defense councils to foster preparedness. As the 1950s passed without any military attack on the continental United States, civil defense groups, seeking to justify their continued existence, expanded their activities to include natural hazards as well as military threats.[38] This move enabled seismic safety advocates, foremost among them Richter, to channel their calls for greater earthquake preparedness through the civil defense apparatus.

Richter, who in the 1940s had still been interested primarily in pure seismology, by the late 1950s had become one of the few seismologists who wanted to "stir up civic organizations and local authorities" to do something about the earthquake problem in California.[39] In the early 1960s, he finally turned to the civil defense apparatus to give his warnings wider circulation. His first talk to a civil defense group came in October 1962, when he addressed the California Women's Civil Defense Conference, in a talk entitled "Earthquake Risks in California." In his speech, he emphasized the earthquake hazards arising from old buildings that did not meet the building codes enacted since 1933 as well as the problems that might be caused by lifelines (such as railroads, aqueducts, and pipelines) severed during fault rupture and by panic spread through rumor-mongering.[40] In talks the following year to the Los Angeles County and Cities Disaster and Civil Defense Commission and the California Governor's Conference on Disaster Preparedness, Richter again urged civil defense officials to be prepared to deal with panic, cut-off lifelines, and collapsed buildings.[41] Finally, in May 1964, Richter addressed the annual regional conference of the United States Civil Defense Council, which was meeting jointly with the Confederation of Northern and Southern California Civil Defense and Disaster Associations in Long Beach. Here Richter found himself on a program that concentrated on responses to nuclear attack, with the keynote address given by Edward Teller, the famous proponent of the hydrogen bomb. Richter targeted his audience by noting that "the effects of a large earthquake are comparable to those of enemy attack in suddenly creating a disaster situation simultaneously over a large area" and focusing particular attention on "general panic and disruption of communications," two effects with which military planners already had considerable experience.[42]

To supplement his talks, in the fall of 1963 Richter drew up a 15-page mimeographed pamphlet, which was also published in a Caltech magazine for general readers. It listed a number of precautions that could be taken by both individuals and communities before an earthquake. In addition to practical advice, Richter offered encouragement to political action: "As a member of a homeowner's or business men's organization you can work to have weak and dangerous buildings removed or strengthened. As a voter, support school bonds for replacing old weak buildings by new safe ones."[43] After publicizing the pamphlet at both the governor's conference and the Long Beach meeting, Richter quickly distributed hundreds of copies to Red Cross and civil defense groups; in 1965, the California Disaster Office reprinted it in order to ensure even wider distribution.[44] Richter worked with the Los Angeles County Disaster and Civil Defense Commission to produce earthquake disaster cards that handily summarized his advice; by mid-1964, the commission had distributed 800,000 of them.[45] Finally, Richter helped local civil defense offices put together realistic scenarios of large earthquakes so that the offices could test and reinforce their preparedness measures through simulation exercises.[46] In all of these ways, Richter made earthquake preparedness an integral part of state-sponsored disaster preparedness activities and enrolled government officials in the promotion of seismic safety while helping the civil defense apparatus justify its continuing existence.

Earthquake Experts and Challenges to Authority

In the postwar era, earthquake experts and the state moved ever closer together. Most earthquake experts were clearly comfortable with this relationship—it provided substantial resources for conducting research and pushing seismic safety, and it was predicated upon the inclusion of experts in elite decision-making circles, where state officials respected and deferred to their technical knowledge and authority. In all of these ways, the relationship reaffirmed the Progressive values that had driven the earthquake research community before the war.

For the most part, the growing connections between earthquake researchers and the state proceeded with few protests or tensions. In one case, though, that of choosing sites for nuclear power plants in California, a strong challenge emerged. In the early 1960s, experts, industry leaders, and government officials had agreed to proceed with constructing two nuclear power plants on the California coast. Grass-roots activists, angered by the closed-door, backroom dealmaking by which these decisions had been made, chose to focus public attention on the seismic hazards inherent in the two plant sites. Their extensive public relations efforts to scare people about the possible consequences of a large earthquake near the sites provided a potential alternative channel through which seismic safety advocates could raise earthquake consciousness in the state. Yet the earthquake experts, disturbed by the activists' lack of deference to expertise and authority, generally recoiled from allying themselves too closely with this new force.

In the late 1940s and '50s, nuclear energy was one of the most insulated policy arenas in the United States, as decisions about nuclear power were made in a tightly confined iron triangle composed of government officials in the Atomic Energy Commission, their supporters in the Joint Committee on Atomic Energy of the U.S. Congress, and top utility officials. These policymakers all strongly favored the development of a commercially viable nuclear power industry in the United States. The use of nuclear fission for civilian power production was an important part of the government's Atoms for Peace program, which was designed to demonstrate that the splitting of atoms, used to such deadly ends at Hiroshima and Nagasaki, could also be employed for beneficial purposes. Toward this end, the Atomic Energy Commission provided substantial economic subsidies to power companies in order to bring the cost of nuclear power plants down to the level of coal-or oil-fired plants. The commission's ceaseless promotion of nuclear energy exacerbated a fundamental tension in its mission, because it was charged with both regulating and promoting civilian uses of the atom. In the 1950s, the commission attempted to satisfy this dual mission by selecting power plant safety experts who could be relied upon to provide safeguards that met minimal standards but did not place onerous burdens on the construction and operation of the plants.[47]

The tightly insulated arena for approving nuclear power plants appeared to work smoothly for Pacific Gas and Electric Company (PG&E), central California's dominant private utility company. In the late 1950s, it decided to build a reactor on Bodega Head, a promontory on the Sonoma County coast about 50 miles northwest of San Francisco. The site was close to the San Andreas Fault zone, but in a submission to the Atomic

Energy Commission, several prominent earthquake experts (including seismologists Don Tocher of the University of California and Hugo Benioff of Caltech and renowned earthquake engineer George Housner) asserted on PG&E's behalf that modern engineering techniques would eliminate any possible hazard arising from a repeat of the 1906 earthquake on the fault. Dangling prospects of economic development and an increase in the tax base before Sonoma County officials, PG&E quickly obtained local government approval for its project in several closed meetings that effectively shut out public input. Through backroom negotiations, PG&E also succeeded in having the California State Division of Beaches and Parks drop its plans to acquire Bodega Head as a state park. The California Public Utilities Commission held a pro forma hearing in early 1962 to give its blessing to the reactor project, and the Atomic Energy Commission appeared unlikely to object to PG&E's plans.[48]

PG&E's Bodega Head project was derailed, however, by a grass-roots coalition of local residents fearful of the disruption of their traditional lifestyles and environmentalists who wanted to preserve the area as a state park. These activists organized in mid-1962 as the Northern California Association to Preserve Bodega Head and Harbor. They were clearly motivated by traditional preservationist concerns for protecting scenic landscapes from economic development, but they were also incensed by the arrogance of PG&E's engineers and the underhanded and antidemocratic ways in which government officials had colluded with PG&E to push forward the Bodega Head project without real input from local citizens. The association's executive director was David Pesonen, a young Berkeley resident and former Sierra Club staff member who moved in the circles that would soon produce the Free Speech Movement and other student demonstrations. Under his leadership, the association engaged in many of the tactics that would be made familiar by subsequent protest movements: letter-writing campaigns, placard-waving demonstrations and rallies headlined by locally popular musical groups, publicity stunts (such as the release of balloons at Bodega Head to dramatize the potential spread of a radioactive cloud along the coastline), distribution of incendiary pamphlets, and incessant lawsuits designed to harass and embarass PG&E. As Pesonen later recalled, the association realized that it could not stop the Bodega Head project through the regulatory process, which was stacked in PG&E's favor. Rather, it turned to fighting the project in the press, seeking to arouse public sentiment against PG&E by any means possible. In the process, it garnered support from many highly placed officials, including California's governor and lieutenant governor, U.S. senator Claire Engle, and most importantly Stewart Udall, the conservation-minded secretary of the interior.[49]

In its campaign, the association seized upon whatever issues might attract public attention—fear of contamination from nuclear fallout, anger at the arrogance of elite decisionmakers, concern for preserving the beauty and ecology of California's coastline. In early 1963, the activists added the issue of seismic safety, raising the specter that, contrary to the company's assertion, a major earthquake might destroy the plant and lead to a nuclear catastrophe. The association was helped in this by Pierre Saint-Amand, a Caltech-trained geologist and seismologist who volunteered his services to the plant's opponents. Far from being a committed environmentalist or grass-roots activist (he worked at the China Lake Naval Weapons Center, where he developed a device used in the Vietnam War to wash out the Ho Chi Minh trail by artificially inducing rainfall), Saint-Amand nevertheless sympathized with the association, for he had witnessed at close hand the catastrophic damage wrought by the 1960 Chilean earthquake on even well-engineered buildings. In a report issued by the association in August 1963, Saint-Amand pointed to various seismic safety problems with the site: the lateral force given off by the greatest possible earthquake on the nearby San Andreas Fault could be much larger than assumed by PG&E consultants; a number of small bedrock faults on Bodega Head near the plant site, which PG&E's experts had declared to be ancient and clearly inactive, could be reactivated as subsidiary faults in a future earthquake and produce fault displacement directly underneath the reactor; and the bedrock was crushed and in places overlain by loose sediment, thus not being as solid a foundation as PG&E claimed.[50]

Even before his report was made public, Saint-Amand had forwarded his conclusions to Harold Gilliam, a newspaper reporter and cofounder of the anti-PG&E association, who had gone to Washington to work under Udall at the Department of the Interior. There, Gilliam and Udall convinced the Atomic Energy Commission to solicit a review of Bodega Head's geology and seismicity from the U.S. Geological Survey, an agency within the Interior Department.[51] Julius Schlocker and Manuel G. Bonilla, the engineering geologists assigned by the survey to investigate Bodega Head, concluded, in a report released in early October 1963, that the PG&E experts had been generally correct in characterizing the head's geology; in particular, the bedrock faults had not moved in a long time and thus posed no real danger of disrupting the reactor. In a separate report, however, the survey's seismologist Jerry P. Eaton, who had been added to the team as an afterthought, came to a significantly different conclusion. He pointed out that signs of recent fault activity might easily have been overlooked in the complex and difficult-to-read rocks on the head. Thus, he argued, "because we cannot prove that the worst situation will not prevail at the site, we must recognize that it might." As a result, he claimed that Bodega Head was "not an adequately safe location for a nuclear power plant." [52]

Both Saint-Amand and the survey geologists maintained a cool, detached style in their reports, with only Eaton indulging in some emotionally charged language. In publicizing the seismic safety issue, though, PG&E's opponents sensationalized it in order to grab wide attention. The association supplemented Saint-Amand's report with numerous photographs designed to highlight the head's proximity to the San Andreas Fault zone, and in a later publication it repeated Saint-Amand's warnings in far more lurid language. In releasing the Geological Survey's reports, Senator Engle and Secretary Udall (who had already used seismic hazards to publicly cast doubt on PG&E's plans) also put a definite spin on the conclusions, dismissing in a single sentence the comments by Schlocker and Bonilla that were favorable to PG&E while quoting at length Eaton's arguments against the site. These efforts had their desired effect, as San Francisco newspapers ran front-page headlines declaring "Bodega is Unsafe." [53] Many earth scientists were clearly made uncomfortable by this explicitly political use of geological expertise. In a rare editorial comment in the magazine of the California Division of Mines and Geology, for example, California State Geologist Ian Campbell, in a stab clearly aimed at the anti-PG&E activists, pleaded for the use of more factual information rather than discussions "founded on ignorance or even on misinformation." [54]

In the end, the seismic safety issue killed PG&E's proposal. In September 1963, during the exploratory excavation of a shaft for the reactor core, PG&E's consulting geologists noticed a particularly prominent bedrock fault as well as an offset in the sediments overlying it. They eventually concluded that the offset did not actually connect with the fault and thus was probably caused by soil compaction rather than an earthquake; moreover, they asserted that the offset disturbed only sediments that were clearly more than 40,000 years old. [55] Schlocker and Bonilla of the Geological Survey, however, reversed their earlier favorable appraisal of the site by concluding that the offset was in fact continuous with the bedrock fault and thus showed that fault rupture caused by an earthquake directly underneath the proposed reactor had occurred within geologically recent times. Thus, "displacement on the order of a few feet, either horizontally or vertically, should be anticipated." [56] PG&E engineers vehemently disagreed with this conclusion, but, in order to assuage even the most remote doubt among the Atomic Energy Commission's regulatory staff, they redesigned the power plant in early 1964 so that it might safely survive even a fault displacement of two feet directly underneath it.[57] The new design incorporated features for isolating the reactor from the surrounding bedrock (essentially by cushioning it with several feet of a special kind of sand) that had never been tried before on large structures. After repeatedly asking for clarifications of the design, the commission's regulatory staff (possibly pressured by the incessant questions raised by PG&E's adversaries) in October 1964 declared itself not completely satisfied that the new design would actually work during a substantial fault rupture. Rather than risking a prolonged fight against this judgment in the glare of publicity directed by Pesonen and other opponents, PG&E decided to abandon its project, much to the dismay of its consulting experts.[58]

A second nuclear power plant proposed for California also foundered in the mid-1960s on questions of seismic safety raised by citizen activists, although in this case the objections were brought to bear through forceful lawyering rather than publicity stunts. Late in 1962, the Los Angeles Department of Water and Power announced plans to construct a nuclear plant near Malibu, on the coast about 30 miles northwest of Los Angeles. Many of the wealthy residents of Malibu, including Bob Hope and Angela Lansbury, banded together with local land developers to oppose the plant, because they objected to this intrusion of industrial technology into their idyllic enclave and feared its effects on real estate values. Unlike Pesonen and his allies, these plant opponents were not community activists working in the milieu from which the protest movements of the late 1960s would emerge; consequently, they did not choose to rally public opinion. Rather, they hired skillful lawyers to oppose the proposed plant in various approval hearings. The effects of this strategy were just as devastating to the insulated nuclear power policy arena as the protests over Bodega Head had been. In a protracted series of hearings before the Atomic Energy Commission's Safety and Licensing Board that extended throughout 1965, the lawyers for the opposition called numerous geologists, mostly unknown local practitioners but including one Caltech professor, to cast doubt on the safety of the Malibu site with regard to both earthquakes and landslides. The lawyers also laid bare the cozy and lucrative working relationships between the Los Angeles Department of Water and Power and its consulting experts, and they focused repeatedly on seemingly innocuous changes of wording in geological reports that favored the department's cause. These attacks, prominently reported in local newspapers, did much to impugn the integrity of the experts testifying in favor of the proposed plant. The strategy was successful; although the board granted approval for the plant, it made the approval contingent on conditions (primarily regarding a redesign of the plant to withstand fault displacement directly underneath it) that the department found itself unable to meet.[59]

To focus public attention on questions of seismic safety, the campaigns against the two power plants used grass-roots organizing and courtroom fireworks, both methods not previously employed by seismic safety advocates. Yet earthquake experts did not rush to embrace these new methods. Rather, leading seismologists, including Richter, Clarence Allen, also of Caltech, and Richard Jahns of Stanford, distanced themselves from the "traumatic experience" of the nuclear plant controversies and steered clear of any continuing involvement with the activists.[60] For one thing, they doubted the sincerity of the activists' concern with seismic safety. Harold Gilliam, the newspaper reporter who had cofounded the association opposing the Bodega Head proposal, did write several columns for the *San Francisco Chronicle* expounding more generally on the seismic hazards facing the Bay area.[61] But in general, the power plant opponents had little interest in highlighting seismic safety issues not directly related to the proposed plant sites. As Jahns perceptively noted, the nuclear power plant controversies were "a prime example of zeroing in on geology as a mechanism available to persons really wishing to block the project for other reasons."[62]

Just as importantly, the methods used by the plants' opponents undermined the very aura of authoritative expertise that seismologists and earthquake engineers sought to maintain. Both the pamphlets distributed in the Bodega Head case and the legal cross-examinations used to stop the Malibu plant portrayed earthquake researchers as consultants with opinions for hire rather than as disinterested experts who could give objective evaluations. Moreover, the activists opposing PG&E raised more general objections to the power that experts held in deciding such issues as land-use decisions. These efforts at deconstructing the authority of experts did not sit well with earthquake experts. Since the 1930s, seismologists and engineers had become accustomed to working with the state in a relationship that relied upon the deference of government officials to their technical expertise. They were not about to give up this deference and the authority it conveyed; they had become too accustomed to being insiders in elite decision-making circles. Consequently, earthquake experts (with a few individual exceptions, such as Pierre Saint-Amand) did not form lasting affiliations with grass-roots activists. Nor, when these experts decided to push more vigorously for expanded seismic safety measures in the late 1960s and '70s, did they turn to litigation or the mobilization of public support. Rather, they worked behind the scenes to enroll allies within government agencies and legislative circles, using their insiders' access and the authority their expertise provided to quietly bring about increased hazard mitigation.

CHAPTER 6

New Initiatives for Earthquake Preparedness, 1964–1971

At 5:36 P.M. on Good Friday, March 27, 1964, an earthquake of magnitude 8.5—the largest earthquake yet recorded in North America—struck Alaska. It was felt over half a million square miles, and its effects reached as far as Oregon and northern California, where a quake-generated tsunami killed 16 people. In Alaska itself, the temblor caused more than $300 million in property damage; in Anchorage, 80 miles west of the epicenter, several buildings, including a department store and an office building, collapsed. Luckily, because many of the buildings were deserted due to the holiday, the death toll was lighter than it might have been, but 115 people lost their lives in Alaska.

The earthquake proved to be a formative experience for California earthquake researchers. Numerous geologists and engineers, traveling on their own impulse or sent by government agencies or private companies like Pacific Gas & Electric, departed for Alaska soon after the quake to examine the damage. Several engineers who had been smug about their ability to design against earthquakes changed their thinking after visiting Alaska. George Housner later recalled that Anchorage's devastation had "opened [his] eyes to the possible disaster that a great earthquake could inflict on a large city." Geologists, meanwhile, noted the extent to which ground conditions had exacerbated the effects of the shaking. In areas underlain by loose soil or mud rather than rock, damage was much higher, as if the shaking had been amplified. Moreover, whole neighborhoods located on bluffs overlooking the ocean near Anchorage had tumbled apart as the bluffs slid toward the ocean; this showed that landslides could also be a significant hazard during an earthquake. The lesson for Californians, as geologist Gordon Oakeshott noted, was that "structural damage is much more closely related to the type of foundation rock or soil than it is to the distance from the trace of an active fault. Location on bedrock is a good insurance."[1]

After the Good Friday earthquake, many earthquake researchers concluded that they needed to do more to prepare California for the next great earthquake. The achievements of the 1930s—the Field Act and the incorporation of basic earthquake provisions in building codes—no longer seemed sufficient in light of the devastation in Alaska. As a result, in the period from 1964 to 1971 scientists and engineers embarked on a number of new initiatives, proposing new means for ensuring seismic safety. A group of nationally prominent seismologists proposed a comprehensive research program on earthquake prediction, hoping that it would significantly reduce the death toll from future earthquakes. The proposal languished, in part because of opposition from those who argued that engineering offered a better means for reducing losses than did prediction. But scientists from the U.S. Geological Survey, the U.S. Coast and Geodetic Survey, and the California Division of Mines and Geology stepped up their research efforts. They increased the mapping and monitoring of fault zones in California and greatly added to knowledge of the state's seismic hazards. Seismologists and earthquake engineers also pressed for increased government regulation of development in particularly hazardous areas, such as fault zones and areas of unstable soil. Their campaign culminated in the establishment of the Joint Committee on Seismic Safety in the state legislature, charged with developing a comprehensive plan for mitigating California's earthquake hazards.

A Proposal for Earthquake Prediction Research

The idea of a comprehensive research program aimed at earthquake prediction was not new in the 1960s. Early in the twentieth century, geologists and seismologists had proposed possible avenues for earthquake prediction. In the early 1920s, for example, Andrew Lawson at the University of California had suggested that a geodetic retriangulation of California might uncover a dangerous buildup of strain, which would be predictive of an earthquake.[2] But by the early 1940s, seismologists had concluded that no viable means existed for predicting earthquakes, especially if such a prediction was expected to include rather narrow limits on the time, place, and magnitude of the earthquake. The methods for measuring accumulated strain along a fault simply were not precise enough to determine when that strain had reached the fault's breaking point. Thus, earthquake researchers left prediction to amateur scientists and publicity seekers, who often resorted to astrology or psychic means for prophesying fault rupture.[3]

By the early 1960s, academic scientists were once again becoming interested in earthquake prediction. In Japan, researchers were encouraged by results from precise leveling of the earth's surface. In this geodetic method, the elevation of various points of the earth's surface relative to a benchmark is measured repeatedly over a period of months or years. In several cases, leveling showed unusually rapid movements of the earth's surface near a fault just before an earthquake. These data suggested that strain buildup changed in character just before an earthquake and that this change could be detected and used to predict an imminent quake. In 1962, a group of Japanese geophysicists presented a plan for further research. Baldly stating that "prediction of earthquakes is one of the most important tasks of seismologists," they proposed that Japan significantly increase its leveling and other geodetic surveys as well as install more seismographs and other instruments in the hope of detecting premonitory signals. The Japanese government finally funded this proposal in 1965.[4]

American scientists also drew inspiration from the Japanese work. In 1963, geophysicists at Columbia University received National Science Foundation funding to hold a joint U.S.-Japanese meeting on earthquake prediction. The dozen Americans who attended the conference, held in Tokyo two weeks before the Alaskan earthquake, came back impressed with the "considerable promise" of earthquake prediction research.[5] Meanwhile, the Caltech Seismological Laboratory obtained $400,000 from the National Science Foundation for a two-year intensive study of the San Andreas Fault that had prediction of the fault's behavior among its goals.[6] Thus, by early 1964, seismologists at Caltech, Columbia, and other leading research universities were eager to pursue earthquake prediction research.

When the Good Friday earthquake struck, they seized the opportunity to lobby the federal government for a comprehensive national research program. The chief lobbyist was Frank Press, who served as the Caltech Seismological Laboratory's director from 1957 until 1965, when he moved to the Massachusetts Institute of Technology. Press was quite familiar with the federal science policy establishment. Since 1958 he had regularly served as technical advisor during nuclear test ban negotiations, and from 1961 to early 1964 he was a member of the President's Science Advisory Committee.[7] On April 2, 1964, less than a week after the Alaskan earthquake, Press contacted Donald F. Hornig, who, as head of the Office of Science and Technology, was President Johnson's chief science advisor. Press suggested that Hornig convene a meeting of prominent seismologists and government officials "to discuss the problem of earthquake prediction . . . and determine if the subject should receive greater emphasis in the U.S."[8] The group of a dozen, which met on April 19, included four seismologists who had attended the U.S.-Japanese meeting on earthquake prediction the previous month. It concluded that "research directed toward understanding of the mechanism of earthquakes is feasible and worthwhile. . . . A program growing deliberately over a 10-year period is appropriate."[9] Press then obtained Hornig's approval for the group to develop a more detailed proposal.[10]

The so-called Ad Hoc Panel on Earthquake Prediction, chaired by Press, delivered its report to the President's Science Advisory Committee in early June 1965. As finally constituted, the panel included prominent researchers from Caltech, MIT, Princeton, Columbia, and the University of California, as well as representatives from the U.S. Geological Survey, the U.S. Coast and Geodetic Survey, the National Science Foundation, and the Advanced Research Projects Agency. It proposed that the earthquake prediction research program focus on fault zones in California, Nevada, and Alaska, as earthquakes could be expected to occur there within the coming ten years. It suggested that instrument clusters containing seismographs, strainmeters, and devices for detecting the tilt of the earth's surface and changes in the local magnetic and gravity fields be distributed along the major fault zones. As it pointed out, such instruments had become much more sensitive during the previous decade because of technological advances produced by the Vela Uniform program. With any luck, they would record some premonitory signal before the next earthquake. In addition, the panel proposed extensive geological mapping and geophysical surveying to produce a better three-dimensional picture of the fault zones, as well as a sizable laboratory program to determine how rocks behave when stressed to their breaking point. All of this research, in the panel's estimate, would cost $137 million, more than half the funding for the Vela Uniform project. This program would thus be a significant boost for federal support of seismology.[11]

The Press panel's report was received coolly by the President's Science Advisory Committee. Hornig wondered whether such an expensive research program was worthwhile, given the low death rates from earthquakes in the United States (an average of only about ten deaths per year). Others also worried about who would carry out the program. Since 1925, the U.S. Coast and Geodetic Survey had been the federal agency responsible by law for seismology, but in recent decades the agency had acquired a reputation for engaging only in routine data collection rather than innovative research. The U.S. Geological Survey had a far better reputation for conducting significant research, and with funding from Vela Uniform it had acquired a small but talented group of geophysicists led by Louis Pakiser to complement its traditional strength in geology. But its legal authority for doing seismological as opposed to geological work was questionable at best.[12] In the end, the Science Advisory Committee could not settle on an agency to head the program, and it also concluded that "the likelihood of success in the reliable prediction of earthquakes" was "so uncertain that the program proposed by the panel should be modified and carried on at a slower pace."[13] This tepid reaction did not, however, kill the proposal. Press had been savvy enough to leak word of his panel's deliberations to the news media even before it had finished its work. California newspapers as well as a newsletter for the geological community had responded enthusiastically to the idea of an all-out federally funded pursuit of earthquake prediction.[14] In the late summer of 1965, Press used this enthusiasm in convincing Hornig to release the proposal to the public even without formal government endorsement.[15]

Empire-building visionaries within both the Geological Survey and the Coast and Geodetic Survey seized upon the published proposal as a means for expanding their agency's turf.[16] On October 8, 1965, two days after the proposal's publication, the Geological Survey announced the establishment of the National Center for Earthquake Research at its West Coast headquarters in Menlo Park, near Stanford University. This new center had long been in the planning. First proposed in 1962, it had received backing from administrators in the Geological Survey and the Department of the Interior in the months after the Good Friday earthquake. By tying announcement of the center to the public release of the Press panel's report, the Geological Survey positioned the new center as the logical place for a national earthquake prediction research program.[17] William T. Pecora, the Geological Survey's ambitious new director, followed up by orchestrating an extensive letter-writing campaign by leading geologists to persuade Hornig to entrust the prediction program to the Geological Survey.[18] The Coast and Geodetic Survey responded in November 1965 by announcing the establishment in San Francisco of its Earthquake Mechanism Laboratory, devoted to research on earthquake prediction.[19] The agency also convened a symposium on earthquake prediction in February 1966 "for the purpose of clarifying [the agency's] technical capability" in the field and to lobby for "an adequate earthquake prediction program."[20] This clamoring by both surveys convinced Hornig in spring 1966 to reopen the question of a coordinated federal research program. He set up a new panel with representatives from the two surveys and other federal agencies to "make recommendations as to what a Federal program in earthquake prediction might consist of and as to how the government might be organized for this function." In a signal victory for Pecora, Hornig asked the Geological Survey to head this panel; Pecora himself ended up serving as its chairman.[21]

In addition to the problems generated by the turf war between the two surveys, the Pecora panel faced considerable opposition, both inside and outside the earthquake research community, to the notion of prediction as a desirable goal. One particularly forceful attack came from Garrett Hardin, a noted population biologist and social critic at the University of California at Santa Barbara. In a piece published in a Stanford University literary magazine, Hardin attacked earthquake prediction as unnecessary "pop research." He broadly challenged the general idea that technological progress was good and should be encouraged by portraying earthquake prediction as an example of progress that would be harmful. In his view, earthquake predictions would be both more uncertain and less constrained than hurricane predictions. While hurricane warnings specified a window of only a few days of danger, an earthquake prediction would likely note a heightened risk over a period of several weeks or even months. Given such a broad window, Hardin argued, little of practical value could be done. Evacuation for such a long period would not be feasible. Real estate values and the psychological state of the general population, however, would suffer greatly. Thus, Hardin concluded, "we need an earthquake-predicting facility like we need a hole in the head."[22]

Hardin's critique attracted notice both in the news media and among seismologists, at least one of whom dismissed it as "shallow thinking."[23] Frank Press responded by arguing in 1968 that worries about "social-psychological aspects" should not be allowed "to obstruct earthquake forecasting research."[24] Pakiser of the U.S. Geological Survey also reacted with exactly the faith in technological progress that Hardin had challenged, by telling a newspaper reporter in 1969 that social considerations should not "stand in the way of increasing knowledge" about earthquake prediction.[25] But other earthquake researchers shared Hardin's specific concerns about earthquake prediction, even if not his broader views on the value of technological progress. As early as 1964, Bruce Bolt, the new director of the seismograph station at the University of California at Berkeley, worried about "the perplexing problem as to the action one would take if calculation yielded 'odds-on' chances of an immediate earthquake near a large city." How could one justify the vast expense and dislocation of evacuation if a predicted earthquake were only likely rather than certain?[26] Similarly, northern California earthquake engineer Karl Steinbrugge asserted repeatedly that a prediction with a reliability of less than 95 percent or a window of more than a few days would do little good.[27]

Charles Richter also opposed efforts at earthquake prediction, although for different reasons. Richter was one of the most publicly visible seismologists in California, both because of his willingness to give interviews and talks in order to promote earthquake preparedness and because his name had come to be attached to the magnitude scale used in reporting earthquakes. As a result, Richter received numerous inquiries from the public concerning predictions made by astrologers, psychics, and miscellaneous cranks. Richter was clearly frustrated by these many "wild-eyed forecasts of earthquakes in California."[28] He became particularly upset in July 1965, when the civil defense director for Santa Barbara County sent a letter to all county department heads informing them that noted psychic Jeane Dixon had predicted an earthquake for the next day. In response, police and beach lifeguards were put on alert and the fire department removed all of its equipment from fire stations that might collapse, thereby giving official credence to the prediction. Richter sent several blistering letters condemning the Santa Barbara civil defense office for having responded so readily to "the vaporings of a crackpot." He pointed out that cranks had already predicted great earthquakes in California for January 17, February 4, February 28, March 17, April 1, April 17, May 30, and June 17 of that year; "of course no earthquake of consequence occurred in California on any of those dates."[29] Given such experiences, it is understandable that Richter would proclaim in a 1968 television interview, "Bah, no one but fools and charlatans try to predict earthquakes!"[30] Establishing an official prediction effort would only further encourage popular belief in cranks, he declared.

Richter also opposed a prediction research program because it might divert attention from what he considered more effective means for reducing earthquake hazards. Throughout the 1960s, Richter vigorously promoted such measures as bringing old schools up to the standards of the Field Act, renovating or replacing other old buildings that might collapse during an earthquake, and strengthening pipelines, highways, high-tension wires, and other lifelines that might be ruptured. In an article published in *Natural History,* he objected to "the pervasive emphasis on prediction" among seismologists. "It directs attention away from the known risks, and the known measures that could be taken to remove them, and fixes it instead on vague, distant hopes."[31] In private, Richter's statements were even stronger. He called prediction research "something between a will-o-the-wisp and a red herring."[32] Others shared this sentiment. Robert A. Frosch of the Advanced Research Projects Agency, for example, asserted, "The most important . . . actions with regard to damage prevention are in the area of building construction and zoning" rather than prediction—a view also expressed by other seismologists, earthquake engineers, and even members of Press's panel on prediction research.[33] The President's Science Advisory Committee likewise concluded, "earthquake engineering to mitigate the consequences of earthquakes is a more fruitful line to follow than earthquake prediction."[34]

Press had not ignored the importance of earthquake engineering. From the beginning, he had actively involved prominent Caltech earthquake engineer George Housner as a member of his panel. At Housner's urging, the panel's final proposal included provisions for research on the behavior of soils during earthquakes and the response of structures to earthquake vibrations. In presenting this proposal to the President's Science Advisory Committee as well as in subsequent discussions, Press made a point of emphasizing the importance of earthquake engineering,[35] but the funding for engineering research suggested by the panel amounted to only $19.6 million of the $137 million proposed for the entire program. Press had kept this amount so low because Housner had implied that the earthquake engineering research community would not be able to absorb a greater increase in funding. To outsiders, though, the small proportion of funding for earthquake engineering suggested that the Press panel saw it as less important than earthquake prediction or other geological research.[36] In retrospect, Housner came to view himself as the panel's "token engineer," whose appointment had been an "afterthought." In the summer of 1965, he convinced the National Academy of Engineering, of which he was a member, to set up a panel of its own, to consider a ten-year program for earthquake engineering research; it would include applied research and design work as well as the basic research recommended by the Press panel.[37]

The interagency group headed by Pecora tried to address all of these criticisms in modifying the Press panel's proposal. For example, it changed the phrase "Earthquake Prediction" in its name to "Earthquake Research," to signal a broadening of its vision. In its report, completed in June 1967, the group also explicitly acknowledged that "the most effective area for action to reduce the damages and loss of life from future earthquakes is in setting guidelines for future construction and land development"; and, despite objections from the Geological Survey's Pakiser that doing so would make the overall program too expensive, the group increased the funding proposed for earthquake engineering to $60 million over ten years.[38] The report also called for research on "the practical uses of earthquake prediction," to soothe those worried about the negative effects of an inexact prediction. The group did, however, insist that there would be an "obvious benefit from the prediction of specific earthquakes . . . in saving lives. Prediction would have significant additional benefits in enabling emergency measures to be taken to limit damage," such as shutting down nuclear reactors and lowering water levels behind threatened dams. Given such benefits, the Pecora report suggested increasing the nonengineering portion of the proposal to $158 million, for a total research program of $218 million over ten years. Finally, despite Pecora's attempt to have a lead agency named for the program, the group sidestepped the turf war between the two Federal surveys by not assigning tasks to any specific agency.[39]

These modifications were still not enough to produce universal support for the research proposal. By 1967, Housner's engineering panel was already considering a ten-year program that would spend $200 million for earthquake engineering research alone; in its 1969 final report, that figure had ballooned to $380 million. The increased funding allotted by the Pecora group was not enough to satisfy the engineers.[40] At the same time, the $218 million overall price tag in that proposal was too much for a federal science policy establishment reeling from the domestic budget cuts required to accommodate the escalating cost of the Vietnam War.[41] The Pecora group's report was not publicly released until December 1968, after Nixon had won the presidential election, and it was promptly ignored by the incoming administration.[42]

New Directions in Earthquake Research

Although the comprehensive program first drafted by the Press panel and later broadened by the Pecora group was not funded, both the federal government and the state government did increase their investment in California earthquake research in the late 1960s. Before 1964, Congress had appropriated only several hundred thousand dollars each year to look at earthquake hazards in California. By the late 1960s, the Geological Survey was receiving about $1.5 million per year for its National Center for Earthquake Research, while the Coast and Geodetic Survey was getting about $2.3 million annually for geodetic surveys, strong-motion seismology, and research at its new Earthquake Mechanism Laboratory. In addition, the National Science Foundation in the late 1960s awarded about $4 million each year in grants for earthquake engineering and geophysical research directly related to understanding earthquakes.[43] The state government, through the California Division of Mines and Geology, also began to increase its funding of earthquake research in the late 1960s. All of this new money led to a significant expansion of scientific knowledge about earthquakes, faults, and seismic hazards.

Both the Geological Survey and the Coast and Geodetic Survey began to implement the Press panel's recommendation that the San Andreas Fault be densely instrumented. In the mid-1960s, researchers at the National Center for Earthquake Research developed methods for automatically recording and transmitting data from seismographs in remote locations. They also wrote computer programs for efficiently analyzing data from large networks of seismographs. These technological advances allowed the Geological Survey to establish an extensive seismograph network along the San Andreas fault system in the San Francisco Bay area and the section of central California just to the south. By 1968, the survey had installed 30 seismograph stations; by early 1971, the network had grown to 83 stations. At the same time, the survey also developed several portable networks of about 10 stations each. After a large earthquake, such networks could be quickly placed near the epicenter to record aftershocks with great accuracy. In addition, the survey installed 12 tiltmeters, sophisticated versions of a carpenter's level that measured tilting of the earth's surface in response to strain accumulation. At 35 sites along the San Andreas and related faults, the survey also installed alignment arrays (groups of markers set up in a long straight line) or other devices across the fault to measure differential movement between the two sides of the fault.[44] Meanwhile, the Coast and Geodetic Survey established a geophysical observatory at Stone Canyon, on the San Andreas Fault south of the bay area, equipped with seismographs, tiltmeters, strainmeters, and instruments for measuring variations in magnetism and electrical resistance along the fault. It also installed 11 seismograph stations, 41 alignment arrays, and 31 other devices for measuring fault displacement in California.[45] Not to be outdone, the University of California set up its own heavily instrumented San Andreas Geophysical Observatory near Hollister, south of the bay area, and Caltech also increased its monitoring of fault displacement.[46]

Increased funding for earthquake research in the late 1960s also served to fulfill the Press panel's hope that laboratory experiments on the mechanics of rock faulting would be accelerated. The Geological Survey established a rock mechanics laboratory at its Menlo Park center, and the National Science Foundation funded new laboratories at Texas A&M University and elsewhere. The most important rock mechanics laboratory was at MIT, where Press's colleague William Brace and his students in the late 1960s observed how tiny cracks accumulate in a rock just before it faults and how these cracks affect the rock's volume, electrical resistance, and other properties.[47]

Despite the profusion of new laboratories and instrumentation along the San Andreas Fault, earthquake researchers in the late 1960s made little progress toward a reliable method for predicting earthquakes. It did appear that magnetic anomalies might occur before earthquakes. Even before the 1964 Alaskan earthquake, Sheldon Breiner, a former student at Stanford, had searched for the changes in the magnetic field near the San Andreas Fault that he reasoned should occur as stress built up along the fault, changing the magnetic properties of rocks near it (the so-called piezomagnetic effect). With funding from the National Science Foundation and the Coast and Geodetic Survey, Breiner and a Stanford colleague in 1965 set up an array of five magnetometers along the fault, and in 1967 they announced that they had indeed found magnetic anomalies just before several seismic events.[48] The excitement produced by this announcement was short-lived, however. The magnetic field is affected not only by the piezomagnetic effect but also by magnetic storms emanating from the sun and by many other effects. Filtering out this noise proved to be far more difficult than anticipated. Moreover, as the Stanford researchers discovered, in many cases seismic events were not preceded by magnetic anomalies, and some clear anomalies were not followed by a seismic event.[49] This problem—that a precursory anomaly would be detected sometimes but not always before earthquakes—would bedevil many other suggested methods of earthquake prediction.

While unsuccessful in producing a reliable method for earthquake prediction, the expanded work of the Geological Survey and the Coast and Geodetic Survey in California did contribute much new knowledge about the behavior of faults. One of the most significant discoveries was fault creep. As early as 1957, structural engineers working for an insurance company noticed cracks in a concrete building at a winery south of Hollister. Subsequent measurements showed that the building was being sheared in half, with the two sides of the building moving past each other by about half an inch per year. The building, it turned out, had been constructed directly atop the San Andreas Fault, and the shearing showed that the fault released strain gradually, through a creeping motion, as well as suddenly, in earthquakes.[50] Observations of displacements along railroad tracks, water pipelines, inside a warehouse, and in a culvert underneath the University of California at Berkeley's football stadium in late 1964 and 1965 revealed a similar creeping motion along the Hayward Fault, which parallels the San Andreas Fault on the east side of the San Francisco Bay.[51] In the late 1960s, researchers from both federal surveys, assisted by local surveyors and even a high school "Fault Finders Club," searched for evidence of small fault displacements. They demonstrated that creep was occurring at numerous sites on the Hayward, Calaveras, Sargent, Owens Valley, and San Andreas faults.[52]

The discovery of fault creep showed that buildings straddling a fault faced the risk not only of being torn apart during an earthquake but also of being inexorably deformed as the underlying fault relaxed through creeping motion. At the same time, the discovery also offered the prospect of dividing the San Andreas Fault into segments of very great and not-so-great hazard. For years, California geologists had noted that the fault's main portion could be divided into three sections. Two had experienced great earthquakes during historic times: a northern section, reaching from the Mendocino County coast to near Hollister and passing through the San Francisco Bay area, had ruptured in 1906, while a southern section, extending from Cholame in San Luis Obispo County to near San Bernardino, had moved in the great 1857 Fort Tejon earthquake that had shaken then-insignificant Los Angeles. The central section had not yet experienced a great earthquake, and as late as 1964 one prominent Caltech seismologist suggested that this section was the most probable site for California's next seismic disaster.[53] In a widely cited 1968 paper, though, Caltech geologist Clarence Allen pointed out that the central section was marked by pervasive creeping motion as well as abundant small and moderate earthquakes, up to magnitude 6, while both creep and small-scale seismicity were absent from the northern and southern sections. This suggested that the northern and southern segments were tightly locked, rupturing only in cataclysmic events when strain had built up to very high levels, while the central section was less tightly locked and able to release strain more continuously in smaller events. By 1970, most California seismologists agreed with Allen that the sections of the San Andreas Fault that had ruptured in 1857 and in 1906 posed the greatest hazard for the future.[54]

Earthquake researchers also provided new information on the width and location of active faults within fault zones. The San Andreas and other important faults are marked not by a single surface break, but rather by fault zones up to several miles wide composed of many intertwining fault strands. Within these fault zones, a single earthquake ruptures only one fault strand. As late as 1964, a U.S. Geological Survey scientist had asserted that it could not be predicted which strand would break in the next earthquake,[55] but field studies after the moderate 1966 Parkfield and 1968 Borrego Mountain earthquakes showed that fault rupture had occurred on the exact same strand that had broken in the previous earthquake.[56] Robert Wallace of the Geological Survey provided independent evidence from a study of ancient stream channels that had been preserved on the Carrizo Plain of central California. These channels had been deflected repeatedly where they crossed the San Andreas fault zone. Within the last 10,000 years, Wallace showed, the changes of stream-course had all occurred on the same strand of the fault. Rupture of a new strand occurred only very infrequently in this fault zone, which had grown gradually over millions of years.[57] Studies of fault creep similarly showed that present-day creeping motion was confined to a single fault strand and that it affected a zone less than 50 feet wide.[58] All of these observations suggested that a narrow zone of active seismic hazard could be identified within the wider inactive fault zone, and by 1968 some California earthquake researchers were arguing that restrictions should be placed on development and construction in these narrow zones of fault activity.[59]

Limiting construction on active faults and in other areas of seismic hazard of course required that maps of such areas be available. From the late 1950s, a small group of engineering geologists at the U. S. Geological Survey had been mapping selected parts of the San Francisco Bay area and Los Angeles County, concentrating on regions of landslide hazards but also noting recently active fault strands.[60] After the establishment of its National Center for Earthquake Research, the survey embarked on a program to map all the active fault strands of the San Andreas and Hayward fault zones. By the early 1970s, it had published maps showing these strands in unprecedented detail.[61] The survey's hazard mapping increased once again in late 1969, when it initiated the six-year Bay Region Environment and Resources Planning Study, funded jointly with the federal Department of Housing and Urban Development. This project was "directed toward learning how man can better accommodate his activities to the physical environment of the earth" by producing various kinds of geological maps of direct use to planners in the nine-county bay area.[62] As part of this project, the survey published maps showing the location of all known active faults and of areas susceptible to soil failure and landsliding during strong shaking. Using new research that quantified how earthquake waves become attenuated with distance from their origin and how they are amplified by loose soil or mud, the Geological Survey also prepared summary maps showing what effects several hypothetical earthquakes might have throughout the bay area.[63]

Earthquake research also increased in the late 1960s at the California Division of Mines and Geology. Before the 1960s, this agency, like the U.S.

Geological Survey, had not been much involved in environmental hazard mapping. Known as the California Division of Mines until 1961, its main work had been "in obtaining and providing information of benefit to the State's mineral industry." It answered to the State Mining Board, which was dominated by representatives of the state's mining industry, ensuring that the division was "a 100% minerals industry oriented organization."[64] This emphasis began to change during the tenure of Ian Campbell, who served as the division's director from 1959 to 1969. Campbell, a former geology professor at Caltech, had ambitions of turning the division into a true state geological survey responsive to the needs of California's growing urban population in addition to its mining industry, which was concentrated in the state's rural counties. In 1960, he initiated a small program of geological mapping in the Palos Verdes Hills and several other rapidly urbanizing portions of Los Angeles County threatened by landsliding. To the State Mining Board he justified this program by pointing out that urban sprawl threatened to pave over vast deposits of sand and gravel, which now were among the state's most valuable mineral resources. Knowledge of where these deposits were located might lead development to be directed elsewhere. But mapping such sedimentary deposits also aided in delineating areas susceptible to landsliding, increased shaking during earthquakes, and other hazards.[65]

After the Good Friday earthquake, Campbell had little difficulty gaining support for expanding his urban hazards mapping program. The State Mining Board agreed "that a continuing study of the San Andreas fault and other active faults in California would be an appropriate undertaking for" the division. A state senate oversight committee also recommended that the division "give primary emphasis to the adequate mapping of geologic hazards."[66] Hugo Fisher, director of the California Resources Agency and thus Campbell's boss in the state administration, likewise endorsed this reorientation. In December 1964, he convened a conference of scientists, engineers, and public officials to generate a consensus on an expanded state geologic hazards research program. In words that echoed those used by G. K. Gilbert in 1909 to define the campaign for seismic safety, Fisher asserted "We need to supply the leadership and funds to encourage our geologists and seismologists to investigate, our engineers to apply this knowledge to construction, and our officials to properly regulate, zone, and exercise necessary controls on construction and developments" in areas subject to seismic and other geologic hazards.[67] Soon after the conference, Fisher appointed two committees—one headed by Caltech geologist Clarence Allen, the other by earthquake engineer George Housner—to draft a program for expanded state-supported research. In a report not completed until 1967, these committees recommended that California fund increased mapping of faults and other hazardous areas, research on fault creep, expanded seismographic networks, some basic earthquake engineering research, and a greater public education effort on seismic hazards. They also urged the state to establish a permanent Earthquake and Geologic Hazards Board to advise the state on earthquake research.[68]

The Allen and Housner committees did not submit their report until after Ronald Reagan had been inaugurated governor in early 1967, and the cost-conscious new administration did not fully implement the committees' ambitious proposal. Nevertheless, earthquake research did expand significantly in the Division of Mines and Geology. In the state legislature, power had shifted decisively in 1965 from rural counties (many still dependent on the mining industry) to urban counties, as the result of a wholesale reapportionment in response to Supreme Court rulings. The new legislature passed a bill making "large-scale geologic mapping to provide timely delineation of geologic hazards in urban areas" an integral part of the division's mission.[69] Instead of establishing a separate Earthquake and Geologic Hazards Board, the legislature added a structural geologist and an engineering seismologist to the State Mining Board and renamed it the State Mining and Geology Board. The two new members, Allen and Steinbrugge, soon allied themselves with Richard Jahns, a Stanford geologist and close friend of Campbell already on the board, to form an active bloc supporting the division's hazards mapping program.[70] In the late 1960s, the legislature also increased the division's funding for this program. By 1969, it had reached $260,000 (out of a total division budget of $1.25 million), supporting 12 full-time geologists. The following year, the program's budget was increased by another $150,000.[71]

This funding allowed the division to add substantially to the seismic hazards work of the two federal surveys. Division geologists mapped fault strands in the San Andreas fault zone near San Bernardino and near Hollister and in the Calaveras fault zone east of the bay area. Division geologists also aided the U.S. Geological Survey in mapping other fault zones, and they contributed measurements of fault creep along the San Andreas. And in 1969, the division took over a geodetic program for monitoring surface displacements near the fault.[72] The division also extended its public outreach activities. It published popularly written articles on faults, historic earthquakes, and seismic hazards in its monthly news magazine, *Mineral Information Service* (renamed *California Geology* in 1971), turning it into a vehicle for public education on seismic safety.[73] At the request of the California Disaster Office, which was responsible for the state's civil defense efforts, the division in 1970 put together a map of expectable damage from earthquakes.[74] And in 1966 the division entered into an agreement with the state Division of Real Estate to review all proposals (about 1,400 a year) filed for land subdivision. Where appropriate, it recommended that the Division of Real Estate add a notice of potential earthquake or other geologic hazards to the subdivision report issued to the public.[75]

New Initiatives in Regulating Seismic Hazards

Another way in which seismologists and earthquake engineers responded to the Good Friday earthquake was to push for increased state regulation of development and construction in areas of seismic hazard. Their efforts first emerged in a debate over whether a new housing development should be allowed to proceed along the shores of San Francisco Bay. While earthquake researchers were not entirely willing to join environmentalists in blocking the development, they did persuade a state-appointed commission to regulate construction on seismically unstable bayshore lands. Later, seismic safety advocates proposed that the state set up a group to develop a comprehensive hazard mitigation program for the bay area. In 1969, they gained the ear of a state senator, and through his efforts the state legislature established the Joint Committee on Seismic Safety, a further enhancement of the regulatory state apparatus for earthquake hazard mitigation.

The future of San Francisco Bay sparked an extensive public debate in the mid-1960s. Since the middle of the nineteenth century, developers had been filling in tidal areas along the bayshore to create new land for harbor facilities, airports, warehouse districts, even whole new cities. As a result, between 1850 and the mid-1960s, the bay's surface area shrank by one-third. This continuing threat to the bay's very existence finally aroused the concern of environmentalists. In 1960, local conservationists founded the Save San Francisco Bay Association. This group convinced the state legislature in 1964 to establish the San Francisco Bay Conservation Study Commission, which was given a year to explore ways of slowing the bay's environmental degradation.[76]

That same year, the Leslie Salt Company, owner of a considerable tract of tidal land near Redwood City, on the west side of the bay about 20 miles south of San Francisco and not far from the Geological Survey's Menlo Park center, proposed a new residential development. Called Redwood Shores, this project was to be a new city, ultimately housing 60,000 residents in single-family homes and 25 high-rise apartment buildings, all erected on bay fill.[77] Conservationists immediately mobilized to oppose the project. In addition to citing the usual concerns about crowded schools and deterioration of the bay's environment, opponents cited seismic hazards. Referring to soil failures and the amplification of shaking in structures placed on muddy ground observed in the recent Alaskan earthquake, they argued that the foundation on which Redwood Shores was to be built—artificial fill dumped on top of bay mud—was about the worst imaginable from a seismic safety standpoint.[78] In December 1964, they received valuable support from G. Brent Dalrymple and Marvin A. Lanphere, two geologists with the U. S. Geological Survey. In testimony to the Bay Conservation Study Commission, they argued at length that "made land and marshland" was "in general, unsafe and should not be used for housing projects, high-rise apartment complexes, or industrial developments" and that the state should impose a moratorium on all bayshore development until proper guidelines for dealing with this seismic hazard were put in place.[79]

Predictably, the Leslie Salt Company strongly disputed Dalrymple and Lanphere's assertions. William Moore, a Caltech-trained engineer whom the company had consulted, issued a withering statement charging that the report and conclusions of Dalrymple and Lanphere were "largely inaccurate and misleading . . . unfounded and quite obviously beyond the competence of its geologist-authors." In his view, the survival of several modern high-rise buildings in the Good Friday earthquake and a 1957 earthquake in Mexico City, another city built on mud, showed that muddy or filled ground was not inherently dangerous—the hazards could be engineered away.[80] The developers went even further by exerting pressure on the Geological Survey through a member of the local Republican congressman's staff who also happened to be a public relations official for bayshore developers. Several days after Dalrymple and Lanphere's testimony, William Pecora, at that time not yet the survey's director but its chief geologist, warned the Menlo Park office that any such statements had to be cleared with the survey's headquarters; Dalrymple and Lanphere's assertions were thus characterized as unauthorized.[81]

Conservationists reacted sharply to these blunt attacks. Newspaper columnists, including Harold Gilliam of the *San Francisco Chronicle*, joined local Democratic party politicians in decrying the "muzzling" of Dalrymple and Lanphere.[82] Meanwhile, the president of the Redwood City Civic Association, which had emerged as the chief opponent to the Redwood Shores project, solicited letters from numerous geologists in support of Dalrymple and Lanphere's statement. A few scientists, such as Charles Richter, shied away from getting involved, having been burned by the acrimony of the debate over Bodega Head.[83] But geologists at the University of California and at Stanford University were quite willing to write letters for public release. One Berkeley geologist insisted that "the Dalrymple-Lanphere report on geologic seismic hazards is accurate, well-reported, carefully documented and scientifically unassailable," while another reiterated that "geologists and seismologists have long known that ground displacements are many times greater on unconsolidated ground (such as marshland, bay mud, and water-saturated alluvium) than they are on bedrock."[84]

Faced with this persistent criticism, the City Council of Redwood City, which had become a partner in the Redwood Shores project, in February 1965 decided to appoint a review panel to assure the project's seismic safety.[85] This panel was dominated by engineers. Like many other California engineers at this time, they had an abiding faith that proper design could overcome the hazards posed by any site. As panel member Henry Degenkolb, echoing William Moore's assertions, later expressed it, "we cannot just accept a geologist's statement that, in effect, says, 'This is bad ground and there is an earthquake hazard, so you should not build there.' . . . In all these locations and situations, the problem is to recognize the hazard, evaluate it and design to resist it." Often, such engineers still denied that sites underlain by mud or alluvium posed special hazards.[86] In keeping with this sentiment, the Redwood City panel concluded, "There is no indication that the site of the Redwood Shores Project will experience any more severe effects from earthquakes than many other areas of California, or that normally used concepts and criteria for earthquake-resistant design in California should be modified." In particular, the panel asserted that the Redwood Shores site resembled in many ways the site of San Francisco's downtown, where nobody would dare restrict construction—neglecting to mention, of course, that that area had suffered severe damage in the 1906 earthquake.[87]

Conservationists, not yet as willing to challenge authority as the opponents of the Bodega Head plant had been, backed away from their objection to Redwood Shores, in the face of the panel's positive report. Construction on the development commenced with backing from the Federal Housing Administration and the Veterans Administration, which approved loans and mortgage guarantees for those buying houses in the project. Not until the summer of 1969, after a hostile congressional oversight committee had publicized continuing doubts by Geological Survey scientists over the project's safety, did the housing agencies drop their support of Redwood Shores, bringing construction to a halt after about 200 homes had been erected.[88] But by this time other efforts to regulate development and construction in seismically hazardous areas had already taken root.

At the end of its one-year life, the Bay Conservation Study Commission recommended that a successor body be established to develop a comprehensive plan for preserving San Francisco Bay. This new body, the Bay Conservation and Development Commission, began work in late 1965 with a membership of 27 representatives of environmental, industrial, and government interests and an annual budget of $240,000. Over the next four years, it constructed a detailed land-use plan for the bayshore and considered a host of regulations for appropriate uses of the bay.[89] Among the issues it considered was the seismic safety of construction on bay fill. One consultant to the commission, soils engineer H. Bolton Seed of the University of California at Berkeley, asserted that if standard engineering procedures were followed—using modern methods of fill compaction, avoiding steep slopes in the bay floor and soils with sand lenses that might liquefy during an earthquake—then "there is no reason why construction on Bay fills should not be considered as satisfactory from a seismic hazard standpoint as other building sites in the Bay Area."[90] But fellow consultant Steinbrugge was less sanguine. He pointed out that loose alluvium strongly amplified long-period earthquake waves, which primarily affected tall structures, such as high-rise apartment buildings, although it had relatively little effect on the short-period waves that primarily affected low structures. Thus, "in a general sense, the seismic risk to buildings on Bay fills . . . is least for one story light mass structures of no more than moderate floor area." Consequently, he recommended that only low-rise development occur on bay fill, and he suggested that the commission appoint a board of consultants to zone bayshore lands according to seismic risk, with more stringent building restrictions placed on higher-risk zones.[91]

The Bay Commission largely agreed with Steinbrugge. In May 1967, it adopted as policy conclusions that "to reduce risk to life and damage to property, special consideration must be given to construction on poor soils throughout the Bay Area," that "the younger Bay mud is the weakest soil and generally requires special engineering to overcome its deficiencies," and that no fill project should proceed unless precautions were taken against seismic hazards.[92] The commission also decided to establish the consulting board recommended by Steinbrugge. This board, appointed in February 1968 and chaired by Steinbrugge, at first did not map out the seismic risk zones along the bay's shore but did define four zones of increasing risk and set restrictions on the height of buildings in each. In the zone of highest risk, it prohibited all residences and allowed only low-rise industrial buildings. It also recommended that a permanent fill review board be established to review all projects for compliance with these zoning provisions.[93] Such a body, named the Engineering Criteria Review Board, was indeed established once the state legislature accepted the commission's overall Bay Plan in 1969 and made the commission into a permanent regulatory agency. The board did not follow the strict rules of zoning proposed by Steinbrugge. Instead, it relied more on case-by-case review of engineering designs. Nevertheless, it still had a considerable impact in encouraging special construction precautions to accommodate the seismic hazards around San Francisco Bay.[94]

By the time Steinbrugge presented his suggestions for risk zoning along the bayshore in 1967, he was already working on an even broader proposal for earthquake hazard mitigation in the bay area, to appear as a monograph published by the University of California at Berkeley's Institute of Governmental Studies. Stan Scott, a public policy researcher with the institute, had become interested in seismic hazards after the Good Friday earthquake. In late 1964, he had attended the conference convened by Hugo Fisher. Scott was one of the few policy analysts among the scientists and engineers at that conference, and he became convinced that effective earthquake hazard mitigation required the input not only of technical experts but of policy people, who could help translate research results into effective hazard regulations. As a first step, he sought a technical expert interested in policy to write on seismic hazards for the institute's series of monographs on policy issues facing the bay area. Through contacts at the University of California's geology department, Scott found Steinbrugge.[95]

Steinbrugge's treatise appeared in the fall of 1968. After briefly recounting bay area earthquake history, he concluded that "for planning purposes it is reasonable to anticipate a major or great earthquake in the San Francisco Bay Area once every 60 to 100 years"—an ominous prediction, given that the last one had struck in 1906. He then discussed mitigation measures that could be taken to prepare for such an earthquake: risk zoning along the bayshore as already proposed to the Bay Conservation and Decelopment Commission, restricting development in fault zones and prohibiting construction across active fault strands, strengthening structures erected before modern building codes had been passed, and reinforcing dams. Finally, Steinbrugge recommended establishment of a regional government organization to address the land-use problems related to earthquake hazards; existing town and county governments were simply not willing enough to restrict development and construction in hazardous areas, because of their reliance on property taxes for revenue.[96]

Reviewing Steinbrugge's monograph for his institute's newsletter, Scott elaborated on this last suggestion. He proposed that a San Francisco Bay Area Earthquake Commission be established analogous to the Bay Conservation and Development Commission, which was then finishing up work on its land-use plan for the bay. The earthquake commission "should be charged with (1) investigating the nature and extent of the hazard, and reporting to the Bay Area public and to the Legislature, (2) drafting a plan for reducing the hazards, and (3) recommending the organization or organizations best able to keep the plan up-to-date and to implement it." Scott argued that this commission should be comprehensive in its scope, looking not only at building code provisions to minimize structural damage but also at "zoning and land-use control, subdivision design and location, transportation planning and design, land drainage and utility operation, provisions for land conservation and open space, and finally, plans for appropriate responses to a disaster after it happens."[97]

Scott and Steinbrugge were not content merely to issue written policy proposals. With the support of the institute's director, they convened a meeting on March 17, 1969, at the University of California Faculty Club to consider strategies for effective earthquake hazard mitigation in the San Francisco Bay area. The meeting was attended by geologists, engineers, architects, and other technical experts interested in seismic hazards, representatives from local government bodies and state and federal agencies, and policy analysts and public policy researchers. They quickly agreed that a bay area earthquake commission as outlined by Scott should be established. On April 1, the institute's director sent a letter, endorsed by the 18 meeting attendees, to all bay area state legislators, urging them to create this commission, which would produce much needed "new policies and intensified earthquake research to guide future decisions on land use and urban development, housing and industrial location, and construction design." The institute also released the proposal to the local press.[98]

The timing of the proposal could not have been better, for the bay area was in the grips of near-hysteria over a prediction of imminent seismic catastrophe. One source of the frenzy was Elizabeth Steen, a 28-year-old housewife in San Francisco who had a reputation as a clairvoyant. In late 1968, she had had a vision that California would be destroyed in a gigantic earthquake, and she and her family promptly moved to Spokane, Washington. The anxiety was fed by followers of Edward Cayce, the so-called Sleeping Prophet, who before his death in 1945 had predicted the destruction of major metropolises, proclaiming, "Los Angeles, San Francisco, most all of these will be among those that will be destroyed, before New York even." Cayce had only suggested that this cataclysm would occur sometime before 2000, but his followers began to proclaim in late 1968 that the destruction would come by earthquake in April 1969. A literary event stoked San Franciscans' worry even further, albeit unintentionally. In late 1968, Curt Gentry published *The Last Days of the Late, Great State of California,* a brilliant examination of the social mores and political life of California under Governor Reagan. Gentry wrote his account as a fictional obituary, lamenting what America would lose if California were to disappear. He opened his book with a graphic account of California falling into the sea during a cataclysmic earthquake at 3:13 P.M. one Friday afternoon in 1969. Many readers did not recognize this as a literary device and added Gentry's book to their reasons for believing that a Friday in April 1969 would see the end of California.[99]

An incensed Charles Richter called this prediction "the most terrific explosion of pedigreed bunk" he'd ever seen in the earthquake field, but reputable earthquake researchers were not entirely innocent of fanning the fears.[100] In a statement publicized by the University of Michigan press relations office and elaborated in a sensationalistic *Esquire* magazine piece, Michigan professor Peter Franken, acting director of the Advanced Research Projects Agency, emphasized that San Francisco would experience a great earthquake within the next 20 years and predicted that fatalities from such an event might well reach into the millions.[101] The Geological Survey's William Pecora, in releasing his panel's report on a federal earthquake prediction research program in late 1968 and again in defending his agency's budget in a congressional hearing in early 1969, also asserted that the Bay area faced a repeat of the 1906 earthquake definitely by the end of the century and probably by 1980.[102] In internal survey correspondence and in interviews with the science correspondent of the *New York Times,* survey geologists even worried that anomalous creep measurements along the San Andreas Fault near San Francisco suggested that the catastrophe might happen within the next two to four years.[103] Even Steinbrugge's monograph on seismic hazards in the bay area, published in late 1968, added to the public concern with its suggestion that a serious earthquake had to be expected within the coming 40 years.[104]

By spring of 1969, the prediction had become firmly entrenched in popular culture. A Calypso song playing frequently on local radio stations asked: "Where can we go, when there's no San Francisco? Better get ready to tie up the boat in Idaho." Another tune warned: "Atlantis will rise. Sunset Boulevard will fall. And where the beach used to be won't be nothing at all." [105] KSAN radio in San Francisco announced in February that it would air feature reports on how to prepare for the earthquake every two hours "until the day the earthquake occurs." [106] Worried residents inundated seismological laboratories around the state with questions on the impending catastrophe. The Division of Mines and Geology handled 52,335 public queries from July 1968 to June 1969, nearly twice its normal rate. At the same time, the National Center for Earthquake Research distributed "90,000 nontechnical leaflets about the San Andreas fault and 20,000 leaflets about earthquake hazards." [107] The San Francisco Chronicle ran an extended series of articles on seismic hazards, quoting many scientists who discounted the apocalyptic rumors and pointed instead to reasonable measures that could be taken to guard against realistic seismic hazards. Meanwhile, demand for earthquake insurance soared in the bay area.[108] Still not reassured, hundreds, if not thousands, of Californians left the state to escape the calamity; preachers led entire congregations away, while a hippie leader struck fear in the hearts of the conservative residents of Pocatello, Idaho, by announcing that San Francisco's hippies would decamp there.[109]

In this context, Democratic state senator Alfred Alquist of San Jose decided to sponsor a bay area earthquake commission. His new administrative assistant Steve Larson, who had been fascinated by Steinbrugge's monograph on seismic hazards, had attended the March 17, 1969, meeting called by Scott and Steinbrugge. Soon thereafter, he obtained Alquist's approval to draft a bill establishing the commission. It followed the model of the Bay Conservation and Development Commission, down to its membership of 27, its life-span of four years, and its initial-year funding of about $100,000. Alquist, noting the "recent interest in earthquakes" generated by what the news media were calling the "Great California Earthquake Scare," submitted this bill in the California Senate on April 8 and committed himself to promoting it among his colleagues.[110]

Outside observers gave the bill little hope for passage. For one thing, the Reagan administration was firmly opposed to funding any new governmental bodies, no matter how meritorious.[111] There was also little popular support for the bill. Conservation groups had loudly backed the original Bay Conservation and Development Commission, and environmentalists collected 250,000 signatures to lobby for passage of the bay plan it released in 1969.[112] By contrast, there were no grass-roots activists pushing for the earthquake commission. While the earthquake prediction had put seismic hazards on many people's minds, it had generally produced a sense of fatalism, as the prophesied cataclysm was so great that little could be done to survive it. Then, too, as the last Friday afternoon in April passed without California's sinking into the sea, Californians relaxed and once again put earthquake hazards out of mind. This indifference extended to legislators. In mid-May, Alquist invited his fellow lawmakers to a special showing of a 30-minute film on earthquakes. Only one other legislator showed up, joining two staff members and four reporters. A film on the dangers of pornography shown the next day drew a far larger crowd.[113] As expected, Alquist's bill was killed by the fiscally conservative Senate Governmental Efficiency Committee on May 27.[114]

Alquist did not give up, however. On June 9, 1969, he introduced a Senate concurrent resolution establishing a seismic hazards study group not as a separate commission but as a joint committee of the legislature. This approach had several advantages: a concurrent resolution required only a majority vote in both houses of the legislature and did not have to be voted on by the Senate Governmental Efficiency Committee; it also did not require Governor Reagan's signature and thus was veto-proof. To secure passage for this resolution, though, he would have to modify it. Southern California legislators were concerned over undue attention for the northern part of the state, so he had to extend the scope of the earthquake study from just the San Francisco Bay area to all of California. He also had to cut the funding for the new committee from his initial proposal of $99,103 to just $5,000. Using his political influence (which was considerable—in 1970 he would run as Democratic candidate for lieutenant governor), Alquist finally won passage for the resolution in late August 1969. Established thereby was the Joint Committee on Seismic Safety, composed of eight legislators representing both houses of the legislature, both political parties, and both northern and southern California. The committee had five years to "develop seismic safety plans and policies and recommend to the Legislature any needed legislation to minimize the catastrophic effects upon the people, property, and the operation of our economy should a major earthquake strike any portion of the State of California." The resolution also created five advisory groups, comprising 74 experts, in seismology, engineering, land-use planning, policy analysis, governmental studies, and other disciplines, to carry out the actual work of the committee.[115] During its first 18 months, the committee had a precarious existence, as it struggled to find funds to carry out its mandate.[116] Nevertheless, it provided a convenient vehicle for conveying earthquake researchers' recommendations on seismic hazard mitigation to state legislators.

The 1964 Alaskan earthquake thus galvanized California earthquake researchers in much the same way that the 1906 San Francisco shock had done. Once again scientists and engineers scrambled to learn more about the nature of seismic hazards and ways of mitigating them, and they renewed their calls to the California public to pay more attention to earthquake risks. With far greater resources at their disposal in the 1960s than they had had half a century before, they were far more effective in mapping dangerous fault zones and identifying other hazardous areas, but they could also project an ambitious program for mitigating hazards through prediction as well as engineering. In addition, they had more governmental resources with which to push for the implementation of mitigation measures: the two federal surveys and the state Division of Mines and Geology were now major channels for public education, and seismic safety advocates could turn to government committees—the Bay Commission and the Joint Committee on Seismic Safety—to develop new means for earthquake protection. This ever-increasing interpenetration of science and the state would lead to a flowering of the regulatory-state apparatus after a destructive earthquake struck northern Los Angeles in 1971.

CHAPTER 7

Seismic Politics

Responses to the San Fernando Earthquake of 1971

At 6:01 A.M. on February 9, 1971, violent shaking rudely awakened most Los Angeles residents. This earthquake was not the long-expected Big One; its magnitude on the Richter scale was only 6.6. But it struck directly underneath the San Fernando Valley, a vast bedroom community that comprises the northern part of Los Angeles. As a result, it was strongly felt throughout the metropolitan area. Once the shaking had ended, 58 people were dead and 30,000 buildings had been damaged, leading to a property loss of more than half a billion dollars. Yet the hazard to San Fernando's residents was not yet over. In the earthquake, the crest of the Lower Van Norman Dam, a structure built in 1915 to store water from the new Owens Valley aqueduct, had collapsed, leaving the top of the dam only four feet above the reservoir level. With the possibility looming that aftershocks might further rupture the dam, the city evacuated 80,000 residents living below the reservoir. Not until late on February 11, after seismic activity had subsided and the reservoir level had been lowered, were the evacuees allowed to return.[1]

Other kinds of structural failure also resulted from the earthquake. Forty-nine of the deaths occurred at the San Fernando Veterans Administration Hospital, where two buildings erected in 1925 and not designed to resist earthquakes collapsed. At the nearby Olive View Medical Center, structures dedicated only a month before the earthquake collapsed; in the psychiatric building (which luckily stood empty at the time of the temblor), the entire first story disappeared as the second story settled to the ground. Other hospitals sustained serious damage as well, disrupting the supply of medical care just as an estimated 2,400 injured people sought treatment. In general, buildings constructed before 1933 fared poorly; in downtown Los Angeles, one person was killed when an old brick wall tumbled apart. The freeway system, which had expanded rapidly in car-happy southern California in the 1950s and '60s, also suffered a blow; five freeway overpasses collapsed, disrupting the main traffic artery north from Los Angeles.[2]

In the months after the San Fernando earthquake, seismologists and earthquake engineers issued a number of reports drawing lessons from the devastation. These reports—from such organizations as the National Academy of Sciences, the Structural Engineers Association of California, the Earthquake Engineering Research Institute, the California Institute of Technology, and a specially convened Los Angeles County Earthquake Commission—offered the same recommendations again and again: Old buildings, designed before 1933, should be renovated or demolished. Building codes should be strengthened. Dams and highway bridges should be made more resistant to earthquakes. Hospitals and other critical emergency facilities should be designed to remain functional even after an earthquake. Development in areas of obvious earthquake hazard, such as fault zones, should be restricted. And further earthquake research should be stimulated, particularly through the placement of more strong-motion instruments to record the exact nature of the forces that produced such devastation.[3]

Over the next four years, earthquake researchers saw many of these recommendations acted upon. In order to produce this increased level of seismic safety, however, they did not, as they had after the 1933 Long Beach earthquake, rely on channeling the energies of an outraged public to demand stronger protective measures. Rather, they worked quietly behind the scenes by serving on technical advisory committees that produced stronger building codes at the local level and by using the legislature's Joint Committee on Seismic Safety to push new earthquake legislation for the state as a whole. This strategy of expert policymaking away from the public eye meant that earthquake researchers frequently had to make compromises, since they could not use public pressure to generate political support. Nevertheless, by being persistent and patient, earthquake researchers did succeed in having most of their agenda implemented.

The San Fernando Earthquake and the General Public

In the weeks after the 1933 Long Beach earthquake, many southern Californians had been aghast at the destruction wrought on buildings, particularly schools. Earthquake experts—through their testimony at coroner's inquests, in editorial columns in newspapers, and in talks at public meetings—directed the public's anger toward the inadequacy of current building codes. The resultant public outcry, fanned by newspaper editorials and by skillful lobbying in Sacramento, resulted in the passage of the Field Act just a month after the earthquake. After the San Fernando earthquake of 1971, though, earthquake researchers did not foster a public campaign for increased seismic safety. They were wary of generating too much public attention, perhaps because they were dismayed by the distortions and acrimony produced by the Bodega Head controversy and other environmental grassroots campaigns. The structural engineer who headed the Los Angeles City Department of Building and Safety, for example, refused to submit building code revisions to the city council in the month after the San Fernando earthquake because he wanted to steer clear of "panic legislation"; in his view, improved seismic safety would be the result of calm deliberation among experts rather than public pressure.[4]

The news media at first did strongly support the efforts of earthquake researchers to improve seismic safety. In the first week after the San Fernando earthquake, editorials in the *Los Angeles Times* called for strengthened building codes, seismically safer designs for hospitals, freeways, and dams, and the consideration of land-use restrictions in areas of great seismic hazard. Radio and television stations in both Los Angeles and San Francisco also broadcast editorials urging better hospital construction, upgrading of old schools, and improved building codes. Such editorial comments appeared sporadically over the next year.[5]

These expressions of support did not, however, coalesce into a sustained public campaign for improved seismic safety. Instead, public concern shifted to bureaucratic failures and incompetence. In mid-March 1971, for example, the Sylmar–Northeast Valley Disaster Committee emerged as a political pressure group that was concerned not with the lack of structural safety that had produced so much damage but rather with the ineptitude with which federal government agencies, such as the Small Business Administration, had distributed disaster aid and loans among victims. Over the summer of 1971, these "complaints from local residents about insufficient amounts of aid, unnecessary red tape and delays, lack of administrative personnel, slowness of inspecting damaged buildings, and, in some cases, failure to clear debris quickly" dominated both congressional hearings and the editorial page of the *Los Angeles Times*.[6]

Meanwhile, the near-collapse of the Lower Van Norman Dam had led to the formation of Homeowners Emergency Life Protection, a vocal group of homeowners residing below the dam. This group also was not interested in the general issue of the seismic safety of dams. Rather, it used petition drives and repeated appearances at legislative committee hearings to try to block the Los Angeles Department of Water and Power from building a replacement dam above their heads, even one that would be far more modern and earthquake-resistant than the one that had failed. Charles Richter, frustrated that this group ignored the city engineers' attempts to design a safer dam, dismissed the group as "largely motivated . . . by a few persons who are more interested in attracting the public eye than in giving due attention to the facts."[7]

Grassroots campaigns for greater seismic safety also failed to emerge in response to a widely available film about seismic hazards. In 1970, the British Broadcasting Corporation released an hour-long film entitled "San Francisco: The City That Waits to Die." It featured gory footage from the 1906 San Francisco and 1964 Good Friday earthquakes as well as a 1967 Venezuelan earthquake in which several reinforced concrete high-rise buildings had completely collapsed. The film also showed Louis Pakiser of the U.S. Geological Survey asserting that the next great earthquake in the bay area could produce carnage on the same scale as the Vietnam War, with fatalities reaching as high as 100,000. The film's narrator closed with the warning, "Political blindness, expediency, and complacency may be preventing the avoidance of a greater disaster than that which overtook San Francisco on April 18, 1906." In the film, the only means of hazard mitigation discussed at length were earthquake prediction and the remote possibility that geologists might one day prevent the occurrence of earthquakes. The film made little mention of more mundane measures such as improved building codes. Thus, the viewer was left with the impression that, barring a miraculous technological breakthrough, the next great earthquake in California would be a cataclysmic disaster.[8]

The BBC film was not distributed in the United States until the summer of 1971. Once it was shown in California, though, it quickly became the centerpiece of campaigns by several grassroots organizations—albeit campaigns that once again did not incorporate a true concern for seismic safety. One grassroots organizer was Alvin Duskin, a San Francisco neighborhood activist who led an ultimately unsuccessful campaign in 1971 to prohibit further construction of high-rise buildings. In June, he showed the film, with its dramatic scenes of collapsed high-rises, four times a day in his North Beach neighborhood; as he admitted, his intention was to scare people about the safety of skyscrapers.[9] That September, 200 miles to the south, the local chapter of the Sierra Club screened the film as part of a campaign against plans to develop several new cities near faults in northern San Luis Obispo County. Here, too, the organizers sought to scare viewers about the possible effects of an earthquake, in order to halt further development.[10]

Earthquake experts did not rush to form alliances with these grassroots groups. Rather, with the exception of Pakiser, they sought to distance themselves from what they saw as the film's distortions and counterproductive scare tactics. On July 25, 1971, San Francisco's KPIX-TV aired the film in prime time, followed by a panel discussion among experts. Here, Henry Degenkolb and other northern California earthquake engineers and seismologists countered what one called its "factually unsound and nearly hysterical utterances" that "grossly exaggerated the dangers to the area." In a review circulated to insurance agents across the state, Karl Steinbrugge also charged that underneath the film's "documentary facade is a biased viewpoint which presents a highly colored and partly untrue story." These earthquake experts asserted that, even under the worst circumstances, a great earthquake in the bay area would cause only about 10,000 rather than 100,000 fatalities; moreover, they emphasized that proper engineering could ensure that even the strongest shaking would not lead to the collapse of high-rise buildings.[11] Hence, earthquake researchers chose not to mobilize public opinion in order to promote their agenda. Relying on public pressure would have meant surrendering to the exaggerations and scare tactics that had become associated with so many environmental grassroots campaigns since the Bodega Head fight. Seismologists and engineers instead preferred to work in the insulated sphere of technical advisory committees, where experts and bureaucrats could work with due deliberation to produce what they considered rational measures for hazard mitigation.

Bureaucratic Politics: Updating Building Codes

A few of the recommendations made by earthquake researchers in 1971 required action in specialized bureaucratic arenas where the power to make decisions had already been ceded to technical experts. This was particularly the case with building codes. Even before 1971, the modification of these highly arcane technical documents had become the province of structural engineers. After the San Fernando earthquake, engineers could go about strengthening the building codes without having to lobby legislators or the general public.

At their most basic, building codes are political documents; they are ordinances passed by city councils or county boards of supervisors that regulate how buildings are to be constructed and vest enforcement power in local building departments. Establishing the first building codes in the late nineteenth and early twentieth centuries, and then adding seismic design provisions to the codes in the 1930s, had involved considerable political struggle to determine the proper scope of government regulatory power. Once the basic principle of government control over seismic design had been established, though, building codes passed from the realm of overt public politics. The codes became very technical documents, often hundreds of pages in length, specifying in great detail the strength and other characteristics that approved building materials had to possess and the precise methods to be used in erecting structures. Changes in these specifications were often arcane and beyond the comprehension of local politicians. Consequently, in large jurisdictions, such as the city of Los Angeles, the local building department, staffed by expert structural engineers, was entrusted with developing changes to the building code, which were then rubber-stamped by the city council. Smaller jurisdictions, which did not have the necessary expertise in their building departments, simply adopted model building codes developed by outside agencies. The one most commonly used in California was the Uniform Building Code of the International Conference of Building Officials. The conference updated this code every three years by voting on suggested changes submitted by engineering groups, representatives of the building material producers, and other interested organizations. In its seismic design provisions, the conference generally followed the recommendations of the Structural Engineers Association of California. Thus, structural engineers exerted a great deal of control over modifications to building codes.[12]

Even before the San Fernando earthquake, many structural engineers had become convinced that the seismic design provisions needed strengthening. In the 1960s, most codes called for designing structures to withstand lateral forces of 0.10g. Strong-motion instruments had measured forces as high as 0.33g during the Imperial Valley earthquake of 1940, but engineers argued that buildings did not have to be designed for such high forces: some of the energy contained in earthquake waves did not propagate from the ground into buildings, and internal friction within the structural frame of the building dissipated even more of the energy. Examination of buildings after earthquakes showed that those designed to withstand a force of 0.10g indeed suffered little structural damage, but several developments in the late 1960s and early 1970s cast doubt on the adequacy of this design standard. Strong-motion records from a 1966 earthquake showed a spike at 0.50g, and one instrument recorded a force of 1.25g during the

San Fernando earthquake. Thus, earthquake forces could be much larger than structural engineers had assumed. Moreover, the ability of structures to resist these forces gradually decreased. Before the 1960s, buildings had been generally overdesigned. Because the strengthening effects of internal partition walls and other elements could not be easily calculated, they were ignored in the design. As a result, many buildings were actually far stronger than design calculations showed. The widespread use of computers for solving complex engineering equations, which began in the 1960s, allowed computation of the contribution of all building elements to structural strength, and the practice developed of designing buildings to be no stronger than had to be. This meant that the gap between design standards and actual earthquake forces became much more important.[13]

Structural engineers also became concerned about the treatment of reinforced concrete in the building codes. In 1958, the Structural Engineers Association had decided to require ductility in the structural frames of high-rise buildings. When ductile materials are stressed beyond their strength, they stretch or bend rather than break. Thus, even if they sustain structural damage, buildings made of ductile materials are less likely to collapse and risk injury to people inside and near them. But reinforced concrete, as it was used in the late 1950s, was usually brittle rather than ductile. The Portland Cement Association undertook an extensive research effort to determine how the steel reinforcing might be redesigned to make reinforced concrete more ductile, and in the mid-1960s it convinced the Structural Engineers Association to permit the use of reinforced concrete in high-rises.[14] Builders did not, however, immediately adopt all of the new design guidelines for reinforced concrete, and as a result many reinforced concrete buildings erected in the mid- and late 1960s provided only a limited degree of ductility. Several of them collapsed in a Venezuelan earthquake in 1967, in the moderate Santa Rosa earthquake of 1969, and in the San Fernando earthquake. Clearly, so engineers concluded, the means for making reinforced concrete ductile had to be spelled out in detail.[15]

After the San Fernando earthquake, structural engineers took the lead in proposing building code changes that would address these problems. The changes fell into two groups. The first group, generally adopted in early 1972, concerned details of construction methods that could improve earthquake resistance. In this group fell, for example, requirements for increasing the ductility of reinforced concrete columns by inserting more steel ties into the concrete. In timber-frame houses, steel joist anchors were now required instead of nailed wood ledges to tie the roof to the walls.[16] The second group of alterations to the code concerned increases in the lateral forces for which structures should be designed. These changes proved to be more controversial among engineers, who thought that they would increase construction costs too much. The new code as eventually agreed to by engineers and adopted in the mid-1970s increased the basic lateral force a building must be designed to withstand by only a small amount from what it had been in the late 1960s. However, this basic force was to be augmented by up to 50 percent if the structure housed medical or emergency facilities or held large numbers of people and by another 50 percent depending on the characteristics of the ground on which the structure was placed. For irregularly shaped structures or tall high-rises, the code went even further and required a dynamic analysis. That is, rather than designing a structure to resist a given force as if it were simply a fixed force applied to the side of the building, engineers now had to model on a computer how the structure would respond to repeated oscillations created by earthquake waves. Such modeling, which required an accurate representation of earthquake ground motion from strong-motion instruments as an input, would produce a structure that would be designed to resist the actual forces encountered in an earthquake rather than ideal representations of them.[17]

Earthquake engineers also influenced seismic design provisions for such structures as highway bridges and dams. Immediately after the San Fernando earthquake had caused the collapse of five freeway overpasses, engineers within the California Division of Highways adopted new design requirements for such bridges. These provisions increased the lateral forces bridges had to resist by 150 percent. They also now required the use of more steel reinforcing, to increase the ductility of reinforced concrete bridge columns. Finally, they insisted that expansion joints on bridges be supplied with cable restraints. As previously designed, highway bridges had expansion joints in which bridge spans were simply left to rest on ledges, free to expand in hot weather. During the earthquake, however, some of these unconstrained spans had slipped off the ledges and fallen to the ground. The division now required that the spans be tied together with cables that still provided room for expansion but prevented them from moving so far as to slip entirely off the ledges. Over the next two years, the highway division's engineers refined these emergency provisions into a detailed new code for bridge design. The division also embarked on a program of supplying existing bridges with cable restraints; over the next two decades, it equipped 1,200 of the state's bridges. All of these actions originated within the division's engineering staff and were implemented with little involvement of the public. By the time the state assembly's Committee on Transportation got around to holding a hearing on the seismic safety of highway bridges in

June 1971, the division could assert that it already had the situation under control and did not need the legislature's help.[18]

In the case of dams, engineers were concerned less with upgrading requirements for new construction than with ensuring the safety of existing structures. The Lower Van Norman Dam was a hydraulic fill dam. In this method of construction, popular in the early twentieth century, the earthen material for the dam was put in place by a jet of silt-laden water from which it settled out to form the dam. While efficient, this method produced a dam in which the material was not fully compacted and pockets of water remained. During severe earthquake shaking, the material could rearrange itself or even liquefy, leading to dam failure. After the San Fernando earthquake, engineers in the California Department of Water Resources decided to check the seismic safety of the state's 28 hydraulic fill dams. This task was entrusted to the Division of Safety of Dams, which had been established in 1929 and already had the power to regulate dam operation. The division ordered all owners of hydraulic fill dams to conduct dynamic analyses of their dams. In almost all cases, these analyses revealed serious earthquake safety problems. Six of the reservoirs were eventually put out of service; others were partially drained while the owners strengthened or rebuilt the dams. Once again, this increase in the state's seismic safety occurred largely out of the public's view, driven as it was mostly by engineers already ensconced in the state's bureaucracy.[19]

Legislative Politics: The Joint Committee on Seismic Safety

Earthquake experts, to achieve other goals, such as increasing research appropriations, ensuring that emergency facilities were designed to remain functional after an earthquake, and introducing land-use planning measures in areas of seismic hazard, had to enter the legislative realm, a more overtly political arena. Yet here, too, they did not proceed by trying to mobilize popular support. Rather, they worked through the Joint Committee on Seismic Safety, quietly garnering support for various pieces of legislation. Taking an approach that avoided grassroots campaigns rendered them politically weak. While legislators might agree that increased seismic safety was desirable in principle, the seismic safety advocates could not apply political pressure to advance specific measures. When faced with entrenched and well-organized interest groups that felt threatened by proposed legislation, earthquake experts often had to back down and significantly weaken their proposals. Nevertheless, they had some success. Senator Alquist, who remained chairman of the Joint Committee throughout its life, skillfully crafted a number of compromises that satisfied the concerns of most involved interest groups, and as a result, state regulation of seismic hazards expanded considerably in the early 1970s.

The Joint Committee on Seismic Safety, as set up in 1969, provided a ready means for earthquake researchers to gain access to state legislators, composed as it was of eight legislators, drawn equally from the Assembly and the Senate, the Democratic and Republican parties, and from northern and southern California. Senator Alquist was by far the committee's most active member and vigorously championed a legislative program put together by its expert advisors. These advisors were organized into five groups, dealing respectively with engineering considerations and earthquake science, disaster preparedness, postearthquake recovery and redevelopment, land-use planning, and governmental organization and performance. The legislation establishing the Joint Committee spelled out in detail the disciplines and professional organizations from which the 82 expert advisors were to be drawn. As a result, engineers from the Structural Engineers Association and the Earthquake Engineering Research Institute, geologists from the California Division of Mines and Geology and the U.S. Geological Survey, academic researchers from Caltech, Stanford, and the University of California, and representatives of local government all found themselves in an official position of advising the Joint Committee, and through it the state legislature, on seismic safety. Of these advisors, earthquake engineers were the most successful in gaining Alquist's ear; and the indefatigable Steinbrugge, who had first promoted the idea of the Joint Committee in the late 1960s, headed the steering committee for the advisory groups.[20]

The Joint Committee's work received a powerful boost from the San Fernando earthquake. It had languished under insufficient funding, receiving only about $40,000 in the first two years. Soon after the earthquake, though, the state legislature approved a special appropriation of $150,000 for the Joint Committee. It certainly helped that one powerful southern California senator had had his car dealership destroyed in the earthquake.[21] Senator Alquist also worked to personalize the issue of seismic safety for other legislators. In March 1971, he had Steinbrugge and several other advisory group members inspect the state's capitol building. They reported that the capitol's nineteenth-century west wing, with its domed rotunda, was in danger of collapsing in an earthquake. In a report released in June 1972, the state architect confirmed this judgment: the capitol was a seismic safety hazard. The legislature did not get around to beginning a renovation until 1976. But in 1972 Alquist did convince the legislature to ban tours of school children in the west wing because of the hazard. This served as a constant reminder to legislators of the earthquake risk, thereby increasing the committee's political leverage.[22]

One of the Joint Committee's first accomplishments was to increase the research resources available to earthquake engineers. Throughout the 1960s, engineers had most desired an increase in the number of strong-motion instruments.[23] These instruments provided accurate records of earthquake motion, which were crucial to testing new methods of making structures earthquake resistant. As dynamic analyses became more widely used in the designing of large structures, strong-motion records, which provided data for such analyses, became even more important. Since the 1930s, the U.S. Coast and Geodetic Survey had operated strong-motion instruments in California, but budget constraints in the late 1960s had limited the expansion of its network of slightly more than 100 instruments.[24] In 1965, the city of Los Angeles added a building code provision requiring owners of new buildings taller than six stories to install strong-motion instruments. Two years later, this provision was adopted as an optional clause in the Uniform Building Code, and about a dozen California cities enacted it. By 1972, this provision had led to the installation of about 250 additional strong-motion instruments. Because they were installed in buildings, however, these instruments recorded the combined motion of the ground and the building; engineers had difficulty in extracting the motion of just the ground (the information of most use to them) from these records.[25]

Within weeks of the San Fernando earthquake, H. Bolton Seed, the soils engineer at Berkeley who had advised the Bay Conservation and Redevelopment Commission and a member of the advisory group on engineering and earthquake science, drafted legislation to establish a statewide strong-motion instrumentation program to be administered by the California Division of Mines and Geology. This program was to be funded by a fee (eventually set at 0.0007 percent of the value of new construction) levied on all applications for a building permit. Lobbyists for the cities of Los Angeles and San Francisco opposed the legislation, arguing that those cities and counties which already required strong-motion instruments as part of their building code should not be forced to levy an additional fee for a state program. Senator Alquist agreed to amend the bill to exempt them, and the resulting bill easily passed the legislature; it was signed by the governor on October 19, 1971.[26] In its first three years, the new program received nearly $1.25 million from building fees, increasing the Division of Mines and Geology's budget by about 25 percent. With this money, the division by the middle of 1975 had installed 196 strong-motion instruments throughout the state. Most of these instruments were so-called free-field instruments, placed on the ground away from buildings or other large structures so they could record directly the motion of the ground during an earthquake. The division also strove to fill in the gaps left by the federal and local programs, so that instruments would be spaced evenly across coastal California, thereby increasing the likelihood that a record would be obtained from a large earthquake anywhere along the coast.[27]

Earthquake engineers also succeeded in having the state legislature increase the seismic design requirements for hospitals. The initial impetus came from Jack Meehan, chief structural engineer of the Schoolhouse Section of the Office of Architecture and Construction, the state agency now responsible for enforcing the Field Act. Under the act, the Schoolhouse Section had the power to promulgate regulations for the seismic design of schools, to check building plans to ensure that they met the regulations, and to inspect building sites to verify that approved plans were being followed. In January 1971, just before the San Fernando earthquake, Meehan suggested to his fellow Joint Committee advisors that the Field Act provisions be extended to hospitals. After the earthquake then graphically demonstrated the vulnerability of hospitals, Senator Alquist rushed to introduce a bill incorporating Meehan's suggestion.[28]

The bill soon ran into opposition. The State Department of Public Health, which already had the general power to approve and supervise hospital construction, objected that giving control over structural aspects to the Schoolhouse Section would fragment the process of approving hospital building plans. As structural engineers pointed out, however, the Department of Public Health did not have the expertise in structural engineering to properly enforce earthquake-resistant design regulations. After consulting with the Joint Committee's expert advisors, Alquist sought to meet the department's objection by amending his bill so that the department retained ultimate approval of plans and oversight of construction, but subcontracted with the Schoolhouse Section to take care of all structural engineering aspects.[29] This compromise, reached in May 1971, still did not satisfy the department, for Alquist's language still gave the Schoolhouse Section too much power for its liking. Consequently, the department and the California Hospital Association (which represented the state's hospital administrators) opposed the bill in a hearing before the state's Senate Finance Committee. The city of San Francisco opposed the bill as being too costly and needlessly adding a layer of bureaucracy. Bowing to this opposition, the Finance Committee killed the bill.[30]

After reintroducing the bill during the 1972 legislative session, Alquist once again amended it to meet the Department of Public Health's objections. The revised bill placed the Hospital Building Safety Board, to be set up by the legislation, more firmly under the control of the department's director than had the defeated bill, and added a hospital administrator to its membership. With this language adopted, the department and the California Hospital Association finally supported the bill. Their endorsement, added to the vigorous lobbying of the Structural Engineers Association and the Construction Inspectors Association of Northern California, overcame the opposition voiced by the cities of San Francisco and Los Angeles over costs and red tape. The bill was finally signed into law by Governor Reagan on November 21, 1972.[31]

The Hospital Seismic Safety Act proved to be far more difficult to implement than its authors had anticipated. It assigned the Schoolhouse Section to review building plans and inspect construction but did not actually specify the seismic design provisions for hospitals. The only guidance came from a "legislative intent" clause, which stated that "hospitals, . . . which must be completely functional to perform all necessary services to the public after a disaster, shall be designed and constructed to resist, insofar as practicable, the forces generated by earthquakes." This statement was open to differing interpretations: did the phrase "must be completely functional" establish an absolute mandate that the most advanced state-of-the-art methods of seismic design must be used irrespective of cost (a standard previously applied only to nuclear reactors), or did the phrase "insofar as practicable" overrule this and allow designers to use only those methods which were not prohibitively costly? The California Division of Mines and Geology, which the Department of Public Health called upon to review geologic and seismologic reports for hospital sites, leaned toward the first interpretation. The division's geologists argued that the act required hospitals to remain functional even after the maximum credible earthquake, that is, the worst earthquake that could conceivably strike the site. In some cases, the division would not approve geologic reports until they specified the lateral forces produced by the maximum credible earthquake as a design criterion. Engineers on the Hospital Building Safety Board, however, insisted that cost should be taken into consideration as well. They argued that hospitals should be designed to remain functional only for the maximum probable earthquake—the worst earthquake that had a better than 50 percent probability of striking during the hospital's assumed lifespan of a century. Not until California's attorney general intervened in July 1974 with a ruling on the legislature's intent did the engineers' interpretation win out. The Hospital Building Safety Board eventually decided to adopt the requirements of the 1976 Uniform Building Code for hospital design.[32] By this time, though, the review process had already gained a reputation among engineers and contractors of being costly, drawn-out, and without clear guidelines. Many designers now believed that the Hospital Seismic Safety Act imposed unreasonably strict design standards and thereby increased construction costs by 15 percent or more.[33]

The dispute over seismic design rules for hospitals probably doomed the chances that Field Act provisions would be extended to other emergency service facilities. In early 1973, Alquist, after consulting with Meehan and several other engineers and building officials advising the Joint Committee, introduced legislation to ensure that all public emergency service facilities would continue to function after the maximum probable earthquake. The facilities covered by this bill included fire stations, police stations, and government buildings housing emergency operating centers and base radio stations. The bill called upon the Schoolhouse Section to set the seismic design provisions, but, in a nod to the increasing protests against the spread of state government, it left plan approval and inspection to local building departments. Nevertheless, the League of California Cities and the County Supervisors Association of California, two lobbying groups for local governments that were quite powerful in Sacramento, opposed the bill. They argued vociferously that it would impose an unreasonable burden, by adding anywhere from 5 to 25 percent to the cost of their emergency facilities. In testimony before a state Assembly committee, the city manager of Sebastopol, a small northern California city, backed up this assertion by detailing his city's negative experiences under the Hospital Seismic Safety Act. Alquist, despite several attempts to reassure local governments that added costs would not be that high, failed to overcome the opposition, and the bill died in committee.[34]

In spite of this defeat, earthquake engineers had accomplished a lot through their service on the Joint Committee's advisory groups. They had seen through the establishment of the State Strong-Motion Instrumentation Program, which provided valuable new data for engineering research, and they had helped pass the Hospital Seismic Safety Act. But the Joint Committee also followed the advice of its consulting geologists and planners, who argued that more use should be made of geologic information in planning appropriate land uses for seismically hazardous areas. As a result of the committee's efforts, the legislature passed several measures restricting construction in fault zones and requiring attention to seismic hazards in both city planning and disaster preparation.

One of the earliest of these measures was the requirement that a seismic safety element be included in the "general plan" of California's cities and counties. In 1947, the state legislature had first mandated that local governments adopt a general plan. By the early 1970s, it required that the plan discuss the impact on development policies of such problems as population increases, transportation and housing needs, and environmental protection. In late 1970, the Joint Committee's Advisory Group on Land Use Planning proposed that the general plan also be required to include a seismic safety element "consisting of an identification and appraisal of seismic hazards such as susceptibility to surface ruptures from faulting, to ground shaking, to ground failures, or to effects of seismically induced waves such as tsunamis." This proposal, which received only token opposition from the city of San Francisco, for adding another burdensome state regulation, quickly passed the state legislature and was signed by Governor Reagan in June 1971, four months after the San Fernando earthquake.[35]

Many local jurisdictions were at first unsure how to prepare the seismic safety element of their general plan. In response to their concerns, the Division of Mines and Geology, in cooperation with the Council on Intergovernmental Relations and other state agencies, in the summer of 1972 developed appropriate guidelines. Ideally, a locality's seismic safety element was to include a basic statement of policy, an emergency response plan for earthquakes, a plan for abating the hazards posed by structures erected before the adoption of modern building codes, a map or maps showing areas of local seismic hazards, and a discussion of how zoning, grading, building, and other codes would be used to reduce seismic risk in new developments.[36]

By 1974, the deadline set by the legislature, many (but by no means all) cities and counties had adopted seismic safety elements meeting these guidelines. In doing so, four urban counties had added geologists to their staff to delineate seismically hazardous areas. Seven other counties called upon the Division of Mines and Geology for assistance in hazard mapping, while still more jurisdictions engaged the services of consulting geologists and engineers. Many of the adopted plans were quite ambitious, calling for the wholesale removal of hazardous old buildings and for restrictions on development in fault zones and landslide areas.[37] Yet the state legislation did not include any mechanism for ensuring that the policies set out in the seismic safety element would be implemented. As a result, many local governments simply prepared them to satisfy the state mandate but did little to put them into actual use. Thus, while the legislation served to increase the technical expertise and knowledge of seismic hazards among local planning department staffs, it did little toward ensuring that geological information would actually be used for modifying land use in seismically hazardous areas.[38]

After the San Fernando earthquake, the Joint Committee and its geological experts sought to provide more effective land-use planning legislation by introducing bills to restrict construction in fault zones. Surface rupture of faults actually produces only a tiny percentage of the damage in earthquakes, because few buildings directly straddle faults. Most damage comes from the shaking produced by earthquake waves and from ground failures when poor soils, such as the mud along the shores of San Francisco Bay, slide or liquefy. But delineating boundaries around areas susceptible to landsliding or liquefaction is generally difficult, while fault zones can be mapped more easily. Moreover, during the rapid suburbanization of the 1960s, residential developments were being placed in fault zones. In Daly City, just south of San Francisco, new tract houses were built directly atop the San Andreas Fault, while on the east side of San Francisco Bay new schools and hospitals studded the Hayward Fault. Many geologists in the late 1960s cited these examples of what they considered flagrant stupidity to urge restrictions on fault zone development as a first step toward proper land-use planning for seismic hazards.[39]

Even after the San Fernando earthquake, it took nearly two years of steady compromising before the legislature passed a bill on fault zoning. In early April of 1971, Alquist introduced a fault zone bill based primarily on suggestions made by Caltech geologist Clarence Allen. It called upon the state geologist to delineate a one-quarter-mile-wide zone centered on the San Andreas Fault and other "sufficiently well-defined and active" faults, and it gave him the power to veto any new real estate development or structures for human occupancy in these zones. Members of the Joint Committee's Advisory Committee on Government Organization, however, thought this bill gave far too much power to a state official over matters traditionally handled at the local level.[40] Assemblyman Paul Priolo, a Republican from Los Angeles who had a deep interest in land-use issues, introduced a bill providing an alternative approach: it called upon the Division of Mines and Geology to transmit maps of fault zones to local jurisdictions, which then were required to adopt zoning ordinances for the fault zones using guidelines established by the division. As Stanford geologist and Joint Committee advisor Richard Jahns pointed out, though, Priolo's bill assumed that the division had already mapped in detail all active faults in the state; this was far from the case.[41] Both bills died in committee at the end of 1971.

In early 1972, both Alquist and Priolo introduced revised versions of their bills. Alquist depended heavily on a draft by Gordon Oakeshott, deputy director of the Division of Mines and Geology, which still called upon the division's head to delineate fault zones but then left it to local jurisdictions to approve new structures or subdivisions in these zones according to criteria set by the State Mining and Geology Board.[42] Priolo's new bill, written after consultation with UCLA engineer C. Martin Duke, now offered a similar approach. It called on the division to provide land-use criteria to guide local governments in adopting zoning ordinances but required that such ordinances be adopted only for the major fault zones that were already well mapped.[43] Among the Joint Committee's advisors, some planners and representatives of local government still thought that the two bills were not sufficiently well developed. They vigorously objected to the absence of any appeals process for local jurisdictions and pointed out that neither the division nor the state board had the necessary expertise to set land-use criteria.[44]

Alquist met these objections by providing that cities and counties be consulted both in the delineation of fault zones and the setting of land-use criteria and by directing that an urban planner be added to the State Mining and Geology Board. He also added a representative of county government to the board to win the support of the County Supervisors Association. Finally, he persuaded Priolo, who was given a seat on the Joint Committee, to join forces with him and back his bill, which he officially renamed the Alquist-Priolo Geologic Hazard Zones Act. It was one of the few Joint Committee bills to attract public attention; at least one major newspaper, the *Oakland Tribune,* endorsed the concept, as did the Sierra Club and a Kern County environmentalist organization that hoped that restrictions on land use in fault zones would lead to preservation of more open space. This outside support, when combined with vigorous lobbying from geologists and Joint Committee members, was sufficient to carry the bill in the legislature. Governor Reagan signed it into law on December 22, 1972.[45]

With the Alquist-Priolo Act passed, both the State Mining and Geology Board and the Division of Mines and Geology set about implementing its provisions. In August 1973, the newly expanded board issued a draft of its guidelines for land-use in fault zones. They stipulated that any fault showing evidence of movement within the past 11,000 years should be deemed active and that no structure for human occupancy should be built on such a fault. Moreover, any site within 50 feet of a known active fault should also be assumed to be underlain by an active fault and thus off-limits for construction unless a detailed geologic investigation conclusively demonstrated that no active fault was actually present. Any application for subdivision or for construction within the fault zone should be accompanied by a geologic report on the presence of faults. These guidelines elicited little opposition from local jurisdictions, and the board formally adopted them in November of 1973.[46]

Meanwhile, the Division of Mines and Geology assembled a team of three geologists to produce the fault zone maps required by the act. In December of 1973, the division released preliminary maps for the four fault zones specifically mentioned in the act—the San Andreas, Hayward, Calaveras, and San Jacinto fault zones.[47] These maps produced vehement objections from speculators who owned land in still sparsely settled northern Los Angeles County; one of them asserted that the maps amounted to "libel of title to the lands inclosed" in the zones. Residents of rural San Benito County also protested, with one going so far as to file a suit charging that the Alquist-Priolo Act "deprive[d] land owners of their property rights without due process of law."[48] By contrast, staff members from planning departments in urban counties, many of whom had worked on the seismic safety element of their general plans, suggested that the division had not gone far enough, overlooking several active fault traces in delineating the fault zones.[49]

But even many of those who supported the Alquist-Priolo Act pointed out that it placed a significant burden on those who simply wanted to build a single-family home. The geologic report required by the act cost anywhere from $1,000 to $3,000. Moreover, in some areas engineering geologists were unwilling to prepare reports for single homesites even at this price, because of the liability they would incur if a fault should subsequently rupture the site. Once informed of these complaints, Alquist moved quickly to amend the act to exclude single-family homes that were not part of a subdivision of more than four homes. He also renamed the act the "Alquist-Priolo Special Study Zones Act," to answer those who considered "Geologic Hazard Zones" too frightening.[50] With these amendments in place, enough of the opposition had been neutralized so that the process of fault zoning could proceed. Over the next decade, the Division of Mines and Geology would continue to delineate more fault zones, and the prohibition on erecting subdivisions or large buildings across active faults within the zones would remain in place, providing an effective land-use measure for dealing with seismic hazards.[51]

Besides working to strengthen earthquake research, building codes for critical facilities, and land-use measures, the Joint Committee and its advisors also concerned themselves with improving emergency preparedness. At the suggestion of Donald Nichols of the U.S. Geological Survey, Alquist in early 1972 introduced legislation requiring dam owners to prepare maps showing the area that would be inundated if the dam should fail in an earthquake. The bill also required local emergency agencies, such as police and sheriff's departments, to draw up evacuation plans for the inundation areas. After Alquist made some technical changes to satisfy the County Supervisors Association and the Department of Water Resources, the bill passed, in late 1972. By 1976, the California Office of Emergency Services had received inundation maps for about three-quarters of the dams that threatened people or property as well as evacuation plans from 12 counties and 11 cities.[52] In 1973, the Joint Committee also introduced a bill requiring cities and counties to establish local disaster councils that would set up and annually review a disaster plan detailing how local agencies would respond in case of an earthquake. This local planning was to complement the work already being done by the Office of Emergency Services and the federal Office of Emergency Planning. The bill died in committee, however, after the influential League of California Cities objected to the cost it would impose on local governments.[53]

Perhaps the most significant accomplishment of the Joint Committee and its advisors, though, was the establishment of a new state agency charged with being a watchdog for seismic safety. As early as 1971, some of the committee's advisors realized that they would not be able to enact a comprehensive plan for earthquake safety by the time the committee's specified duration expired in 1974. To ensure that attention to seismic hazards did not flag after 1974, Karl Steinbrugge and George Gates, a geologist with the U.S. Geological Survey, in early 1972 suggested "a high level executive branch effort" in the form of a commission with "a small, highly qualified staff directly responsible to the governor" that would "have control and responsibility over pre-earthquake hazard reduction policies at State and local levels."[54] In its final report, released in early 1974, the Joint Committee proposed a California commission on seismic safety, whose role would be "to develop seismic safety goals and programs, help evaluate and integrate the work of State and local agencies concerned with earthquake safety, and see that the programs are carried out effectively and the objectives accomplished." The committee also noted that this was its "single most important" recommendation.[55] On February 14, 1974, Alquist introduced legislation, based on a draft apparently written by Steinbrugge, to establish this commission.[56]

Getting the bill passed required considerable compromising. As initially written, the bill proposed a very strong agency: a commission appointed by the governor but independent of the executive branch and given the power to review state agency budgets for proper attention to earthquake matters, to devise "criteria and standards" for earthquake hazard mitigation, and to compel all state agencies and local governments to abide by these standards. To centralize this power even further, several boards set up in the early 1970s—the Building Safety Board established by the Hospital Seismic Safety Act and the Strong-Motion Instrumentation Board that supervised the instrumentation program in the Division of Mines and Geology—would be abolished and their responsibilities transferred to the new commission. These provisions evoked strong opposition both from other state agencies, fearing an intrusion on their turf, and from engineering associations worried that the new commission would take over their power to write building codes. Responding to these criticisms, Alquist amended the bill to continue the two boards in existence and to change the Seismic Safety Commission from a rule-setting to a merely advisory body.[57] In order to get the bill past the fiscally conservative Senate Finance Committee, Alquist had to amend it further to provide that the commission would expire in four years and would have a budget of only $85,000, rather than $195,000 as originally proposed.[58] Winning approval from the full legislature required reducing the commission's life-span even further, to only two years. Even in this weakened form the commission drew opposition; the state Department of Finance, Governor Reagan's fiscal watchdog, recommended the governor veto the bill, calling it "a costly and unnecessary expansion of governmental entities." But as a result of the compromises that had been made, most other government agencies had no objections, and engineering organizations strongly urged Reagan to sign the bill, which he did on September 26, 1974. The Seismic Safety Commission was formally inaugurated on May 27, 1975, with the swearing-in of 12 members.[59]

Thus, in the four years after the San Fernando earthquake, the Joint Committee on Seismic Safety and its expert advisors oversaw the enactment of numerous measures designed to increase earthquake hazard mitigation. They convinced legislators to concern themselves with the seismic safety of the state capitol. They provided earthquake engineers with more strong-motion data with which to improve building codes. They extended the Field Act process to hospitals. By requiring a seismic safety element in general plans, they increased awareness of seismic hazards among local planning department staffs. They restricted the construction of buildings for human occupancy over active faults. They required local emergency agencies to be prepared for dam failures due to earthquakes. They significantly expanded the funding and regulatory powers of the Division of Mines and Geology. And in the Seismic Safety Commission, whose life was repeatedly extended after 1976, they established a continuing voice for seismic safety in state government. In many of these cases, the final legislation was considerably weaker than originally envisioned by earthquake researchers, but, because seismic safety advocates chose not to mobilize large-scale pressure from the general public for stronger safety measures, they had to make compromises in order to get the bills passed. In case after case, Alquist moved deftly to defuse the opposition from entrenched professional organizations, local government groups, and rival state agencies. This skillful maneuvering, while not always successful in bringing about the passage of desired legislation, did enable the Joint Committee to add considerably to the regulatory apparatus for earthquake hazard mitigation in California.

Local Politics: Fights to Upgrade Schools and Other Buildings

In the absence of broad popular support, earthquake experts implemented much of their agenda after the San Fernando earthquake by working in restricted arenas in which they already had a measure of influence, such as the bureaucratic realm in which building codes were written or the legislative arena in which Alquist's political support could bring about numerous victories. One of the most important parts of their agenda, however, could not be handled in these insulated arenas. Experts considered the continuing existence of old buildings that did not meet even the relatively weak building codes of the 1930s to be one of the greatest seismic hazards. Richter in particular decried the continuing presence of "obsolete and weakly built" structures that were an "inheritance from the years when it was considered very bad form even to speak of earthquakes in California." As he and others pointed out, in the next major earthquake "the majority of deaths and most of the destruction" would come from these "unsafe relics of the pioneer past." [60] But reconstructing these buildings to meet modern building codes (a process for which the term "seismic retrofitting" was coined in the 1980s) was far too costly to be done simply by bureaucratic or legislative fiat. While the additional costs imposed on new buildings by seismic design provisions ranged from less than 1 percent to perhaps 15 percent, the cost of retrofitting an old building in some cases exceeded its entire value. To bring about retrofitting, then, required the mobilization of considerable public support for the necessary funding. Earthquake researchers succeeded in having old school buildings retrofitted by 1977, mostly by getting the state legislature to make financing easier and to threaten sanctions if local school districts did not retrofit. But there was no similar state aid to retrofit privately owned buildings, and earthquake researchers failed to overcome the strong opposition of well-organized building owners who objected to bearing the cost of retrofitting.

The initial impetus that got retrofitting of schools rolling came from a decision of California's attorney general on the duties of local school board trustees. The Field Act had not been retroactive, but it had given school boards the option of requesting a structural inspection of old school buildings. If this inspection showed the structures to be unable to withstand an earthquake, the school board was then obligated to renovate or demolish them. Failure to do so would open the trustees to personal liability if anyone was injured or killed in a subsequent earthquake. In the 1930s, many trustees had protested that voters did not approve the bonds necessary to finance retrofitting. After passage of the Garrison Act in 1939, trustees were no longer personally liable if their good-faith efforts to raise funds were defeated. In 1963, the legislature quietly removed this waiver of liability. Most school boards avoided the whole issue by simply not requesting a structural inspection. But in 1966, the attorney general ruled that failure to request an inspection constituted negligence and also exposed school board trustees to personal liability.[61]

After this ruling, school board trustees once again clamored for legislative relief from personal liability. The California legislature obliged in 1967, reinstating the waiver of liability originally provided by the Garrison Act—but only if school boards requested a structural inspection of all their old buildings by 1970 and thereafter sought either bonds or a tax increase to finance the retrofitting at least once every five years until approved by local voters.[62] This bill was drafted by Leroy Greene, the chairman of the Assembly Committee on Education, who was a civil engineer and strongly believed in the necessity for school retrofitting. The following year, he introduced a companion measure requiring that all school buildings not meeting Field Act standards be abandoned by 1975. The measure, which squeaked through the Senate on a 21–17 vote and was signed by Governor Reagan on July 18, 1968, put additional pressure on school board trustees to find money for retrofitting; if they failed, they would have to put students into tents, temporary buildings, or double sessions in newer buildings once the 1975 deadline passed.[63]

School board trustees faced an enormous task. A state Department of Education official estimated in 1966 that about 20,000 classrooms, 15 to 20 percent of the state's total inventory, needed retrofitting and that the cost would be $1.16 billion at a minimum. Other estimates ranged as high as three billion dollars.[64] School retrofitting did enjoy considerable support in influential circles. The pronouncements of Richter and other earthquake researchers convinced many metropolitan newspapers, chambers of commerce, politicians, and school parent-teacher associations that retrofitting was necessary to avoid a disaster. Architectural firms, contractors, and labor unions that stood to profit from retrofitting projects also endorsed trustees' efforts to raise the money.[65] But many voters balked at approving the bond issues, which would be repayed through higher property taxes. In the late 1960s and early 1970s, property taxes were a potent political issue in California. As property values had increased dramatically in rapidly growing California in the 1960s, property tax bills had skyrocketed, causing property owners to seek any kind of relief from the burden. In 1978, their protests culminated in the successful campaign to limit property taxes through the statewide Proposition 13. These tax protesters argued that school retrofitting should be paid for by the state, perhaps through a sales tax, rather than by local property tax payers.[66]

Some property owners went further, denying that there was a need for retrofitting. As one charged, "never has one of our 'Sweet Kiddies' even been hurt [by earthquake damage] in a school." Destructive earthquakes had never shaken California during school hours, always striking early in the morning or in the evening. Another argued that he was "more concerned with our children being hit by a car, molested, assaulted, subjected to narcotics, permissiveness, brain-washing, and academic deficiencies"; seismic safety to him was the least of the problems facing schools. These vociferous protesters insinuated that the earthquake hazard was simply a ploy by cynical school officials to grab more money. As one put it, "official proponents of new schools . . . [who use the] rare possibility of a killer quake as a reason and vibrate with glee when the earth trembles a little remind me of a slobbering rabid dog."[67] Unfortunately for proponents of retrofitting, some of these charges were true. In proposing bond issues, many school districts sought money not only for retrofitting, but also for other measures to bring old schools up to what they considered modern educational requirements. In some cases, this doubled or even tripled the size of the bond issue. In Pasadena, the padding of the bond issue was so blatant in 1968 that even Richter voted against it.[68]

Supporters of retrofitting faced yet another hurdle. A clause of the California state constitution, upheld in 1971 by the U.S. Supreme Court, required *local* bond issues to receive a two-thirds majority for passage. This meant that opponents needed to muster only one-third of the votes to block funding.[69] With the growth of the property tax revolt in the late 1960s and early 1970s, this was increasingly easy to do. While 67 percent of school bond issues passed in 1961, only 35 percent did in 1968. By 1970–71 the situation had become even worse; according to the *Los Angeles Times,* only $50 million worth of bond issues out of $450 million worth that were proposed managed to obtain a two-thirds majority, while another $350 million worth received a simple majority but failed to reach the higher bar. During these years, retrofitting issues failed three times in San Diego and at least twice in San Francisco and Los Angeles.[70]

Blocked at the local level by the opposition of tax protesters, supporters of school retrofitting worked at the state level to find funding. Shortly after the San Fernando earthquake, Assemblyman Greene introduced legislation calling for a $350 million statewide bond referendum in June 1972, with $250 million of the money to be earmarked as loans to local school districts for retrofitting. This bill, which was supported by the Reagan administration because it provided aid in the form of loans rather than grants, easily passed the legislature.[71] In a companion measure, Greene both allowed school districts to raise property taxes to repay the loans and required them to accept state loans if local means of funding failed at the ballot box. This bill, too, squeaked through the legislature.[72] The bond issue, which was heavily supported by the state's major newspapers, received 53.7 percent of the votes; this was enough for passage, as *state* bond issues required only a simple majority.[73] After this success, Greene obtained legislative approval for further state loans for retrofitting. In 1972, when the state for the first time imposed withholding for state income taxes, the legislature agreed to earmark $30 million of the windfall from the earlier collection of taxes for retrofitting. The legislature also agreed to use $15 million of unspent money from a 1966 state bond issue for a loan fund for poor school districts facing problems with retrofitting. Put together, this state funding covered nearly half of the bare-bones cost of retrofitting, which a revised estimate in 1972 put at $634 million.[74]

In order to make local funding of retrofitting easier, supporters also changed the rules of the game. In March of 1972, Senator George Moscone of San Francisco introduced a bill for a state referendum to amend the state constitution so that local bond issues for school retrofitting would need only a simple majority to pass. Governor Reagan and other conservative Republicans opposed this measure, fearing it would put the state on a slippery slope toward easing the requirements for all bond issues and thus driving up property taxes. But the Democratic-controlled legislature passed the measure, and Reagan allowed it to become law without his signature, perhaps fearful of the political fallout if he were to be tagged as opposing increased safety for schools.[75] The ballot proposal, once again strongly supported by the state's major newspapers, won 54.5 percent of the votes in November of 1972, enough for passage.[76]

As most retrofitting bond issues had been receiving simple but not two-thirds majorities, this change significantly increased the number of issues being passed. In 1973, school districts also were far more careful in submitting bond issues strictly for retrofitting, not padded with money for other goals, and they pointed to the looming 1975 state deadline (which in 1974 was extended to 1977 for districts with extenuating circumstances) as yet another incentive for voters to approve funding.[77] Taken together, these measures finally brought about wholesale retrofitting of the state's old schools. While 1,593 pre–Field Act school buildings were still being used in California in 1972, only 19 remained, primarily in small rural districts, after 1977.[78] One more significant earthquake hazard had thus been removed. Achieving this victory, however, had required extraordinary measures by seismic safety advocates. They had needed a state mandate for retrofitting, state funding, and a state-approved change in the rules for bond issues to overcome the opposition of taxpayers who had balked at paying the high cost of school retrofitting.

Retrofitting private buildings proved to be even more difficult than renovating schools. Schools, after all, were public institutions charged with housing children, whose safety could be made into a potent political symbol. With the passage of the Field Act and its interpretation by the state attorney general, the state of California had acquired the power to require school retrofitting. Moreover, the cost of retrofitting could be spread across property owners, because schools were publicly owned. These advantages did not exist for private buildings. Estimates in the early 1970s showed that about 10,000 pre-1933 structures still existed in San Francisco and 17,000 in the city of Los Angeles, with another 11,000 in the remainder of Los Angeles County. Retrofitting all of these buildings would cost perhaps $3 billion each in San Francisco and in Los Angeles, and in the absence of tax credits or other state financing, these costs would have to be born by the buildings' owners. Furthermore, many of the hazardous structures were apartment buildings or welfare hotels in the poorer sections of town, often housing low-income residents. If building owners were to raise rents to cover the cost of retrofitting, much affordable housing for the most vulnerable members of society would disappear.[79]

A few local governments did proceed with retrofitting despite the overwhelming economic and social problems. They usually were prodded by stubborn structural engineers in their building departments. In Long Beach, for example, Edward O'Connor, head of the city's building department, in the late 1960s began using the city's nuisance ordinance to declare hazardous old buildings a public nuisance that had to be abated. By 1970, he had condemned 118 buildings using this procedure. The legality of his tactic was upheld in general by the California Supreme Court in a case involving fire hazards; nevertheless, O'Connor found himself subject to lawsuits from outraged building owners, who also showered the city council with numerous protests. The council responded in 1971 by passing a retrofitting ordinance that reined in O'Connor but nevertheless affirmed retrofitting of private buildings as a policy goal. In 1976, the council amended the complex ordinance to make it easier to enforce, thereby placing Long Beach at the forefront of retrofitting efforts in California.[80] Santa Rosa, a small city north of San Francisco that had been damaged in a 1969 earthquake, also passed an ordinance requiring certain old buildings in its downtown to be retrofitted. Here, too, the driving force was an energetic building department official. It also helped that Santa Rosa received substantial federal urban renewal funding for renovations in its downtown, thereby lessening the economic impact on building owners.[81]

More typical than these isolated success stories, though, was the experience of the city of Los Angeles. Beginning in 1973, the city's Building and Safety Department, dominated by structural engineers, worked on a retrofitting ordinance. By mid-1976, after agonizing over the political and financial difficulties of retrofitting, the department finally proposed an ordinance that would require all seismically hazardous old buildings to be posted as a first step toward retrofitting. A council member added a requirement that all retrofitting be completed within the next ten years. When the city council as a whole considered the ordinance in late 1976, though, it encountered considerable opposition. The most vocal and best organized of the opponents was the Apartment Association of Los Angeles County. The association's executive director, Howard Jarvis, was a well-known tax protester and demagogue; later, he would head the successful Proposition 13 campaign to limit property taxes. Before the city council hearing on the retrofitting ordinance, Jarvis's organization placed a large ad in the *Los Angeles Times* addressed to "all tenants, businesses and churches" occupying old buildings, warning them in bold letters that they would be "EVICTED if a so-called seismic safety ordinance, which requires the virtual re-building of these structures, passes" and urging them to attend the hearing. More than 400 worried residents, many of them low-income tenants, heeded the call. Faced with their vocal opposition, the city council backed down, replacing the ordinance with a simple call for further study of the situation.[82] Seismic safety advocates, who had relied for so long on behind-the-scenes negotiation rather than public pressure, simply did not have the means to mobilize an effective counter to this organized grassroots opposition to their agenda. Not until the late 1980s, after many more years of patient negotiation and persistent coalition-building, did they succeed in beginning large-scale retrofitting of private buildings.

CHAPTER 8

Pushing Prediction

Establishment of the National Earthquake Hazards Reduction Program

After the San Fernando earthquake, seismologists and engineers in California succeeded in having building codes tightened, land-use provisions introduced, a new watchdog agency established, and retrofitting of at least school buildings carried out. Earthquake experts had advocated all of these measures with increasing vigor in the 1960s, but they had also urged another strategy for hazard reduction: research toward producing an operational earthquake prediction system. In the days after the San Fernando earthquake, such prediction advocates as Frank Press of MIT and Louis Pakiser of the U.S. Geological Survey repeated these calls. Pakiser asserted that experts would "be able to predict earthquakes within five years"— if research funds were forthcoming. Some members of the press agreed that research on prediction should be pushed forward. KNX Radio of Los Angeles editorialized in 1971 that "the February quake cost millions of dollars and many lives. These losses might have been prevented if we could have predicted the earthquake."[1] In early 1972, high-ranking Nixon administration officials approved an $8 million increase in the budget of the Geological Survey and other agencies for earthquake research, of which about $3 million would go to prediction research.[2] But to prediction enthusiasts, who in the late 1960s had developed a blueprint for a $220 million program, this was barely a beginning.[3]

Prediction advocates gained an important ally in Alan Cranston, the Democratic senator from California, who decided that prediction research would be an appropriate target of federal largesse for his state. In February 1972, after consulting with the Geological Survey's Jerry Eaton and other prediction advocates, Cranston introduced several bills providing $91 million over five years for a comprehensive earthquake prediction and engineering research program in California and neighboring Nevada.[4] But the bills died in committee after the Office of Management and Budget and other administration officials voiced opposition. To them, throwing so much money at prediction research seemed like a waste of funds, especially since earthquake researchers had little to show for the millions already spent on it. The prediction program also continued to suffer from other problems that had plagued it in the late 1960s: the political handicap of interagency squabbling between the Geological Survey and the Coast and Geodetic Survey, opposition from earthquake engineers who saw prediction as a threat to their own agenda for seismic safety, and a general feeling among policymakers that, in a time of budget stringency, earthquake hazards were one of the less important problems facing the nation.[5]

In the mid-1970s, prediction advocates overcame all of these obstacles. To do so, they used many of the same strategies employed at the state level by fellow earthquake experts. Earthquake prediction attracted much more media interest than other mitigation measures, and prediction researchers sometimes played to this interest by making unguarded promises about the capability of future prediction systems which turned out to be unsupportable. The state's major newspapers, intrigued by the dramatic potential of prediction for solving the state's earthquake problems in one fell swoop, in turn lent their editorial support to prediction advocates. But the advocates did not seek grassroots support beyond these editorials for their drive to establish a well-funded research program. Rather, like their colleagues who advised the Joint Committee on Seismic Safety, they played the political insider game of patient compromising and alliance building.

Prediction advocates' first step was to gain scientific credibility. In 1973, researchers developed the dilatancy-diffusion hypothesis, a promising theory for understanding earthquakes. Its subsequent use to predict several small earthquakes finally enabled them to demonstrate that prediction research might pay off. At the same time, Geological Survey officials worked with Cranston's advisors to remove political opposition to the prediction program. They developed a proposal for a comprehensive National Earthquake Hazards Reduction Program that would sponsor research on engineering and hazard mapping as well as prediction. Finally, in 1976, advocates skillfully used several vague indications that a catastrophic earthquake was imminent in southern California to persuade both Congress and the president to acknowledge earthquake hazards as a significant problem. As a result, the federal government in 1977 finally established the National Earthquake Hazards Reduction Program, of which prediction research was an important and well-funded component. In this manner, prediction advocates added yet another important component to the growing science-state alliance for seismic safety.

The Dilatancy-Diffusion Hypothesis

Despite the millions of dollars that had been devoted to earthquake prediction research by the Geological Survey and the Coast and Geodetic Survey since 1964, researchers had obtained few results by the early 1970s. Although the San Andreas Fault, especially its central section near Hollister, now bristled with seismographs, tiltmeters, strainmeters, and other tools for catching possible earthquake precursors, scientists had not identified any kind of anomaly that reliably preceded a temblor. But the fortunes of prediction research changed after 1971. Following up on Soviet work, American scientists determined that anomalies in seismic wave velocities appeared to precede some earthquakes. Researchers at Stanford and at Columbia University also developed a persuasive theory, the dilatancy-diffusion theory, to explain these anomalies. Using it, seismologists in 1973 and '74 were able to predict small earthquakes in upstate New York and in California. These results elated researchers and captured media interest: the era of earthquake prediction seemed finally at hand.

The initial breakthrough for prediction research came from Soviet work in the early 1960s on the variation of seismic wave velocities before earthquakes. The velocity of seismic waves changes as a function of the density and elasticity of the materials they encounter. Since the turn of the century, seismologists had used spatial variations in wave velocities to map variations in the density and other physical characteristics of the earth. The Soviet researchers noted, however, that there were temporal as well as spatial variations. They focused on the ratio of the velocities of the two main body waves given off in earthquakes, the primary (P) and the secondary (S) wave. This ratio, designated V_P/V_S, is usually about 1.75 for materials close to the earth's surface. Soviet seismologists monitoring the earthquake-prone Garm region of Tajikistan reported, though, that from time to time the V_P/V_S ratio of waves passing through small areas would decrease to 1.60 or even 1.55. The ratio would be abnormally low for one to three months, then return to normal. Within days, the areas would be stricken by earthquakes of Richter magnitude 3 to 5. From this empirical observation, it appeared that abnormally low V_P/V_S ratios were a reliable earthquake precursor in the Garm region.[6]

The Soviet researchers had first presented their observation to an international audience in 1962 and had referred to their results repeatedly in print afterward, albeit usually only in Russian-language publications. American seismologists did not become generally aware of the Soviet work until the summer of 1971, when many leading researchers attended the quadrennial meeting of the International Union of Geodesy and Geophysics, in Moscow. After learning about the Soviet observations at the meeting, Lynn Sykes, a prediction research advocate at Columbia University's Lamont-Doherty Geological Observatory near New York City, wangled an invitation to Tajikistan. There he inspected some of the original data gathered by the Soviet seismologists. Convinced that they were on to something, Sykes spread the word about V_P/V_S anomalies on his return to the United States, as did others who had attended the Moscow meeting.[7]

One researcher inspired by Sykes's report was Yash Aggarwal, a young graduate student who had just arrived at Lamont-Doherty. In the summer of 1971, one of Aggarwal's colleagues had set up a seismograph network near Blue Mountain Lake in the Adirondack Mountains of upstate New York to monitor the aftershocks of a magnitude 3.6 earthquake that had just occurred there. Aggarwal decided to search for velocity anomalies in the data collected by this network. He indeed found that the V_P/V_S ratio for seismic waves traveling through the area had decreased by up to 13 percent for several days and then returned to normal just before two magnitude 3 earthquakes in July 1971.[8]

Further confirmation soon came from California. At Caltech, James Whitcomb, a young researcher with a deep interest in earthquake prediction, searched through the data gathered by Caltech's network before the San Fernando earthquake. His analysis indicated that near the Pasadena and Riverside seismograph stations there had been a V_P/V_S anomaly for three-and-a-half years before the magnitude 6 earthquake. Whitcomb and his coauthors also pointed out that there was a direct relationship between the duration of the anomaly and the magnitude of the subsequent earthquake. This meant that the velocity anomaly method could be used to predict not only the time of the next earthquake (it would happen soon after velocities returned to normal), but also its size.[9]

Belief in this prediction method increased even further among seismologists after Amos Nur, a young Stanford professor, in late 1972 suggested a theoretical explanation for why the V_P/V_S ratio would decrease and then rebound just before an earthquake. As a graduate student at MIT in the late 1960s, Nur had conducted experimental and theoretical investigations on the phenomenon of rock dilatancy. When rock is squeezed (as it is when stress builds up near an earthquake fault), it reaches a point where it does not decrease in volume, but rather increases or dilates, as numerous tiny cracks form and open up. In a dry rock, these cracks impede the propagation of P-waves, leading to a decrease in P-wave velocity and thus in the V_P/V_S ratio. In a fluid-saturated rock, however, P-wave velocity increases during dilatancy. Nur proposed that the buildup of stress prior to an earthquake led to dilatancy in dry rocks, resulting in the decreased V_P/V_S ratio. The opening of the cracks, however, would then allow infiltration of groundwater, leading to fluid saturation and a rebound in the V_P/V_S ratio. Moreover, the groundwater would counteract the forces clamping together rocks on opposite sides of the fault. Thus, the rocks would be free to slip, producing an earthquake, just after diffusion of the groundwater was complete, that is, just after the V_P/V_S ratio had fully rebounded.[10]

In the spring of 1973, the dilatancy-diffusion hypothesis was further developed by Lamont-Doherty's Christopher Scholz, who had also worked on dilatancy as a graduate student at MIT. Scholz pointed out that if rocks were dilating before an earthquake, other effects besides just velocity anomalies should be observable. As the rock's volume increased, the ground surface should bulge slightly, an effect that could be measured using ground leveling. The density of the rock should decrease, producing a slight reduction in the force of gravity. The opening of cracks and subsequent introduction of groundwater should affect the electrical resistivity of crustal rocks. And crack opening should make it easier for radon, produced by radioactive decay in subsurface rocks, to escape to the surface, producing a spike in surface radon measurements. Scholz and his coauthors provided numerous examples that at least suggested the presence of such precursors before earthquakes in Japan, Tajikistan, and Alaska. Moreover, according to Scholz, all of these precursors fit the relationship between the duration of the anomaly and the magnitude of the subsequent earthquake that had already been established for seismic velocity anomalies.[11]

At the spring 1973 meeting of the American Geophysical Union in Washington, D.C., prediction advocates greeted Scholz's work with exuberance. Scholz himself proclaimed that "earthquake prediction . . . appears to be on the verge of practical reality," an ebullient assertion frequently repeated by seismologists and journalists.[12] Scholz's colleague Sykes, in reviewing his paper several weeks later for a National Academy of Sciences committee, asserted that "prediction of some classes of earthquakes in some tectonic provinces is virtually a reality." Likewise, Senator Cranston told Los Angeles reporters that "we have the technology to develop a reliable prediction system already in hand." [13]

Thus far, however, the velocity anomaly method had predicted earthquakes only in hindsight. Seismologists exulted even more after it produced its first forward-looking prediction in the summer of 1973. After a couple of magnitude 3.6 earthquakes near Blue Mountain Lake on July 15, 1973, Aggarwal and his colleagues reestablished their seismograph network there. On August 2, Aggarwal noted an anomaly in the V_P/V_S ratio; based on its duration, he predicted an earthquake of magnitude 2.5 to 3.0 for the following day. Indeed, on the evening of August 3, a magnitude 2.6 earthquake struck. Both the *New York Times* and *Time* magazine noted this successful prediction. After Aggarwal published a detailed paper in the *Journal of Geophysical Research, Time* once again wrote up the case, making it the cover story for its September 1, 1975, issue. Even the *American Legion Magazine* featured an article on the prediction, boldly proclaiming that "they said it couldn't be done, but it is now close to a sure thing that earthquakes will be routinely predicted." [14]

Not all seismologists immediately joined the jubilation. After all, the earthquakes in Tajikistan, New York State, and southern California which had provided evidence for the dilatancy-diffusion hypothesis represented only one type of earthquake. In these thrust earthquakes, one portion of the earth's crust slides under another portion along an inclined fault. Most destructive earthquakes in California, though, were of the strike-slip type, in which one portion of the crust slides horizontally past the other portion along a vertical fault. Did the dilatancy-diffusion hypothesis apply to strike-slip earthquakes as well? After analyzing data going back to 1961, Thomas McEvilly of the University of California at Berkeley announced in the summer of 1973 that there had been no velocity anomalies before any of the 20 magnitude 5 earthquakes to hit the central San Andreas Fault during that period. Teams of researchers at Caltech, Stanford, and the U.S. Geological Survey also failed to find anomalies before several strike-slip earthquakes.[15] But supporters of the hypothesis refused to be discouraged. Perhaps the seismograph stations used in these studies were simply not spaced close enough together to pick up the anomalous signals. Or perhaps the cracks produced by dilatancy before a strike-slip earthquake were oriented in such a way that they impeded only seismic waves traveling in a certain direction, making the anomaly more difficult to pick up.[16]

Further discoveries reaffirmed the value of the dilatancy-diffusion hypothesis in late 1973 and 1974. Anomalies of the duration required by the theory were discovered before thrust earthquakes that had struck New Zealand in 1966 and southern California in 1973.[17] More importantly, researchers now began reporting anomalies (albeit smaller and of shorter duration than expected) before strike-slip earthquakes as well. It was suggested, for example, that the Matsushiro swarm of earthquakes that had plagued central Japan in the late 1960s had featured velocity anomalies as well as strain and gravity anomalies consistent with the dilatancy-diffusion model.[18] Even more convincing to the seismological community, one of the teams that had failed to find anomalies before strike-slip earthquakes along the San Andreas Fault now asserted that there actually had been one before the 1972 Bear Valley earthquake south of Hollister, a finding also announced by U.S. Geological Survey scientists. It had been missed earlier, they claimed, because of its small geographic extent and short duration.[19] All of these discoveries, presented in rapid succession at the 1974 American Geophysical Union's spring meeting, made this "a time of considerable excitement for geophysicists," in the words of one science reporter.[20]

The growing excitement was fueled even further by the announcement of what appeared to be two more successful predictions. In the summer of 1973, Caltech's Hiroo Kanamori noted that a P-wave velocity anomaly had persisted at the Riverside station from 1969 to earlier that year. Kanamori refused to attach any definite significance to it.[21] His colleague Whitcomb, however, was less reticent. In a letter sent to the Geological Survey in December 1973 and to the American Geophysical Union early in January 1974, Whitcomb predicted that a magnitude 5.5 earthquake would strike near Riverside in early 1974. There was indeed a strike-slip earthquake just east of Riverside on January 30, albeit only of magnitude 4.1. Nevertheless, Whitcomb claimed success for his prediction, and fellow seismologists and journalists were willing to grant the claim.[22]

Another successful but less formal prediction was made by scientists at the U.S. Geological Survey in late 1974. At a meeting of the Pick and Hammer Club, a social group of northern California earth scientists, on November 27 (the day before Thanksgiving), survey researchers were discussing preliminary data taken from tiltmeters and magnetometers near Hollister, in central California. These data suggested that there had been both a tilting of the earth's surface and a short-lived anomaly in the magnetic field near the San Andreas Fault. One scientist commented that these anomalies were "the sort one would expect to see before a quake." Another survey geophysicist then suggested that an earthquake would strike "maybe tomorrow." Indeed, a magnitude 5.2 earthquake occurred near Hollister the next afternoon. The prediction had been only informal, as the tilt and magnetic data had not been fully processed. Later analysis indicated that there had also been a seismic velocity anomaly before the Thanksgiving Day earthquake. If data-processing had been speedier, the Geological Survey might have been in a position to issue a formal prediction.[23] Even in the absence of one, the survey saw its success as evidence "that some earthquakes can be predicted by presently available geophysical instrumentation and analytical techniques."[24]

Seismologists' optimism reached its peak when word reached the West that Chinese scientists had saved thousands of lives through a successful earthquake prediction in February 1975. American researchers were already envious of the Chinese prediction program, which enjoyed considerable government support. Hundreds of trained scientists, assisted by tens of thousands of interested amateurs, monitored 250 seismic stations and 5,000 other instruments there, looking for indicators of a coming earthquake—a scale of effort that dwarfed the research program in the United States. In the early 1970s, these observers noted tilting of the earth's surface and an increased number of small earthquakes in Liaoning Province, northeast of Beijing. As these precursors, together with unusual variations in well-water levels and increasing unrest among animals, became concentrated near the large industrial city of Haicheng, local authorities issued an earthquake alert and finally lured the city's residents from their houses on February 4 by showing outdoor movies. Hours later, a magnitude 7.3 earthquake devastated the city. Thanks to the authorities' actions, however, few people died.[25]

In the wake of the Thanksgiving Day and Haicheng earthquake predictions, a National Academy of Sciences committee prophesied that "a scientific prediction will probably be made within the next five years for an earthquake of magnitude 5 or greater in California. With appropriate commitment [of funds and other resources], the routine announcement of reliable predictions may be possible within 10 years." The U.S. Geological Survey was even more optimistic, claiming that "the Chinese success in saving lives during the Liaoning earthquake signals that the age of earthquake prediction is upon us." As the survey's director, Vincent E. McKelvey, put it, "we are close enough to having an established earthquake prediction capability that an effort should be made to seek funding for greatly increased deployment of monitoring networks." McKelvey pressed this point home by beginning work on formal guidelines for how predictions and forecasts should be formulated by the survey and conveyed to local government officials.[26] It seemed only a matter of time before the public would have to respond to formal predictions.

In the early 1970s, then, earthquake prediction researchers had greatly increased the scientific credibility of their field. No longer were they simply lining faults with instruments in the hope of detecting some kind of anomaly. With the dilatancy-diffusion hypothesis, they had a definite idea of what earthquake precursors to look for. Each year, more and more data seemed to be falling into place, confirming the hypothesis. Buoyed by these successes, prediction advocates in the mid-1970s felt it was time to call for a significant expansion of the prediction research program.

Lining Up the Ducks: Removing Opposition to Prediction Research

If they were to advance their agenda, prediction advocates needed to surmount considerable political difficulties. Interagency squabbling between the Geological Survey and the Coast and Geodetic Survey over who would lead a national prediction program alienated many potential supporters. Earthquake engineers strongly opposed devoting more resources to prediction, arguing that other mitigation measures (most prominent among them, of course, engineering) would be far more effective and cost-efficient in reducing seismic risks. These opponents also continued, as they had since the mid-1960s, to argue that an earthquake prediction, by producing panic and disrupting the economy in the affected area, might actually do more harm than good. Throughout the mid-1970s, advocates of prediction research worked to overcome these obstacles by eliminating interagency disputes and by winning over their opponents.

By the early 1970s, the turf battle between the two federal surveys over earthquake prediction had become politically embarrassing. As early as 1970, agents of the General Accounting Office, the investigative arm of Congress, had begun to examine the two surveys' research activities in California for possible waste and duplication of work. Two years later, they issued a highly critical report charging that the "two agencies conduct duplicative and overlapping earthquake research studies along the same portions of the major California earthquake faults. Neither the studies nor the placement of the research equipment have been coordinated. . . . Failure of the two agencies to coordinate their research activities resulted in competition for funding for similar programs."[27] Concerned by the evidence of duplication, the Office of Management and Budget in late 1972 decided to freeze some of the recently granted increase in funding for earthquake research until the two surveys could coordinate their work. Highly placed officials in both surveys scrambled to form interagency committees so as to improve cooperation. At least one Office of Management and Budget examiner was not impressed with their efforts, commenting that "the examples of . . . cooperation . . . are pretty disheartening if that is the best that can be produced."[28]

In the end, the interagency disputes were resolved when the Coast and Geodetic Survey relinquished the field. Since 1970, it had been part of the National Oceanic and Atmospheric Administration (NOAA), whose top leaders were less keen on earthquake research than the survey's earlier administrators had been. In late 1972, the Office of Management and Budget released word that agency budgets for fiscal year 1973 would have to be cut back in order to avoid an increasing budget deficit. These cuts endangered NOAA's plans for a new research ship and a satellite. Faced with this threat, NOAA's top administrator decided in early 1973 to jettison his agency's earthquake work, thereby freeing up room in the budget for what he considered higher-priority items. The Geological Survey quickly agreed to take over much of this work, incorporating NOAA's seismologists and equipment. As a result, the Geological Survey now stood undisputed as the sole federal agency involved in earthquake prediction research.[29]

Prediction research still faced considerable opposition from earthquake engineers. In early 1973, Senator Cranston held a congressional subcommittee hearing on two bills he had just reintroduced, which provided $60 million over five years to the Geological Survey for prediction research and $30 million to the National Science Foundation for engineering research. Eight prominent earthquake engineers testified against the bills, arguing that engineering research needed a funding boost more than did prediction research. Henry Degenkolb, representing the Structural Engineers Association of California, bemoaned the "diversion" of money to the "will-o'-the-wisp goal" of prediction. Even a geophysicist, Thomas McEvilly of the University of California at Berkeley, argued that engineering "will save orders of magnitude more lives and property losses in this century" than prediction and thus should receive more funding. Similarly, at a hearing in late 1974 of the California state legislature's Joint Committee on Seismic Safety, a San Francisco structural engineer asserted "that more attention should be given" to retrofitting of buildings than to "the exotic approach of earthquake predictions." He added, "I see no useful purpose to this type of research."[30]

Advocates of prediction research worked to overcome this opposition by emphasizing their support of a broad program of earthquake hazard mitigation of which prediction was only a part. In a general overview of its earthquake hazard reduction program published in 1974, the U.S. Geological Survey argued that a balanced program of hazard mitigation should include land-use planning, engineering, earthquake prediction and control, insurance, emergency preparedness, and public information. The Geological Survey sought funds to increase not only its prediction research program, but also to produce more seismic hazard maps for land-use planning and to gather more data for engineering use through the strong-motion seismology program it had taken over from NOAA. Moreover, it asserted that its work should be only part of a broader national program that also involved the National Science Foundation, emergency preparedness organizations, and other federal agencies.[31]

Senator Cranston similarly sought to build a coalition with earthquake engineers. When he reintroduced his earthquake bills in 1975, they provided for $250 million over ten years to the Geological Survey and $240 million to the National Science Foundation, putting earthquake engineering on essentially equal footing with the Geological Survey's work in prediction and hazard mapping. Indeed, in a congressional committee hearing in early 1976, numerous engineers strongly endorsed Cranston's revised bills.[32] Cranston also solicited the support of his fellow legislators by emphasizing that earthquakes were a problem not just for California and other Pacific Coast states. As he repeatedly pointed out, 39 states faced the threat of a moderate or severe earthquake. Moreover, he estimated that, over the next 30 years, earthquakes across the country would cost the nation's economy on average one billion dollars a year. Thus, reducing the damage, so Cranston argued, was a task for the national government.[33]

Finally, advocates of earthquake prediction research moved to answer critics such as Garrett Hardin, Karl Steinbrugge, and Charles Richter, who had argued since the mid-1960s that a prediction, by producing panic and reducing real estate values, would do more economic harm than good. In 1973, the National Academy of Sciences appointed a panel to consider the public policy implications of earthquake prediction. The panel, headed by sociologist Ralph Turner of UCLA, considered the lessons drawn from hurricane warnings, flood watches, and other predictions of natural catastrophe. In a report issued in 1975, it acknowledged that "there will be disadvantages as well as advantages to earthquake prediction. Under the worst combination of an inaccurate prediction and an inappropriate public response, the prediction and quake together might even be more costly than an unpredicted quake would have been." However, it asserted that "earthquake prediction *can* be a means for substantially reducing the losses from earthquakes if appropriate social, economic, engineering, and legal actions are taken prior to the quake. Even in the case of a false alarm, some of the costs of a well-planned hazard-reduction program will contribute to the long-term seismic safety of the community."[34] Prediction advocates in the Geological Survey and in the press repeatedly cited this favorable assessment that the benefits of such research would outweigh its costs.[35]

Forcing the Issue: The Palmdale Bulge, the Whitcomb Prediction, and the National Earthquake Hazards Reduction Program

By 1975, advocates of earthquake prediction research had reason for considerable optimism. The success of the dilatancy-diffusion hypothesis, as well as the Thanksgiving Day and Haicheng earthquake predictions, indicated that prediction research would soon lead to the establishment of a routine earthquake prediction system. Moreover, the advocates had answered charges that prediction research was mired in bureaucratic infighting, was a diversion of resources from more effective means of hazard reduction, and would produce more harm than good. But they still faced one last hurdle: lack of interest from high administration officials and Congress. In fact, as the federal budget deficit mounted in the mid-1970s, the Office of Management and Budget saw earthquake research as an easy target for budget-cutting. In early 1975, it cut the National Science Foundation's budget for earthquake engineering, which had grown to $8 million in fiscal year 1974, by $2.6 million. Later that year, it announced that it would reduce the Geological Survey's fiscal year 1976 funding for earthquake research from $11 million to $9 million.[36] Meanwhile, Senator Cranston could not garner sufficient political support to move his earthquake bills out of committee and onto the floor of Congress for a vote. Clearly, advocates of prediction research stood in need of a dramatic event to highlight the importance of their work.

Two such events occurred early in 1976: geodetic data suggested an ominous bulge along the San Andreas Fault north of Los Angeles, and a Caltech seismologist suggested that a strong earthquake might strike southern California within the next year.

The first opportunity for prediction advocates came from a comprehensive review by Geological Survey scientists of leveling data for southern California. Throughout the 1950s, '60s, and early '70s, the Coast and Geodetic Survey and surveyors for both the city and the county of Los Angeles had run leveling surveys along five routes across the southern San Andreas Fault. Each survey began at sea level near the Los Angeles harbor, which was assumed to be a fixed point, and measured the elevation of various points along the routes radiating north from there. Comparison of the results from surveys done in different years showed whether there had been any elevation changes in the intervening time. A preliminary analysis published in 1974 revealed that in the 1960s there had been upwards of eight inches of uplift along one route due north of Los Angeles. The final results were even more dramatic: along all five routes, the earth's surface had bulged up by six to ten inches since about 1960, with most of the uplift coming in surges in the early '60s, late '60s, and around 1973. The greatest amount of uplift occurred near the town of Palmdale in northern Los Angeles County; hence, the uplift soon was called the "Palmdale Bulge." It extended along a hundred-mile-long segment of the San Andreas Fault that had ruptured during the last great southern California earthquake, in 1857, and had been ominously quiet and free of even small earthquakes since at least 1932.[37]

Frank Press, still one of the leading advocates of prediction research, first heard of these results in early December 1975 at a meeting of the Geological Survey's Earthquake Studies Advisory Panel, which he headed. At this time, the survey's scientists did not attach any particular interpretation to the bulge. But Press knew that a similar uplift had occurred before the destructive 1964 Niigata earthquake in Japan. He thus immediately recognized the importance of the Palmdale Bulge as both an ominous harbinger of a catastrophic earthquake about to strike and as an opportunity for pushing prediction research. As he remarked in his notes taken at the advisory panel meeting, "the vertical uplift reported on SA [San Andreas] north of LA is to me the most significant data and the most dangerous indication of a killer earthquake (m > 7) I have seen in U.S."[38]

Five weeks later, Press saw his chance for bringing the Palmdale Bulge to the Ford administration's attention. Together with several Geological Survey officials, he was invited to give a presentation on January 14, 1976, to the White House Advisory Panel on Anticipated Advances in Science and Technology. They made their regular pitch for an increased prediction research program, suggesting that funding be increased by anywhere from $15 to $50 million a year.[39] After the panel's meeting, Press briefly mentioned the as-yet-unpublished discovery of the Palmdale Bulge to Vice President Nelson Rockefeller. Intrigued, Rockefeller asked for further details. In his reply, on January 21, Press noted that this "discovery, which will soon be released publicly, is most disturbing because such uplifts in the past have preceded earthquakes of great destructive power." He further asserted that "in Japan, a geophysical anomaly of this magnitude would trigger an intensive study or a public alert." After urging an increase in the funding for prediction research, he then pointed to the Chinese success in predicting the Haicheng earthquake and warned that "because of insufficient resources a similar achievement may not be possible in this country."[40]

In a follow-up conversation with Rockefeller's advisors in early February, Press noted that the data on the Palmdale Bulge would soon be published in *Science*. The implied warning was not lost on the advisors; they wrote to the Vice President, the administration was "likely to incur a political liability" if news of the bulge should be coupled by the media with the fact that the Office of Management and Budget had cut the Geological Survey's budget for earthquake research.[41] When Rockefeller's staff asked Press for his recommendations, he suggested that $2 million be restored to the Geological Survey's budget for monitoring the uplift and that an additional $20 million per year be provided for earthquake prediction research. The director of the National Science Foundation and the Secretary of the Interior concurred that the survey's budget should be increased by at least $16.6 million.[42]

The president's top domestic advisors, including Rockefeller, agreed that something had to be done in response to the Palmdale Bulge, but they balked at adding $16.6 million for earthquake prediction research, instead preferring to establish an ad hoc panel to produce yet another study on the most cost-effective design for an earthquake mitigation program. The administration did agree to restore $2.6 million to the Geological Survey's earthquake research budget for monitoring the bulge.[43] This increased funding, however, was a disaster for the survey. Half of the money had come from internal reprogramming within the survey, meaning that no funds were available for adding new staff; rather, unqualified staff were transferred from other divisions. The other $1.3 million came from the National Science Foundation rather than through a direct budget increase; as a result, the survey had to go through the formal procedure of applying to the foundation for the money and did not have it available for use until June.[44] Rather than taking any effective steps in response to the Palmdale Bulge, then, the Ford administration had merely done enough to avoid a public relations disaster. However, the study panel set up to consider an earthquake mitigation plan would later play a crucial role in the establishment of the National Earthquake Hazards Reduction Program.

Although the Geological Survey's article presenting the evidence for the Palmdale Bulge, submitted on January 26, 1976, did not appear in *Science* until April 16, the survey released word of the uplift to the press on February 12, just days after a magnitude 7.5 earthquake had killed more than 20,000 in Guatemala.[45] Over the next several months, survey scientists and other researchers repeatedly publicized the bulge, using it to warn southern Californians to be prepared for seismic disaster and incidentally highlighting the need for further prediction research. Survey officials held briefings for the California congressional delegation on March 4 and for Governor Brown's staff on March 17. In them, survey director McKelvey expressed his opinion that a "great earthquake" would occur "in the area presently affected by the . . . 'Palmdale Bulge' . . . possibly within the next decade." He noted that it might result in up to 12,000 fatalities, up to 48,000 people hospitalized, 40,000 buildings severely damaged, and up to $25 billion in property damage.[46] On March 11, the state Seismic Safety Commission also held a hearing on the bulge, at which a Geological Survey representative testified. A month later, the commission passed two resolutions stating that "the uplift should be considered a possible threat to public safety and welfare in the greater Los Angeles metropolitan area" and urged both state and federal agencies to take all measures possible to prepare for a potential catastrophe, a call endorsed by the *Los Angeles Times*.[47] A couple of weeks later, the California Earthquake Prediction Evaluation Council, a new state board composed of leading academic and government seismologists, likewise concluded that the bulge "should serve to give us a renewed sense of urgency in preparing for the large earthquake that some day inevitably will occur."[48] Meanwhile, the director of the state's Office of Emergency Services sent several letters to local and state officials, alerting them to the bulge and noting, "[this news] should stimulate each of us to review and update our earthquake emergency response plans." High officials within the state Department of Transportation and Department of Water Resources told their engineers to focus their bridge and dam retrofitting efforts on the uplifted area. And at least one county decided to hold more frequent earthquake drills in its schools.[49]

In all of this activity, the need to lobby for earthquake prediction was not forgotten. On March 16, Senator Cranston inserted the text of the survey's briefing on the bulge into the *Congressional Record* as part of his argument for increased funding for earthquake prediction research. The following month, Frank Press, in a special address at the American Geophysical Union's spring meeting, once again contrasted the wholehearted government support that prediction research received in China with the more modest effort that had followed announcement of the uplift in the United States.[50]

In April 1976, a tentative prediction issued by Caltech's Whitcomb underscored the argument that southern California was under imminent threat of a seismic catastrophe and that prediction research should receive more support. Whitcomb had already made a name for himself by uncovering evidence for a V_P/V_S anomaly preceding the San Fernando earthquake and by predicting an earthquake near Riverside in early 1974 (albeit missing its magnitude). Now, in a presentation at the American Geophysical Union's meeting, he predicted, based on an observed V_P/V_S anomaly, that there would be an earthquake of magnitude 5.5 to 6.5 in the Los Angeles basin within the next twelve months. Whitcomb emphasized that this was merely a test of a scientific hypothesis, not a true prediction, as the dilatancy-diffusion theory was not yet sufficiently confirmed to serve as the basis for formal predictions. The California Earthquake Prediction Evaluation Council agreed; after a meeting on April 30 to discuss Whitcomb's evidence, it concluded that his data were not sufficiently unambiguous to show that there had indeed been a V_P/V_S anomaly.[51] The southern California media, however, treated Whitcomb's announcement as a full-blown and scientifically credible prediction. The *Los Angeles Times,* for example, interviewed Whitcomb repeatedly, explained the dilatancy-diffusion hypothesis, and gave advice on how locals should prepare themselves for the coming quake. The publicity went so far that a Los Angeles city councilman threatened to sue Whitcomb for hurting property values in the city.[52] Journalists and editorial columnists ridiculed this threat. It would be far more worthwhile, so the *Times* argued, for funding for prediction research to be increased significantly, so that a more precise warning of the coming earthquake might be given. As a result, the *Times* joined the *San Francisco Chronicle* in repeatedly publishing editorials that strongly supported the prediction research program.[53]

During most of 1976, advocates of earthquake prediction research pinned their hopes primarily on Senator Cranston's earthquake research bill. As originally written, this bill authorized nearly $500 million over ten years for earthquake prediction and engineering research and hazard mapping by both the U.S. Geological Survey and the National Science Foundation. On February 19, two weeks after the disastrous Guatemalan earthquake and only days after the survey publicly announced its discovery of the Palmdale Bulge, Cranston held a Senate Commerce Committee hearing at which both geophysicists and engineers strongly endorsed his bill. Later, in order to make his bill more palatable politically, Cranston amended it to limit the program's duration to three years while retaining the annual authorization of $50 million. He also included a 15-member national advisory committee to provide scientific oversight and accountability for what he now called the National Earthquake Hazard Mitigation Program. This amended bill easily passed the Senate Commerce Committee on May 4 and received a unanimous vote on the Senate floor on May 24. Its passage was due in part to what Cranston called the "disturbing discovery" of the Palmdale Bulge and the emotional impact of recent seismic disasters in Italy and the Soviet Union as well as Guatemala.[54]

As the bill moved to the House of Representatives, it encountered some opposition. Charles Mosher, a Republican congressman from Ohio, introduced an alternate bill that established a 29-member National Earthquake Hazards Reduction Conference to coordinate the work of all federal agencies involved in earthquake hazard mitigation without providing them any additional funding. These agencies included not only research agencies, such as the Geological Survey and the National Science Foundation, but also emergency response agencies, such as the Federal Disaster Assistance Administration, and agencies owning buildings, such as the Veterans Administration. Fellow Republican representative Barry Goldwater, Jr., of California supported Mosher's bill, it being less expensive and offering a more comprehensive management of the federal earthquake response. At a House committee hearing in late June, representatives of the Ford administration also endorsed Mosher's approach and argued that a legislatively mandated national program, as provided for in the Cranston bill, was not needed. Simple coordination among executive branch agencies, as outlined in Mosher's bill, was all that was required.[55]

After working with Mosher's staff, Cranston significantly revised his bill, reducing the authorization to $80 million over three years and setting up an Office of Earthquake Hazards Reduction to manage the national program and involve all concerned federal agencies. This bill finally emerged from committee in late August and came to a vote on the House floor on September 20. At this late date, just before congressional adjournment for the election season, the bill required a two-thirds majority for passage. Cranston's advisors expected little opposition. In the debate on the bill, however, several fiscally conservative Republicans voiced great concern over setting up a new program that would further increase the budget deficit and lead to the establishment of a new bureaucracy. Congressman William Ketchum, whose district included the Palmdale Bulge area, even argued that there was no need for a federal earthquake program at all, as sufficient resources already existed at the state level. The bill failed on a 192-192 vote.[56]

All was not lost, however, for prediction research advocates. The cause was now taken up by the Ford administration. To be sure, President Ford's domestic advisors had opposed the Cranston bill. Some had even talked about Ford's vetoing the bill if it should pass Congress, even though such a veto might be difficult to explain in vote-rich California just before the presidential election. But they did not oppose earthquake hazard mitigation—who could be against such a worthy objective? They simply did not want the credit for establishing a national earthquake program to go to a Democratic senator, and they did not want Congress to curtail administrative prerogative by prescribing the program's goals and how they were to be accomplished.[57]

As the Cranston bill had been making its way through Congress in the summer and fall of 1976, the Ford administration had been conducting its own study "to define an overall plan for earthquake hazard reduction." This study was primarily the task of the panel first authorized in February in response to the Palmdale Bulge and finally appointed in May. Composed of representatives from fields as diverse as architecture, economics, and disaster mitigation, this panel was headed by Nathan Newmark, a distinguished earthquake engineer at the University of Illinois. In its report, released on September 15, the panel recommended a balanced program of federally funded research that would aid in reducing seismic hazards. In many ways, the recommended program was similar to that of the Cranston bill, reflecting the laboriously achieved consensus among geophysicists and earthquake engineers about research priorities. The Newmark panel report recommended that funding of earthquake research by the Geological Survey and the National Science Foundation be increased from the current $20 million to anywhere from $75 to $105 million in fiscal year 1980. Funding for prediction and engineering research should each be increased by $15 to $23 million; support for hazard mapping, fundamental earth science research, and social science studies of how to make mitigation more effective should also be boosted. The report differed from the Cranston-Mosher bill only in suggesting that overall coordination of the hazard mitigation program be left to the discretion of the Executive Office of the President rather than entrusted to a newly created office or advisory panel.[58]

The Newmark panel's recommendations soon found acceptance among President Ford's advisors. One important supporter was H. Guyford Stever, director of the National Science Foundation and the president's chief science advisor. Stever had been instrumental in putting together the first draft of the Newmark panel's report. Once the final report had been issued, Stever several times recommended to the Office of Management and Budget that funding for earthquake research be increased to $54 million for fiscal year 1978, as a first step toward ramping up to the recommended level of $90 million in 1980. The Office of Management and Budget, happy to support an earthquake program over which the executive branch retained control, endorsed this figure. President Ford did likewise in the budget he submitted to Congress on January 17, 1977.[59] By this time, of course, defense of the budget proposal was already out of the hands of the Ford administration, as Jimmy Carter took office on January 20. But in the spring of 1977, Carter appointed none other than Frank Press as his science advisor, so earthquake research retained a strong advocate among the president's staff. The budget request for earthquake research remained unchanged, and Congress, in appropriating the money, even added $2 million more for the Geological Survey. As the House Appropriations Committee noted in its report on the item, it considered the earthquake research program to be "very important."[60]

The formal establishment by Congress of a National Earthquake Hazards Reduction Program now followed quite easily. Soon after Congress opened its new session in early January 1977, Senator Cranston once again introduced his earthquake research bill, which now authorized $220 million over three years. In April, at the request of the Carter administration, he amended it to remove all references to a National Advisory Committee on Earthquake Hazard Reduction and to leave implementation of the program entirely to the administration's discretion. With the level of authorization reduced slightly to $205 million, the bill passed the Senate on May 12 and the House, on a 229-125 vote, on September 9. President Carter signed it into law four weeks later.[61]

Thirteen years of hard organizational and political work had finally paid off for the advocates of earthquake prediction research. A National Earthquake Hazards Reduction Program, with prediction research a well-funded component, was now officially part of federal government policy. The final success had come about because of the concern raised by the Palmdale Bulge and the Whitcomb prediction. Earthquake researchers had skillfully used these events to remind both politicians and the general public of the severe threat looming over southern California. As a result, the president's advisors and congressional leaders finally agreed that the earthquake problem was worth paying attention to. After another year of compromising and alliance-building to resolve political squabbling over how best to address the problem, these leaders accepted the program that prediction advocates had over the years endowed with both political acceptability and scientific credibility.

The Demise of the Dilatancy-Diffusion Hypothesis

It was well that the National Earthquake Hazards Reduction Program was established in 1977, for by that year optimism about earthquake prediction had already waned considerably among earth scientists. On July 28, 1976, a magnitude 7.6 earthquake hit the Chinese city of Tangshan, about 100 miles east of Beijing, without any effective warning. The Chinese government did not release an official death toll, but Western analysts quickly concluded that the quake had killed at least 100,000 and perhaps as many as 600,000. This staggering catastrophe showed that even the vaunted Chinese prediction effort, supported at a much higher level than the American program, could not prevent seismic disaster.[62] Meanwhile, the U.S. Geological Survey by 1977 had to admit that it did not know what the Palmdale Bulge portended for the seismic future of southern California. Repeated leveling across the bulge showed that the earth's surface in places rose and sank quite rapidly, but without any correlation with earthquakes. By 1979 most of the bulge had deflated without any release of strain in a great earthquake. In fact, a number of geophysicists came to argue that the bulge had never existed and that the appearance of an uplift near Palmdale had been due to measurement errors rather than real movement of the earth's surface.[63] The Whitcomb prediction fared even worse. In December 1976, Whitcomb quietly withdrew his prediction, because the V_P/V_S data on which he had relied refused to conform to the pattern required by the dilatancy-diffusion hypothesis. By April 1977, there had indeed been no earthquake of magnitude 6 in the area covered by his "hypothesis test."[64]

Support also waned for the dilatancy-diffusion theory as such. In early 1978, Allan Lindh of the U. S. Geological Survey and several colleagues published a very careful and devastating paper showing that the velocity anomaly that had supposedly occurred just before the Thanksgiving Day earthquake was an artefact of interpretation. The original investigators, eager to find confirmation for the dilatancy-diffusion theory, had selected only favorable points from a data series that contained a lot of random noise, and thereby created an apparent anomaly in the V_P/V_S ratio. The same was true of the supposed anomaly preceding the 1972 Bear Valley earthquake. Likewise, on reconsideration, the data from the Matsushiro earthquake swarm in Japan, which supporters of the dilatancy-diffusion hypothesis had cited as a confirmation, turned out not really to fit the theory.[65] Meanwhile, Hiroo Kanamori and his colleagues at Caltech continued their systematic and thorough search for velocity anomalies in southern California. In 1977, they concluded that there had been none before several moderate earthquakes in the region—and that their network of observations had been dense enough that, if an anomaly had existed, it would have been detected. Teams from the University of California at Berkeley and Stanford University also failed to detect any velocity anomaly before a magnitude 5.7 earthquake near Oroville, north of Sacramento, in 1975.[66]

As science journalist Richard A. Kerr noted in 1978, the dilatancy-diffusion hypothesis had "fallen on hard times, the result of too few observations consistent with the theory." As a result, he reported, the "euphoria" about earthquake prediction was gone from the scientific community. The last remnants of that exuberance evaporated a year later, when the magnitude 5.7 Coyote Lake earthquake struck 15 miles north of Hollister, within the portion of the San Andreas fault zone most heavily instrumented by the Geological Survey. Despite an intense search, no anomalies of any sort could be found in the seismograph, tiltmeter, strainmeter, magnetometer, and other instrumental data recorded before that earthquake. An effective and reliable earthquake prediction system would be far more difficult to establish in California than its advocates had assumed.[67]

Although the dilatancy-diffusion hypothesis faded away, the National Earthquake Hazards Reduction Program remained, a legacy of the optimism over earthquake prediction of the mid-1970s and the concern over seismic disaster engendered by the Palmdale Bulge and the Whitcomb prediction. In establishing the program, Congress had officially declared it to be the federal government's role "to reduce the risks of life and property from future earthquakes in the United States."[68] And both Congress and the executive branch had committed themselves to supporting a comprehensive program of earthquake research and emergency preparedness at an overall annual level of about $75 million—a level of support equivalent to that of the Cold War–era Vela Uniform project of the early 1960s. Thus, earthquake hazard mitigation became entrenched within the federal government apparatus, as it had already done at the state level in California.

CHAPTER 9

The Regulatory-State Apparatus in Action

With the establishment of the California Seismic Safety Commission in 1975 and the National Earthquake Hazards Reduction Program in 1977, the framework for the government regulation of earthquake hazards was complete. Both the state and the federal government had formally acknowledged a duty to mitigate seismic hazards, and they had established formal institutions through which this duty might be exercised: a state commission that would urge the adoption of improved building codes, land-use regulations, retrofitting ordinances, and other means for lessening the risks from earthquakes, and a federal research program that would provide the knowledge necessary for making these measures effective. Over the next two decades, both the state commission and the federal program would produce new initiatives for making California safer from earthquakes. In doing so they would flesh out the regulatory-state apparatus that had been emplaced in the mid-1970s.

The National Earthquake Hazards Reduction Program

Of the two institutional frameworks, the National Earthquake Hazards Reduction Program has fared less well. It has been riven by internal management problems, conflicts over its overall mission, and cuts in funding, and it has suffered from disappointing results in prediction research. Nevertheless, the program has also produced some significant achievements, such as the initiation of several earthquake preparedness projects and the refinement of earthquake engineering guidelines.

For all practical purposes, the National Earthquake Hazards Reduction Program as established in 1977 was a research program. The only groups to receive budget authorizations under the founding act were research agencies: the U. S. Geological Survey and the National Science Foundation. To be sure, the act did assert that the program should foster both research and the implementation of effective mitigation measures, but many of the researchers who had lobbied so hard for the program firmly believed that the key to increased seismic safety lay in more research. Once scientists had increased their knowledge about how earthquakes occur and engineers had identified better methods for building against them, so they asserted, the implementation of new mitigation measures would follow naturally from the availability of new information.[1]

When it came time for the administration to designate a lead agency for the program, however, President Carter and his advisors (foremost among them Frank Press) chose not one of the research agencies but rather the Federal Emergency Management Agency. This agency had just been created, in early 1979, by amalgamating various civil defense and emergency preparedness offices within the federal government. It was a mission-oriented agency, charged with preparing the nation for both natural and military disasters, and it worked closely with state and local governments to reduce the risks from various hazards. When the National Earthquake Hazards Reduction Program was renewed in 1980, Congress gave the Federal Emergency Management Agency a budget authorization of nearly $9 million to lead it and make sure that research results were widely disseminated and used. The National Bureau of Standards (now the National Institute of Standards and Technology) also received some funding for applied work in earthquake engineering.[2]

The program has failed to achieve coordination among its four primary members and to define a cohesive hazards reduction strategy. When made lead agency, the Federal Emergency Management Agency was not given the power to review budgets for earthquake work among other agencies; instead, each agency still negotiates its earthquake budget individually with the Office of Management and Budget and with its congressional appropriations committees. Moreover, throughout the 1980s the Federal Emergency Management Agency was a generally weak body, struggling to define its mission and stake out its turf in the federal bureaucracy. Within the agency, which was dominated by officials concerned with civil defense against nuclear attack, the earthquake threat had relatively low priority. As a result, the Geological Survey and the National Science Foundation, both of which already had well-established earthquake programs, continued to pursue their individual research objectives with little regard for overall co-ordination in a national program. Congressional oversight committees have repeatedly voiced concern over the lack of a unified strategy with clearly stated goals and objectives for achieving a reduction in earthquake hazards. Despite the introduction of new management tools, such as annual reports to Congress and the development of five-year plans, the inherent institutional weaknesses of the multiagency national program have prevented any effective coordination.[3]

The program has also suffered from anemic funding. In 1978, a year after concern over the Palmdale Bulge and the Whitcomb prediction had led to a tripling of funding for earthquake research, congressional appropriations committees started trimming budgets again—a process that accelerated in 1979 when a sociologist studying the effectiveness of earthquake prediction and other mitigation measures was caught using funds for personal expenses. In each year from 1978 to 1989, funding for the program dropped in real terms (adjusting for inflation). Even with some sizable increases in the three years after the Loma Prieta earthquake of 1989, the program's budget in the mid-1990s had not yet returned to the level of 1977. This tight budget situation has led to conflicts between various agencies over the relative weight that should be given to research and implementation. The Federal Emergency Management Agency increased its share of the program's budget, used for implementing mitigation measures in cooperation with state and local governments, from less than 5 percent in the early 1980s to about 20 percent in the mid-1990s. The shrinking share given to research has led seismologists and engineers to complain about the starving of important earthquake research. Others, however, are still concerned that so little money is earmarked for using the knowledge that is already in hand for more effective hazards reduction, a problem that has come to be known as the "implementation gap."[4]

Individual components have also had their problems. Earthquake prediction research, for example, has still not produced the results its advocates initially expected. With the loss of faith in the dilatancy-diffusion hypothesis in the late 1970s, earthquake researchers once again shifted to an empirical search for any kind of signal that might be a reliable precursor to earthquakes. They looked at radon emissions, foreshock patterns (that is, series of small earthquakes that occur just before a larger one), even the behavior of animals. But this search produced few encouraging results. In part this was due to a Catch-22 situation: because researchers could not predict where earthquakes would occur in the near future, they could not set up their instruments and experiments close enough to the epicenter to record signals that might turn out to be precursors. Hence, in many cases analysis was done in hindsight on sparse and inadequate data.[5]

In the early 1980s, the U.S. Geological Survey decided to concentrate its prediction efforts on one 20-mile-long stretch of the San Andreas Fault near the small town of Parkfield, halfway between San Francisco and Los Angeles. Survey geologists discovered that earthquakes of magnitude 5.5 had struck this segment in 1857, 1881, 1901, 1922, 1934, and 1966. If the 1934 earthquake is assumed to have occurred ten years prematurely, this series suggests that an earthquake strikes approximately every 22 years. Moreover, the seismographic records produced by the 1922, 1934, and 1966 earthquakes look identical, suggesting that the mechanism responsible for the shocks was the same. From this, the geologists concluded that another magnitude 5.5 earthquake would occur in 1988, or, to be on the safe side, within the interval 1983–93. This expected earthquake would be especially valuable for prediction researchers, because both the 1934 and 1966 earthquakes had been preceded by a magnitude 5.0 foreshock 17 minutes before the main shock, and in 1966 there had been some unusual tectonic creep along the fault in the days before the earthquake. Thus, the coming Parkfield earthquake promised to provide valuable data on earthquake precursors. In order to record these data in sufficient detail, earthquake researchers turned the Parkfield fault segment into the most densely instrumented segment in the country. To their great chagrin, however, the expected Parkfield earthquake by the end of 2000 still had not occurred. As a result, few new leads have emerged that might allow the short-term prediction of earthquakes; moreover, the survey has come under considerable attack for concentrating its resources at Parkfield, on the basis of what critics called an inherently fallacious forecast.[6]

The Geological Survey has claimed greater success in long-term forecasting. In 1988 it issued a report assigning probabilities for earthquakes along various segments of the San Andreas fault zone during the following 30 years. The destructive Loma Prieta earthquake of 1989, which originated on the San Andreas Fault in the Santa Cruz Mountains south of San Francisco, indeed (but perhaps fortuitously) struck along a segment that had been assigned one of the highest probabilities for seismic activity.[7] Yet obstacles still remain even for long-term forecasts couched in terms of probabilities. Three of the more destructive California earthquakes of the past two decades—the Coalinga earthquake of 1983, the Whittier earthquake of 1987, and the Northridge earthquake of 1994—originated on faults that had not been designated as active. In fact, activity on these faults could not have been detected by the standard methods of mapping faults on the earth's surface. These three faults are all blind thrust faults: they produce fault offsets only deep in the earth's crust; the active fault plane peters out before it reaches the earth's surface. The existence of blind thrust faults underneath downtown Los Angeles has been revealed by excavation for subway tunnels, but absent such expensive tunneling, or the patient piecing together of data from oil wells and other drillholes, the activity and even the existence of buried faults is still difficult to determine. And of course no reliable predictions, short-or long-term, can be issued for faults whose existence is unknown.[8] Thus, for the foreseeable future, California must still respond to a generalized earthquake threat rather than a seismic hazard that is tightly circumscribed in space and time. The advice issued by G. K. Gilbert in 1909 — be prepared as if an earthquake could strike at any time — is still valid today.

Despite the disappointments in earthquake prediction research, the National Earthquake Hazards Reduction Program has had its share of successes. Among the most important has been the Southern California Earthquake Preparedness Project. This project, funded with money from the Federal Emergency Management Agency, was established in 1980 under the supervision of the California Seismic Safety Commission. Originally it had been designed to prepare southern Californians for the effects of an earthquake prediction; soon, though, its purpose shifted to preparing for the earthquake itself. In its first three years, the project established partnerships with a southern California county (San Bernardino County), a large city (Los Angeles), a small city (Westminster, in Orange County), and a corporation (Security Pacific Bank). For each of these institutions, project staff members worked together with emergency preparedness personnel to develop detailed response plans and contingency plans for a large earthquake. Once these prototype plans had been worked out, they could then be transferred to similar entities across the region. In this manner, earthquake planning increased signally in southern California, because individual preparedness offices now had detailed models of emergency plans available to them. The project as a whole also was duplicated in other regions: in 1983, the Federal Emergency Management Agency initiated a Bay Area Earthquake Preparedness Project in northern California; later, a similar one was started in the central United States.[9]

More generally, the National Earthquake Hazards Reduction Program has represented an ongoing federal commitment to earthquake research. Even with the funding cutbacks in the past two decades, the federal budget for earthquake research is two to three times greater than it was in the early 1970s and ten to twenty times greater than in the mid-1960s.

With this funding, geologists, seismologists, and earthquake engineers have continued to increase their knowledge about where and how earthquakes occur and how structures respond to earthquake shaking. Some of this new knowledge has also made its way into the practice of earthquake hazard mitigation. Under the program's auspices, for example, engineers, architects, and others have developed a building code that serves as a model for earthquake provisions across the country.[10] Despite its problems, then, the National Earthquake Hazards Reduction Program has served to increase seismic safety in California and elsewhere in the United States.

Earthquake Hazard Mitigation in California

Largely leaving the fostering of earthquake research to the national program, the California Seismic Safety Commission has concentrated on implementing new hazard mitigation methods and improving old methods that are already in place. It has done so in the absence of a sustained groundswell of grassroots support for seismic safety. As was the case in the 1960s, so in the 1980s and '90s the general public and the news media in California have not actively lobbied for greater earthquake protection except for nuclear reactors and large dams. Thus, the Seismic Safety Commission has continued the strategy of using behind-the-scenes consensus building within the existing bureaucratic structure to achieve its goals. This strategy has brought a number of successes: most cities and counties now require retrofitting of old buildings; bridges, highway structures, hospitals, and emergency services buildings are being strengthened; and the hazard zoning process has been extended to include areas of landsliding and liquefaction as well as fault zones.

In the past two decades, public demands for greater seismic safety have continued to focus exclusively on large projects opposed by environmentalist groups. The classic example is the Diablo Canyon nuclear power plant, located on the seashore in San Luis Obispo County, halfway between San Francisco and Los Angeles. The Pacific Gas and Electric Company in the late 1960s began building at this site two reactors that were designed to resist the moderate earthquakes to be expected from the known faults exposed in the nearby Coast Ranges. As the first reactor was nearing completion in late 1975, however, a UCLA geologist suggested in print that an off-shore fault that ran within three miles of Diablo Canyon was much longer than previously suspected and might be capable of generating a magnitude 7.5 earthquake, greater than the reactors could resist. The timing of this article could not have been worse for Pacific Gas and Electric. Within the next several months, revelation of the Palmdale Bulge and the Whitcomb prediction would heighten concern about seismic hazards in southern California. Moreover, in the spring of 1976 antinuclear activists were heavily campaigning for a ballot initiative to impose a moratorium on all new nuclear reactors. The activists of course seized upon the issue of seismic safety as an important argument against nuclear power in California.[11]

As it turned out, the moratorium initiative was defeated, and the Nuclear Regulatory Commission (which had replaced the Atomic Energy Commission as the federal safeguard for reactor safety) allowed Pacific Gas and Electric to continue work on the Diablo Canyon reactors once they had been redesigned to withstand the stronger earthquakes. In late 1981, though, the company suffered a major embarassment when it was discovered that the wrong blueprints had been used for the second reactor. Once again, environmental and antinuclear activists protested that the nuclear industry could not be trusted to ensure the seismic safety of their plants. Such local grassroots groups as the Abalone Alliance and the San Luis Obispo Mothers for Peace kept up their protests until 1985, when the Nuclear Regulatory Commission finally granted a full power license for the plant after it had been retrofitted to conform to the correct blueprint. While not successful in stopping the Diablo Canyon nuclear reactors, the protests and delays did add at least $2 billion to their cost, thereby making the construction of future atomic power plants in California economically unattractive.[12]

Environmental activists had even greater success playing the earthquake card in their opposition to the Auburn Dam. In the mid-1970s, the federal Bureau of Reclamation proposed building a large dam at Auburn, on the American River about 30 miles upriver from Sacramento. This was in the foothills of the Sierra Nevada, a region considered relatively free from damaging earthquakes. But in 1975, a magnitude 5.7 earthquake struck near Oroville, site of another large dam in the foothills north of Sacramento. Subsequent studies by the U. S. Geological Survey and a geotechnical engineering firm suggested that a fault within two miles of the Auburn site was capable of a magnitude 6.5 or perhaps even magnitude 7 earthquake. Activists who opposed the dam on environmental grounds seized upon the seismic hazard as an argument, suggesting that the dam might fail in a large earthquake and unleash a wall of water on defenseless Sacramento. The high cost of a dam redesigned to withstand the earthquake threat eventually led the federal government to abandon plans for the Auburn site. In the mid-1990s, a local congressman revived plans for a flood-control dam at Auburn; once again, environmentalists are pointing to seismic hazards in opposing it.[13]

Concerted opposition from grassroots activists has blocked the building of some large projects in earthquake country. Those nuclear power plants and dams that do get built despite this opposition tend to be designed to very high seismic safety standards, to satisfy the protesters; as a result, they are among the safest structures in California during earthquakes. As was the case in the 1960s, though, environmental and antinuclear activists are not interested in a broader program of seismic safety; rather, they see seismic hazards as an opportunity for stopping projects they oppose for other reasons. Thus, other measures for improving seismic safety, most notably the retrofitting of old buildings, have not enjoyed support from a sustained grassroots movement. Rather, seismologists, earthquake engineers, bureaucrats within the Seismic Safety Commission, and other members of the earthquake community have had to work on their own to implement these measures. And without the political support brought by grassroots pressure, they have usually had to rely on their strategy of quiet and patient consensus building to achieve their goals.

One of the most important victories for earthquake researchers came in 1981, when the patient work of seismic safety advocates brought about the passage of a retrofitting ordinance in Los Angeles, an ordinance that opened the way for similar ordinances in other cities and counties throughout California. Retrofitting old buildings to bring them up to higher standards of earthquake resistance enjoyed considerable support among engineers and geologists, who saw it as the most important step to be taken in California in the late 1970s and '80s to ensure greater seismic safety. In Los Angeles, retrofitting was also vigorously advocated by the *Los Angeles Times* and, after his 1979 election, by new city council member Harold Bernson, who had lived through the Bakersfield earthquakes of 1952.[14] Opposed were realtors and apartment owners as well as tenants afraid that the cost of retrofitting would lead to significant rent increases. Opposition from tenants orchestrated by the Apartment Association of Los Angeles County had already led to a resounding defeat of a retrofitting ordinance in early 1977.

After that debacle, retrofitting proponents in 1979 and 1980 once again sought to build support for an ordinance in the city council. While supporters had their own scare tactics, predicting that the collapse of unretrofitted buildings might lead to 8,500 deaths in a major earthquake, they focused mostly on removing the objections of their opponents. A revised ordinance gave 15 years rather than 10 for building owners to retrofit, and it required them only to bring their buildings up to the earthquake standards in existence before the 1971 San Fernando earthquake, rather than the stricter standards enacted since then. Extensive tests by consulting engineers retained by the city showed that the retrofitting could be done at a cost of $6 to $12 per square foot, leading to a total cost of $750 million for retrofitting rather than the several billion claimed by building owners. Bernson, a conservative who otherwise defended the interests of the business community, also demonstrated considerable energy in seeking out state and federal loans and other financial support for retrofitting projects. By demonstrating a willingness to bend over backwards to meet all objections, the proponents of retrofitting made their opponents seem unreasonable in their continued shrill opposition. As a result, the city council finally approved the ordinance on January 7, 1981, on an 11-3 vote.[15]

The passage of a retrofitting ordinance in Los Angeles, California's largest city, served as an inspiration to retrofitting advocates elsewhere. Urged on by concerned geologists, engineers, and building inspectors, such cities as Santa Ana and Pasadena now also considered what to do about hazardous buildings. This movement accelerated after 1985, when an earthquake that killed more than 7,000 people in Mexico City received graphic coverage in the California news media. The following year, the state legislature, spurred by the Seismic Safety Commission, passed a law requiring all cities and counties to inventory their unreinforced masonry buildings and suggesting that they institute a retrofitting program. By 1989, 34 cities and counties (about a tenth of all local jurisdictions) had enacted retrofitting ordinances, often over considerable opposition from property owners. That year, nine Californians died in the collapse of old unreinforced masonry buildings during the Loma Prieta earthquake. Within 12 months, another 120 local jurisdictions had made retrofitting such buildings mandatory. By 1997, the number of cities and counties with some form of retrofitting program (either mandatory or voluntary) had grown to about 320, and about half of the state's privately owned unreinforced masonry buildings had been either retrofitted or demolished. Retrofitting showed its value in the 1994 Northridge earthquake, which centered in the San Fernando Valley section of Los Angeles. The nearly 6,000 retrofitted buildings in the city sustained considerably less damage than unretrofitted buildings in neighboring towns; moreover, no one was killed by a retrofitted building.[16]

Over the past two decades, the Seismic Safety Commission has also been an important advocate for retrofitting other sorts of structures. During the 1980s, for example, commissioners argued that hospitals built before the Hospital Seismic Safety Act of 1972 took effect should be up-graded. As they pointed out, the pace of hospital construction was rather sluggish; even in the early 1990s, about 90 percent of the state's hospitals still predated the hospital act. A state law enacted in 1994 finally provided that old hospitals should be retrofitted by 2030. In 1997, the target date for retrofitting was moved up to 2008, in part based on evidence from the Northridge earthquake that newer hospitals performed far better than older ones.[17] The Seismic Safety Commission had less success in spurring the retrofitting of state-owned buildings. In the mid-1980s, it proposed an $800 million bond issue for retrofitting such buildings, but the state submitted only a $300 million issue to voters; it passed in 1990 in the wake of the Loma Prieta earthquake. Five years later, none of the money had yet been spent on any actual retrofitting; the state program was still mired in bureaucratic foot-dragging and endless design studies. Meanwhile, a number of the state's public universities had begun retrofitting older campus buildings on their own initiative with money from other sources.[18]

Among the most expensive and publicly noticed retrofitting projects of the 1980s and '90s has been the retrofitting of the state's highway bridges and structures. After the San Fernando earthquake led to the collapse of several highway overpasses, the state Department of Transportation required that more steel reinforcing be used to make reinforced concrete bridges more resistant to earthquake shaking. This rule, however, applied only to new construction. The only retrofitting that the department conducted during the 1970s and early 1980s was to install steel restraining cables to prevent roadway slabs from sliding off their supports during an earthquake. In some cases, these restraining systems were not adequately tested before installation. In the 1989 Loma Prieta earthquake, two slabs on the double-deck Bay Bridge connecting San Francisco with Oakland pulled off their supports and fell down, leading to one death. More tragedy befell motorists on the Cypress Street Viaduct in Oakland. This double-deck elevated highway (constructed partially on filled ground that amplified earthquake motions) had been built in the 1950s, at a time when engineers did not yet realize how much reinforcing was needed to make the structure resist severe shaking. In the Loma Prieta earthquake, 48 sections of the upper deck collapsed as the columns holding them failed; as a result, 41 people died, accounting for two-thirds of the fatalities in the earthquake.[19]

Since the Loma Prieta earthquake, the California Department of Transportation and local highway departments have been diligently retrofitting bridges and other structures. Throughout the state, baskets of steel reinforcing have been introduced around bridge columns to render the reinforced concrete more capable of resisting earthquakes. In San Francisco, several elevated highways built at the same time as the Cypress Street Viaduct have been completely removed. This activity has of course been costly. In 1994, voters rejected a $1 billion bond issue for bridge retrofitting; two years later, though, they approved a $2 billion issue. Included in this amount is $650 million for retrofitting the seven state-owned toll bridges. The most expensive project, however, the replacement of the eastern span of the Bay Bridge, will alone cost more than $1.5 billion. Thus, a complete retrofitting of the state's critical bridges will require still more financing and time.[20]

In the 1980s and '90s, earthquake researchers and their allies in the Seismic Safety Commission concerned themselves with other hazard mitigation strategies as well. In 1985, the California legislature passed an act requiring fire and police stations, sheriff's and highway patrol offices, and emergency dispatch centers to conform to Field Act standards. This act was the final culmination of the efforts begun by Senator Alquist and his advisors in the early 1970s to provide special earthquake resistance to such critical emergency services facilities.[21] Earthquake researchers also continued their interest in using land-use planning for mitigation. Throughout the 1980s, geologists with the California Division of Mines and Geology continued their mapping of active faults under the Alquist-Priolo Geologic Hazard Zones program. Analysts have generally lauded this program for preventing the proliferation of new housing developments across active faults. Some geologists in the late 1980s were unhappy, though, that no special provisions existed for areas subject to landsliding, ground amplification, or liquefaction. In 1990, after the Loma Prieta earthquake, the state legislature finally passed an act directing the Division of Mines and Geology to delineate such areas in a manner similar to the Alquist-Priolo fault zones. The division issued its first map of general earthquake hazard zones in 1996.[22]

By pushing for more comprehensive hazards mapping, improved building practices, and wholesale retrofitting, then, the Seismic Safety Commission over the past two decades has served as a primary advocate for increased earthquake safety in California. Together with the research-oriented National Earthquake Hazards Reduction Program, it has become the core of today's regulatory-state apparatus for confronting seismic hazards.

Conclusion: The Persistence of Progressivism

In the latter part of the twentieth century, the reduction of earthquake hazards in California has been directed and primarily carried out by a regulatory-state apparatus. The primary advocates for greater seismic safety are agents of the state: scientists who are either employed directly by the government (U.S. Geological Survey, California Division of Mines and Geology, state universities) or have their research funded by it; bureaucrats who work for the Seismic Safety Commission, the Office of Emergency Services, and other government agencies concerned with disaster mitigation; building inspectors working for local governments. Even earthquake engineers in private practice often learn their craft at public universities, take advantage of publicly funded research, attend government-sponsored meetings, and serve on government commissions. Furthermore, the measures that these agents invoke to reduce earthquake hazards often rely on the power of the state: building codes that regulate how structures are designed, retrofitting ordinances that require or at least encourage the renovation of old buildings, land-use laws that prohibit construction in fault zones and limit it in other areas of heightened risk. Emergency planning, public education, and earthquake research are also frequently carried out by official bodies using government resources. The public in general has ceded seismic safety to the state: a poll conducted in 1972 indicated that nearly 90 percent of Pacific Coast residents considered earthquake protection to be a governmental rather than a personal responsibility.[23]

The regulatory-state apparatus for earthquake hazard mitigation differs greatly from other fields of environmental protection, because the state and its agents initiate as well as enforce protective measures. To be sure, government power is also very important in achieving other environmental goals: antipollution laws enforced by the Environmental Protection Agency, restrictions on land-use in wilderness areas, prohibitions on ozone-depleting chemicals, and protection of endangered species all rely on state regulation and coercion. But these measures did not originate in quiet, behind-the-scenes negotiation of government bureaucrats with technical experts whom they trust. Rather, they are the result of political pressure exerted by large and vocal nongovernmental organizations, such as the Sierra Club, the Wilderness Society, and Greenpeace. These environmental groups derive their power not from government-sponsored scientific expertise but from sizable membership bases willing to contribute money and make their opinions heard, and they frequently operate by drawing on widely held moral beliefs concerning individuals' duty toward the environment. In the field of seismic safety, by contrast, there are no such broad-based groups; new measures are initiated by advocates wielding technical expertise and good relations with government bureaucrats rather than political muscle and fundraising ability as their main tools. Even in the area of earthquake prediction, which probably enjoyed the most public attention among mitigation measures, new government programs in the 1970s originated not from public pressure but from the access enjoyed by prediction advocates to Vice President Rockefeller, Senator Cranston, and other influential politicians.

The reliance of earthquake hazard mitigation on state action that became the *modus operandi* for the cause in the 1970s differs greatly from the approach at the beginning of the century. In 1906, responding to an earthquake was widely held to be an individual test of moral character, requiring fortitude and a willingness to rebuild whenever disaster struck. There was no regulatory-state apparatus to direct attention to seismic hazards, and even many scientists and engineers, as well as the boosters and journalists, denied that there were significant earthquake threats in California. Earthquakes were seen simply as law-bound but unpredictable events against which humans could do little; the idea that man could do something to limit earthquake damage, and even more that the government should ensure that these measures were undertaken, would have seemed foreign.

The regulatory-state apparatus that exists today clearly had its roots in the Progressive movement that flourished at the turn of the century. Disgusted with the excesses of laissez-faire capitalism and of unchecked urbanization and industrialization at the end of the nineteenth century, Progressive scientists, engineers, and other professionals sought reforms that would produce a more efficient and sustainable society. Many of them saw science, with its emphasis on objectivity and rationality, as the main source for reform ideas—if only its values could be impressed on businesses and other institutions, then society would become far better regulated and just. Thus, Progressives championed the application of technical expertise under state supervision through, for example, the establishment of the National Forest Service, with its college-trained foresters, and the proliferation of expert commissions to regulate public utilities and other functions of local government. The Progressive ideal of the marriage of science and the state, which persisted through the New Deal years, found its fullest expression in Lyndon B. Johnson's Great Society program in the 1960s: expert social scientists working for the federal government would engineer a just society in which poverty and social discord no longer existed.

The efforts of seismic safety advocates grew out of this Progressive re-

form impulse. A few scientists, awakened by the 1906 San Francisco earthquake to the existence of seismic hazards in California, recognized that the
continued denial of these hazards by boosters and the general public was inefficient. In the long run, so they held, a society that built with earthquake-
resistance in mind would be more sustainable than one that had to repair
flimsy structures after every moderate temblor. These scientists soon developed an infrastructure capable of accumulating new knowledge on California's seismic hazards, and by the 1930s they had involved engineers in the
search for the principles of earthquake-resistant construction. In addition
to generating new knowledge about seismic hazards they wanted to see it
applied to their mitigation. Toward this end, they sought the aid of state
power in the form of building codes that compelled attention to earthquake
hazards. In 1933, they succeeded in establishing an embryonic regulatory-
state apparatus. With the passage of the Field Act that year, a state agency
was entrusted with the enforcement of building regulations drawn up by
engineering experts. After World War II, the apparatus flowered. Scientists and engineers, who had formerly depended on their own meager resources to fund their research, came to rely more and more on federal largesse. In turn, they gained access to governmental power structures, such
as civil defense agencies, the California Joint Committee on Seismic Safety,
and Senator Cranston's office, and they used them to advance their campaign by having more and more mitigation measures imposed by law. The
establishment of the California Seismic Safety Commission in 1975 and the
National Earthquake Hazards Reduction Program in 1977 then confirmed
the interpenetration of science and the state, marking the fulfillment of the
Progressive dream for seismic safety.

The campaigns for resource conservation and wilderness protection
emerged from the same Progressive milieu as that for seismic safety. Many
of the Seismological Society's early members also belonged to the Sierra
Club, and the two groups pursued similar strategies for bringing about a
more efficient and harmonious adaptation to nature. In one case, the campaign of John Muir and his followers against the damming of Hetch Hetchy
Valley in Yosemite National Park, environmentalists in the Progressive Era
did engage in an emotional battle dependent on the mobilization of public
opinion, but this was an aberration; for the most part, environmentalists
used the same means as seismic safety advocates—accumulation of expertise and alliance building with like-minded professionals and public officials.[24]

Beginning in the early 1960s, though, many environmental groups
abandoned these Progressive roots and became increasingly radical, argu-

ing that the preservation of environmental values required a principled stand against the forces of despoliation rather than accommodation with business and government leaders. As they became increasingly moralistic and confrontational in tone, these groups turned to large-scale grassroots mobilization to pursue their goals. Earthquake experts did not participate in this radical turn; rather, they developed an increasing distaste for the distortions of facts and emphasis on scare scenarios that became a hallmark of environmentalist campaigns. This is not to say that seismic safety advocates sacrificed their objectives in deciding to continue their cooperation with the establishment. Indeed, they have probably been as successful in promoting earthquake resistance as other activists have been in building an environmentally sustainable society. They have simply used different strategies in pursuing their goals.

What brought about this divergence? At this time, only speculations can be offered. One factor that may have played a role is a subtle difference in the nature of reformist organizations before World War II. The Sierra Club and similar environmental groups, while peopled with engineers, biologists, foresters, and other experts, also attracted large numbers of amateurs. Moreover, through wilderness expeditions, birding trips, and other outdoor adventures, they offered members an opportunity to form an emotional bond with unspoiled nature. Once that nature seemed threatened in the postwar era, these members reacted by turning increasingly intransigent in their demands for environmental protection. By contrast, the Seismological Society remained primarily a scientific organization. Although Bailey Willis before his fall from grace sought to recruit many amateurs, the society continued to be the domain of scientists, engineers, insurance workers, and others with a professional interest in earthquakes. Furthermore, it did not offer many opportunities for emotional bonding or socialization; it confined its activities to the publication of a quarterly bulletin and the holding of an annual meeting dominated by scientific presentations. As a result, environmental groups found it less of a stretch to become mass-based protest groups in the 1960s than did the Seismological Society.

The importance of individual leadership also cannot be overlooked. In California, such charismatic visionaries as David Brower and David Pesonen did much to radicalize the Sierra Club and the nascent antinuclear movement. They also had the advantage of being able to point to the revered John Muir, the decidedly non-Progressive founder of the Sierra Club, as an inspiration for a more radical stance.[25] No similar leadership emerged among seismic safety advocates. Bailey Willis did attempt to infuse a moralistic tone with mass appeal into the campaign in the mid-1920s, but this approach became discredited after he overreached in his earthquake prediction. Those who led the campaign after him were far more skilled at and interested in high-level political negotiation and consensus building than in confrontational rhetoric. Of course, their dependence on the largesse of the Cold War state for their research support also made them far less likely to adopt the stridently antigovernment and antiestablishment rhetoric that percolated among environmental protest groups after 1960.

What, then, are the insights that can be gained from this history of earthquake hazard mitigation? One important one concerns the relation of mitigation measures to natural disasters. Sociologists have long pointed out that the occurrence of a disaster provides a "window of opportunity," at most a few months in length, for producing public action before the memory of catastrophe fades again. The best example of this in the present history is, of course, the enactment of the Field Act a month after the Long Beach earthquake, a result of the skillful manipulation of public outrage by seismologists and engineers while the memory of collapsed school buildings was still fresh in the public mind. But it must be kept in mind that mitigation measures do not necessarily follow upon disasters. In fact, two of the largest earthquakes to strike California this century, the 1906 San Francisco earthquake and the 1952 Kern County temblor, did not produce noteworthy increases in seismic safety. In the former case, scientists and engineers were simply unprepared to take advantage of the opportunity, being themselves surprised by the magnitude of the destruction. In the latter case, earthquake experts were satisfied with the level of seismic safety they had achieved and merely used the earthquake to reinforce the value of the Field Act and the building codes that had already been enacted. The most significant growth in hazard mitigation measures has occurred after more moderate earthquakes, such as the Long Beach temblor of 1933 and the San Fernando earthquake of 1971. It appears that the San Francisco earthquake and the great Alaska earthquake of 1964 served mainly to awaken concerns about seismic safety among scientists and engineers, who then prepared themselves to be ready to push mitigation measures once another opportunity arose in the form of a more moderate earthquake.

The importance of these windows of opportunity also seems to have changed over time. In 1933, the Field Act and building code changes were rushed into law within a few months of the Long Beach earthquake. In the 1970s, however, new measures trickled out over the course of years. The December 1972 Alquist-Priolo Act and the November 1972 Hospital Seismic Safety Act were not enacted in the immediate aftermath of the

February 1971 San Fernando earthquake. Neither the Seismic Safety Commission created in 1975 nor the National Earthquake Hazards Reduction Program inaugurated in 1977 was established in direct response to a particular seismic event. In these years, because seismic safety advocates were well represented among government agency employees and had direct access to legislators through the Joint Committee on Seismic Safety, they did not have to rely on public opinion to achieve their goals.

In the 1960s and '70s, earthquake experts were able to open their own windows of opportunity, ones not directly related to seismic disasters. The Joint Committee was established in the wake of the Great California Earthquake Scare of 1969, conjured up by psychics and other cranks. Prediction advocates had an even more direct hand in 1976, when they used announcement of the Palmdale Bulge and the Whitcomb prediction to pry open a window for getting a national earthquake program enacted. Broadly speaking, then, the evolution of hazard mitigation is not a simple story of response to natural disasters; also important have been the degree to which mitigation advocates have been organized and have had the resources necessary to mobilize public opinion or the political process in pursuit of their goals.

Abbreviations

The following abbreviations of frequently cited sources are used throughout the notes.

A&E	*Architect and Engineer*
A&EC	*Architect and Engineer of California*
ALP	Andrew Lawson papers (Bancroft Library, University of California, Berkeley)
BSSA	*Bulletin of the Seismological Society of America*
BWP	Bailey Willis papers (Huntington Library, San Marino, California)
CIWR	Carnegie Institution of Washington records (Washington, D.C.)
CRP	Charles Richter papers (California Institute of Technology Archives, Pasadena, California)
EERC	Earthquake Engineering Research Center
EERIN	*Earthquake Engineering Research Institute Newsletter*
EMLR	Earthquake Mechanism Laboratory records (RG 370, National Archives and Records Administration, Pacific Region, San Bruno, California)
EN	*Engineering News*
ENR	*Engineering News-Record*
ESABR	*Engineering Supplement to the American Builders' Review*
FPP	Frank Press papers (MC 159, Massachusetts Institute of Technology Archives, Cambridge, Massachusetts)
HWP	Harry Wood papers (California Institute of Technology Archives)
JBP	John Branner papers (SC 34, Department of Special

	Collections, Stanford University Libraries, Stanford, California)
LAE	*Los Angeles Examiner*
LAT	*Los Angeles Times*
LBPT	*Long Beach Press-Telegram*
NYT	*New York Times*
RAP	Ralph Arnold papers (Huntington Library)
RCT	*Redwood City Tribune*
RJP	Richard Jahns papers (SC 306, Department of Special Collections, Stanford University Libraries)
RMP	Robert Millikan papers (California Institute of Technology Archives)
SB	*Sacramento Bee*
SEB	*Sacramento Evening Bee*
SFC	*San Francisco Chronicle*
SFE	*San Francisco Examiner*
SSAC	Seismological Society of America Correspondence (Bancroft Libary)
SSRG	Seismic Safety Record Group (F 3713, California State Archives, Sacramento)
SUPGS	*Stanford University Publication in Geological Sciences*
USGS	United States Geological Survey

Notes

Introduction

1. See, for example, Samuel P. Hays, *Conservation and the Gospel of Efficiency: The Progressive Conservation Movement, 1890–1920* (Cambridge: Harvard UP, 1959); Brian Balogh, "Reorganizing the Organizational Synthesis: Federal-Professional Relations in Modern America," *Studies in American Political Development* 5 (1991): 119–172; Kendrick A. Clements, "Engineers and Conservationists in the Progressive Era," *California History* 58 (1979): 282–303; Susan R. Schrepfer, *The Fight to Save the Redwoods: A History of Environmental Reform, 1917–1978* (Madison: U of Wisconsin P, 1983), chaps. 1–5; Michael L. Smith, *Pacific Visions: California Scientists and the Environment, 1850–1915* (New Haven: Yale UP, 1987); Anne F. Hyde, "William Kent: The Puzzle of Progressive Conservationists," in William Deverell and Tom Sitton, eds., *California Progressivism Revisited* (Berkeley: U of California P, 1994), 34–56.

2. Samuel P. Hays, *Beauty, Health, and Permanence: Environmental Politics in the United States, 1955–1985* (Cambridge: Cambridge UP, 1987); Michael J. Lacey, ed., *Government and Environmental Politics: Essays on Historical Developments since World War Two* (Washington, DC: Woodrow Wilson Center Press, 1989); George Hoberg, *Pluralism by Design: Environmental Policy and the American Regulatory State* (New York: Praeger, 1992); Schrepfer, *To Save the Redwoods*, chaps. 6–11; Thomas Raymond Wellock, *Critical Masses: Opposition to Nuclear Power in California, 1958–1978* (Madison: U of Wisconsin P, 1998).

3. Charles Edwin Clark, "Science, Reason, and an Angry God: The Literature of an Earthquake," *New England Quarterly* 38 (1965): 340–362, on 355–361; William D. Andrews, "The Literature of the 1727 New England Earthquake," *Early American Literature* 7 (1973): 281–294, on 287–291; Maxine Van de Wetering, "Moralizing in Puritan Natural Science: Mysteriousness in Earthquake Sermons," *Journal of the History of Ideas* 43 (1982): 417–438.

4. A. L. Stone, *The Finger of God* (San Francisco, 1868); "Lessons of the Earthquake," *Daily Alta California,* Nov. 23, 1868, p. 1 (quote); "The Spiritual View," *San Francisco News Letter and California Advertiser,* Nov. 14, 1868, p. 9.

5. "The 'Queen of the West' Laid Low," *Signs of the Times* 32 (1906): 276–77; "Unchangeable LAW Crushed San Francisco," *SFE,* June 10, 1906, p. 17; see also "Foolish Sermon on the San Francisco Calamity," *SEB,* Apr. 26, 1906, p. 4; "Divine Law and Disaster," *SFC,* May 14, 1906, p. 8.

6. Clark, "Science, Reason, and an Angry God," 347–355; Andrews, "Literature," 282–287; Van de Wetering, "Moralizing in Puritan Natural Science."

7. Roy Porter, *The Making of Geology: Earth Science in Britain, 1660–1815* (New York: Cambridge UP, 1977); M. J. S. Rudwick, *The Great Devonian Controversy: The Shaping of Scientific Knowledge among Gentlemanly Specialists* (Chicago: U of Chicago P, 1985); George P. Merrill, *The First One Hundred Years of American Geology* (New Haven: Yale UP, 1924).

8. Charles Davison, *The Founders of Seismology* (Cambridge: Cambridge UP, 1927), 2, 9, 39–40, 47–63, 71–75; Dennis R. Dean, "Robert Mallet and the Founding of Seismology," *Annals of Science* 48 (1991): 50–57.

9. Davison, *Founders of Seismology,* 25–33.

10. Ibid., 75–80; Dean, "Robert Mallet," 58–63.

11. James Dewey and Perry Byerly, "The Early History of Seismometry (to 1900)," *BSSA* 59 (1969): 183–227, on 183–193.

12. Ibid., 195–207; A. L. Herbert-Gustar and P. A. Nott, *John Milne: Father of Modern Seismology* (Tenterden, Kent: Paul Norbury Publications, 1980), chaps. 4 and 5; Davison, *Founders of Seismology,* chap. 10.

13. Alan E. Leviton and Michele L. Aldrich, "John Boardman Trask: Physician-Geologist in California, 1850–1879," in Alan E. Leviton, Peter U. Rodda, Ellis L. Yochelson, and Michele L. Aldrich, eds., *Frontiers of Geological Exploration of Western North America* (San Francisco: Pacific Division, American Association for the Advancement of Science, 1982), 37–69, on 54.

14. Charles Wollenberg, "Life on the Seismic Frontier: The Great San Francisco Earthquake (of 1868)," *California History* 71 (1992): 494–509; William H. Prescott, "Circumstances Surrounding the Preparation and Suppression of a Report on the 1868 California Earthquake," *BSSA* 72 (1982): 2389–2393; Michele L. Aldrich, Bruce A. Bolt, Alan E. Leviton, and Peter U. Rodda, "The 'Report' of the 1868 Haywards Earthquake," *BSSA* 76 (1986): 71–76; W. Frank Stewart, "Earthquake Philosophy," *Daily Alta California,* Nov. 8, 1868, p. 2; "California Academy of Natural Sciences," *Mining and Scientific Press* 17 (1868): 324; "The Electric Theory of Earthquakes," ibid., 17 (1868): 328–329; Thos. Rowlandson, *A Treatise on Earthquake Dangers, Causes and Palliatives* (San Francisco, 1869). For further contemporary debates over earthquake theories, see "Regular Meeting, November 7th, 1870," in *Proceedings of the California Academy of Sciences* 4 (1868–1872): 141–144, on 142–144. For the general weakness of California scientific institutions in California at the time, see Smith, *Pacific Visions,* esp. chap. 2. State geologist Josiah D. Whitney, on a sojourn in California in 1872, did produce a thorough study of the great Owens Valley earthquake of that year; that study, though, was an isolated phenomenon. See [J. D. Whitney,] "The Owen's Valley Earthquake," *Overland Monthly* 9 (1872): 130–140, 266–278.

15. Davison, *Founders of Seismology*, 144–146; Alexis Perrey to C. G. Rockwood, Apr. 11, 1872, C. G. Rockwood Earthquake Scrapbook, General MSS (bound), CO199, Special Collections, Firestone Memorial Library, Princeton University, vol. 1; Rockwood to G. K. Gilbert, Apr. 12, 1890, ibid., vol. 4.

16. Wallace Stegner, *Beyond the Hundredth Meridian: John Wesley Powell and the Second Opening of the West* (Lincoln: U of Nebraska P, 1953), 158–161; G. K. Gilbert, "A Theory of the Earthquakes of the Great Basin, with a Practical Application," *American Journal of Science*, 3rd ser., 27 (1884): 49–53; William Morris Davis, "Earthquakes in New England," *Appalachia* 4 (1886): 190–194.

17. Cleveland Abbe to Chief Signal Officer, Jan. 28, 1875, Records of the U.S. Weather Bureau (RG 27), National Archives II, College Park, MD, entry 7, box 327, file 124; Abbe to Chief Signal Officer, Aug. 11, 1884, ibid., box 460, file 4839; C. F. Marvin to Chief Signal Officer, Oct. 3, 1884, ibid., entry 4 (NC3), box 107, file 1627; T. C. Mendenhall, "Seismoscopes and Seismological Investigations," *American Journal of Science*, 3rd ser., 35 (1888): 97–114; Richard Rubinger, ed., *An American Scientist in Early Meiji Japan: The Autobiographical Notes of Thomas C. Mendenhall* (Honolulu: U of Hawaii P, 1989). As early as the 1850s, Smithsonian Institution secretary Joseph Henry had requested information on earthquakes from his meteorological correspondents; see James Rodger Fleming, *Meteorology in America, 1800-1870* (Baltimore: Johns Hopkins UP, 1990), 82.

18. Robert P. Stockton, *The Great Shock: The Effects of the 1886 Earthquake on the Built Environment of Charleston, South Carolina* (Easley, SC: Southern Historical Press, 1986); Rockwood to J. W. Powell, Sept. 1, 1886, Records of the USGS (RG 57), National Archives, Washington, D.C., microform M590, roll 34, frame 802; W. J. McGee, "Some Features of the Recent Earthquake," *Science* 8 (1886): 271–275; Everett Hayden, "The Charleston Earthquake: Some Further Observations," ibid.: 246–248; T. C. Mendenhall, "Report on the Charleston Earthquake," *Nature* 35 (1886): 31–33; WJ McGee to Earle Sloan, Sept. 9, 1886, WJ McGee Papers, Manuscript Division, Library of Congress, box 21, letterbook "July 1886–January 1887," 158; McGee to Charles E. Correre, Sept. 30, 1886, ibid., 226; W. B. Hazen, untitled memorandum, Nov. 15, 1886, and Dunwoody to Hazen, undated [ca. mid-Oct. 1886], both in Records of the U.S. Weather Bureau, entry 4 (NC3), box 171, file 9046; Mendenhall, "Seismoscopes and Seismological Investigations," 110.

19. C. E. Dutton and Everett Hayden, "Abstract of the Results of the Investigation of the Charleston Earthquake," *Science* 9 (1887): 489–501; T. C. Mendenhall, "The Charleston Earthquake" (letter to editor), ibid.: 584–587; Simon Newcomb and C. E. Dutton, "The Speed of Propagation of the Charleston Earthquake," *American Journal of Science*, 3rd ser., 35 (1888): 1–15; Clarence E. Dutton, "The Charleston Earthquake of August 31, 1886," *Ninth Annual Report of the United States Geological Survey* (Washington, DC: Government Printing Office, 1889), 203–528; Davison, *Founders of Seismology*, 148–151.

20. George D. Louderback, "History of the University of California Seismographic Stations and Related Activities," *BSSA* 32 (1942): 205–229, on 206–212; Edward S.

Holden, "Earthquakes in California (1888)," *American Journal of Science,* 3rd ser., 37 (1889): 392–402; idem, "Earthquake Observations at the Lick Observatory," *Harper's Weekly* 41 (1897): 767; idem, *List of Recorded Earthquakes in California, Lower California, Oregon, and Washington Territory* (Sacramento, 1887); idem, *A Catalogue of Earthquakes on the Pacific Coast, 1769 to 1897,* Smithsonian Miscellaneous Collections, no. 1087 (Washington, DC, 1898); Rockwood to Gilbert, Apr. 12, 1890, Rockwood Earthquake Scrapbook, vol. 4. For an excellent account of Holden's turbulent years at the Lick Observatory, see Donald E. Osterbrock, "The Rise and Fall of Edward S. Holden," *Journal for the History of Astronomy* 15 (1984): 81–127, 151–176.

21. Wollenberg, "Life on the Seismic Frontier," 503–507; "Earthquake Philosophy, and Earthquake-Proof Building," *Mining and Scientific Press* 17 (1868): 296; J. H. White, "How Brick Houses May be Built Earthquake-Proof," ibid., 18 (1869): 26–27; David Farquharson, "Architecture Adapted to Earthquakes," *Daily Alta California,* Feb. 19, 1869, p. 1; Stephen Tobriner, "A History of Reinforced Masonry Construction Designed to Resist Earthquakes: 1755–1907," *Earthquake Spectra* 1 (1984): 125–149, on 130–131; idem, "Bond Iron and the Birth of Anti-Seismic Reinforced Masonry Construction in San Francisco," *Masonry Society Journal* 5 (1986): G12–G18. The use of iron bolts in brick walls was also the major engineering response to the earthquake in Charleston; see Stockton, *The Great Shock,* chap. 8.

22. Clark C. Spence, *Mining Engineers and the American West: The Lace-Boot Brigade, 1849–1933* (New Haven: Yale UP, 1970); Richard Michael Levy, "The Professionalization of American Architects and Civil Engineers, 1865–1917" (Ph.D. diss., University of California, Berkeley, 1980), chap. 1; Carroll W. Pursell, Jr., "The Technical Society of the Pacific Coast, 1884–1914," *Technology and Culture* 17 (1976): 702–717.

23. "Pluck and Push," *Winters Express,* Apr. 23, 1892, quoted in Richard Cowen, Janice Cooper, and Richard Cooper, *The Quakes of '92: The History and Geological Background of the Vacaville-Winters Earthquakes of 1892* (Davis: Seismic Press, 1992), 24; "Forward, March!" *Vacaville Reporter,* Apr. 28, 1892, quoted in ibid., 29.

24. "Some Facts About Earthquakes," *Daily Alta California,* Oct. 24, 1868, p. 2 (quote); untitled editorial, ibid., Oct. 28, 1868, p. 2; ibid., Nov. 3, 1868, p. 2; ibid., Dec. 4, 1868, p. 2; "Editorial Notes," ibid., Dec. 20, 1868, p. 2.

25. "The Great Earthquake of 1868," *Daily Alta California,* Oct. 22, 1868, p. 1; "Earthquake Damages," ibid., Nov. 13, 1868, p. 1.

26. "After the Event," *SFC,* Apr. 21, 1892, quoted in Cowen et al., *The Quakes of '92,* 37; *Sacramento Record-Union,* Apr. 20, 1892, quoted in ibid., 39.

27. Leviton and Aldrich, "John Boardman Trask," 54.

28. Holden, *List of Recorded Earthquakes,* 18; repeated in idem, "Earthquakes in California and Elsewhere," *Overland Monthly* 11 (1888): 49; and idem, *A Catalogue of Earthquakes,* 30.

29. "No Cause for Alarm in Recent Earthquakes," *Daily Californian* [University of California student newspaper], Dec. 12, 1904, p. 1.

Chapter 1

1. Andrew C. Lawson, ed., *The California Earthquake of April 18, 1906: Report of the State Earthquake Investigation Commission,* 2 vols. (Washington, DC: Carnegie Institution of Washington, 1908–1910); Judd Kahn, *Imperial San Francisco: Politics and Planning in an American City, 1897-1906* (Lincoln: U of Nebraska P, 1979), chap. 6; Gladys Hansen and Emmet Condon, *Denial of Disaster* (San Francisco: Cameron and Co., 1989).

2. The reaction in 1906 has already been noted by, among others, Arnold J. Meltsner, "The Communication of Scientific Information to the Wider Public: The Case of Seismology in California," *Minerva* 17 (1979): 331–354, on 333–337; Hansen and Condon, *Denial of Disaster,* 107–111; and James C. Williams, "Earthquake Engineering: Designing Unseen Technology against Invisible Forces," *ICON: Journal of the International Committee for the History of Technology* 1 (1995): 172–194, on 179–181.

3. For a particularly lurid account that employed faked pictures of supposedly destroyed buildings, see James Russel Wilson, *San Francisco's Horror of Earthquake and Fire: Terrible Devastation and Heart-Rending Scenes* (n.p., 1906), esp. illustrations following 128. Such sensation mongers are referred to in, for example, "Small Loss of Life," *SFC,* May 1, 1906, p. 6; "Slandering the State," *SFC,* May 10, 1906, p. 6; "Leave Us Something," *A&EC* 5, no. 3 (July 1906): 74–75; *San Francisco Imperishable,* pamphlet issued by General Passenger Department, Southern Pacific Company, copy in Huntington Library, San Marino, CA.

4. Hansen and Condon, *Denial of Disaster,* 124–127; see also "The Insurance Situation," *SFC,* May 6, 1906, p. 18; "In the Matter of Certain Insurance Companies," *SFE,* May 7, 1906, p. 20; "Blaze Started 399 Hayes Street," *SFE,* May 23, 1906, p. 3; "Questions Have a Bad Look," *SFC,* June 5, 1906, p. 16; "Agents Are Busy Looking for Earthquake Damage," *SFC,* June 22, 1906, p. 1; "To Take Views to Germany," *SFC,* Oct. 17, 1906, p. 12; "Will Lay Fire to Earthquake," *SFC,* Jan. 10, 1907, p. 7.

5. George C. Pardee to Curtis Guild, Jr., Apr. 19, 1906 (first quote), and Sacramento Valley Development Association press release, Apr. 21, 1906 (second quote), both in George Pardee Papers, Bancroft Library, Berkeley, CA, box 30.

6. "Insurance Situation in San Francisco Is Question of the Hour," *SEB,* Apr. 30, 1906, p. 1 (first quote); Horsburgh quoted in Hansen and Condon, *Denial of Disaster,* 108–109.

7. First quote from *San Francisco Imperishable;* second quote from "An Earthquake Advertisement That Isn't Needed," *SFE,* June 6, 1906, p. 14.

8. "Earthquake Will Not Recur," *SEB,* Apr. 22, 1906, p. 1; "Danger of Another Heavy Shock Gone," *SFC,* Apr. 24, 1906, p. 8; "Stability of the Earth," ibid., May 2, 1906, p. 6; "Tall Buildings and Shortsighted Restrictions," *SFE,* May 6, 1906, p. 16.

9. "Fire Losses and Earthquake Losses," *LAT,* Apr. 20, 1906, II:4.

10. "The Torrid East," *SFC,* July 2, 1906, p. 6 (first quote); "The Worst Convulsion," *LAT,* Apr. 25, 1906, II:4; "The Future of the City," *SFC,* May 2, 1906, p. 6 (second quote).

11. "Bad Building in This City," *SFC*, May 29, 1906, p. 6 (first quote); S. B. Christy, "Some Lessons From the Earthquake," in *After Earthquake and Fire: A Reprint of the Articles and Editorial Comment Appearing in the Mining and Scientific Press Immediately After the Disaster at San Francisco, April 18, 1906* (San Francisco: Mining and Scientific Press, 1906), 35–40, second quote on 36; [T. A. Rickard], "After the Disaster," ibid., 78–84, third quote on 81–82; "Honest Construction Work," *SFC*, May 8, 1906, p. 6 (fourth quote).

12. Christine Meisner Rosen, *The Limits of Power: Great Fires and the Process of City Growth in America* (Cambridge: Cambridge UP, 1986), 92, 177–179, 249; A. L. Todd, *A Spark Lighted in Portland: The Record of the National Board of Fire Underwriters* (New York: McGraw-Hill, 1966), 22, 24, 41; F. W. Fitzpatrick, "The Red Monster," *A&EC* 7, no. 3 (1907): 57. All dollar figures are given in terms of actual rather than constant (inflation-adjusted) dollars.

13. Todd, *A Spark Lighted*, 24–29, 33–45, 143, quote on 40; Everett U. Crosby, "Fire Prevention," *Annals of the American Academy of Political and Social Science* 27 (1905): 404–418; "Some Water Supply and Fire Protection Lessons From the San Francisco Earthquake and Fire," *EN* 55 (1906): 471.

14. Quoted in John Stephen Sewell, "The Effects of the Earthquake and Fire on Buildings, Engineering Structures, and Structural Materials," in Grove Karl Gilbert, Richard Lewis Humphreys, John Stephen Sewell, and Frank Soule, *The San Francisco Earthquake and Fire of April 18, 1906 and Their Effects on Structures and Structural Materials*, USGS Bulletin 324 (Washington, DC: GPO, 1907), 64.

15. "The San Francisco Disaster: Earthquake and Fire Ruin in the Bay Counties of California," *EN* 55 (1906): 478–480, on 478 (first quote); "Some Water Supply and Fire Protection Lessons," 472 (second quote); "Water Supply, Fire Protection and Conflagration Hazard at San Francisco, Cal.," *EN* 55 (1906): 474–476; J. L. Van Ornum, "Some Engineering Lessons of the San Francisco Disaster," *Journal of the Association of Engineering Societies* 38 (1907): 147–151, on 147 (third quote).

16. A. L. A. Himmelwright, *The San Francisco Earthquake and Fire* (New York: Roebling Construction Co., 1906), esp. 264–265; S. Albert Reed, *The San Francisco Conflagration of April, 1906: Special Report to the National Board of Fire Underwriters Committee of Twenty* (n.p., 1906), 22–25; Sewell, "The Effects of the Earthquake and Fire," 119–124; Van Ornum, "Some Engineering Lessons," 148–151.

17. "Informal Discussion, Exterior and Floor Openings," *ESABR* 1, no. 2 (1906): (unpaginated); "Metal Sash and Window Frames and How They Saved a Building," ibid., no. 4: 16; A. McBean, "Exterior and Interior Openings and Partitions," ibid., no. 5: 11–12; W. W. Thurston, "Exterior and Floor Openings," ibid., no. 6: 5–6; "Transcript of Notes, Meeting of July 5, 1906," ibid., no. 6: 7–9; Ralph R. Reed, "The Problem of Fireproofing Openings," ibid., no. 11: 8–11; "Transcript of Notes, Meeting of August 16, 1906," ibid., no. 12: 8–11; "Transcript of Notes, Meeting of September 20, 1906," ibid., 2, no. 3 (1906): 11–12.

18. F. H. Meyer, "Fireproofing of the Structural Parts or Skeleton Frame of a First-Class Building," *ESABR* 1, no. 3 (1906): 7–9; E. V. Johnson, "The Fireproofing Problem,

With Special Reference to the Fireproofing of Columns," ibid., no. 4: 5-6; "Notes on the Fireproofing of Columns," ibid.: 7-10; George J. Wellington, "Fire Protection For Buildings; Such As Automatic Sprinklers, Roof Tanks, Cellar Reservoirs, Etc.," ibid., 2, no. 1 (1906): 5-6; H. S. Tittle, "Advantages of Conduit Systems Over Knob and Bushing Work; and Circuit Breakers on Service Boards Over Enclosed Fuses," ibid., no. 5: 8-10; F. H. Meyer, "Report of the Committee on Fireproofing of Steel, Partitions and Interior Finish," ibid., no. 8: 5-13, and no. 9: 5-10.

19. "Local Station for Fire Tests," *SFC,* June 24, 1906, p. 36; "Executive and Editing Committee Meeting of June 30, 1906," *ESABR* 1, no. 3 (1906): 15; C. B. Wing, Jr., and C. Derleth, "The Fire Testing Station," ibid., 2, no. 4 (1906): 10.

20. Luther Wagoner, "Safe Structural Treatment of Building Fronts," *ESABR* 1, no. 10 (1906):9-12, on 10 (first quote); "Transcript of Notes, Meeting of August 16, 1906," ibid., no. 12: 8-11, on 9 (second quote).

21. "New Year's Announcement," *A&EC* 7, no. 2 (1906): 84-85. Fitzpatrick had earlier reported on the San Francisco fire for the magazine; see F. W. Fitzpatrick, "Notes on the San Francisco Fire," *A&EC* 5, no. 2 (1906): 35-42.

22. F. W. Fitzpatrick, "Architects Must Do Better," *A&EC* 6, no. 3 (1906): 23-25; "Wanted: Fire Insurance," ibid.: 29; F. W. Fitzpatrick, "The Municipal Encouragement of Better Building," ibid., 7, no. 1 (1906): 26-28; idem, "The Red Monster," ibid., no. 3 (1907); idem, "Fears Another Wooden City," ibid.: 77; "More Fire-Prevention Needed," ibid., 8, no. 2 (1907): 95; "The Fire Insurance Habit," ibid., no. 3: 90; "The Cities in Dangerous Condition," ibid.: 91; "Better Fire-Prevention," ibid., 9, no. 1 (1907): 62; "Less Water nd Safer Buildings," ibid.: 88; F. W. Fitzpatrick, "The Confessions of an Architectural Reformer," ibid., no. 2: 64-67; "A Plea for Better Building Inspection," ibid.: 81; F. W. Fitzpatrick, "The Folly of Fire," ibid., no. 3: 86; "Unprotected Windows the Gateway of Destruction by Fire," ibid.: 98.

23. Reed, *The San Francisco Conflagration,* 4-11; Sewell, "The Effects of the Earthquake and Fire," 66-67; Hansen and Condon, *Denial of Disaster,* 49-53, 64.

24. Reed, *The San Francisco Conflagration,* 27-28; Hermann Schussler, *The Water Supply of San Francisco, California Before, During and After the Earthquake of April 18th, 1906, and the Subsequent Conflagration* (n.p., 1906), 8, 34; "No Decision Reached Regarding Fire Limits," *SFC,* May 12, 1906, p. 9; "Reports City Should Own Water Supply," ibid., May 29, 1906, p. 9; "Salt Water Discussed by Down-Town Owners," ibid., June 13, 1906, p. 8; Wm. H. Hall, "Municipal Engineering—Its Relation to City Building," *A&EC* 7, no. 2 (1906): 23; Richard L. Humphreys, "The Effects of the Earthquake and Fire on Various Structures and Structural Materials," in Gilbert et al., *San Francisco Earthquake and Fire,* 55; Frank Soule, "The Earthquake and Fire and Their Effects on Structural Steel and Steel-Frame Buildings," in ibid., 153; Sewell, "Effects of the Earthquake and Fire," 117-118.

25. L. B. Cheminant, "High-Pressure Fire Protection in San Francisco," *BSSA,* 19 (1929): 80-85, on pp. 80-81.

26. Andrew C. Lawson to T. W. Ransom, May 11, and June 17, 1907, both in ALP, box 15, folder "Apr.-June 1907."

27. "Some Water Supply and Fire Protection Lessons."

28. [T. A. Rickard], "The Earthquake," in *After Earthquake and Fire*, 26-27; "Report by Prof. C. Derleth, Jr.," *EN* 55 (1906): 503-504; Charles Derleth, Jr., "The Destructive Extent of the San Francisco Earthquake of 1906," ibid.: 707-713, on 709; Wm. Ham Hall, "Some Lessons of the Earthquake and Fire, I: Reminiscences of the City's Site as Accounting for Earthquake Effects," *SFC*, May 19, 1906, p. 3; idem, "Some Lessons of the Earthquake and Fire, II: A Record of Earthquake Disturbances," ibid., May 20, 1906, p. 32; Frank Soule, "Foundations as Affected by Earthquakes," *ESABR* 1, no. 7 (1906), 5-8, on 8; Octavius Morgan, "A Los Angeles Architect's Impressions of the San Francisco Earthquake and Fire," *A&EC* 5, no. 1 (1906), unpaginated; William Ham. Hall, "Earthquakes and Earth Faults," ibid., 7, no. 3 (1907): 20-25, on 23-24; Reed, *The San Francisco Conflagration*, 1-2; Humphreys, "The Effects of the Earthquake and Fire," 15, 21, 25-26; Sewell, "The Effects of the Earthquake and Fire," 118; Soule, "The Earthquake and Fire," 133-136.

29. Himmelwright, *The San Francisco Earthquake and Fire*, 268-269; Reed, *The San Francisco Conflagration*, 2; C. Derleth Jr., "A Supplementary Report on the San Francisco Buildings by Prof. Derleth," *EN* 55 (1906): 525-526; Wm. Ham Hall, "Some Lessons of the Earthquake and Fire: III. Integrity of Buildings Under Earthquake," *SFC*, May 24, 1906, p. 14; "William Sooy Smith on Building Problem," ibid., June 5, 1906, p. 10; "Omori Advises Good Foundation," ibid., Aug. 6, 1906, p. 5; William F. Scott, "Earthquakes and Earthquake Proof-Buildings [*sic*]," *A&EC* 5, no. 2 (1906): 33-34, on 34; F. W. Fitzpatrick, "Notes on the San Francisco Fire," 41; Robert Morgeneier, "Cause of Earthquake Shock and Its Effect on Building Foundations," *A&EC* 8, no. 1 (1907): 35-39, on 38-39; Soule, "The Earthquake and Fire," 143; idem, "Foundations as Affected by Earthquakes"; Carl Uhlig, "The Foundation of the Union Depot and Ferry House of San Francisco," *ESABR* 1, no. 7 (1906): 8-10; L. M. Scofield and M. De Palo, "Notes on Foundations," ibid.: 10-12.

30. J. D. Galloway, "The Recent Earthquake in Central California and the Resulting Fire in San Francisco," *EN* 55 (1906): 523-525, on 524; idem, "The Possible Uses for Reinforced Concrete in Earthquake Proof, Fire Resisting Building Construction" (letter to editor), ibid.: 700; Fitzpatrick, "Notes on the San Francisco Fire," 37; "Effects of Fire and Earthquake," *SFC*, May 17, 1906, p. 11; Wm. Ham Hall, "Some Lessons of the Earthquake and Fire: IV. Effect of the Quake on Wooden and Steel Construction," ibid., May 28, 1906, p. 3; Scott, "Earthquakes and Earthquake Proof-Buildings"; Himmelwright, *The San Francisco Earthquake and Fire*, 247, 269; [C. Derleth, Jr.], "Illustrations: Column Design," *ESABR* 1, no. 4 (1906), p. 10; Sewell, "Effects of the Earthquake and Fire," 124-125; Soule, "Earthquake and Fire," 143-144; J. D. Galloway, M. C. Couchot, C. H. Snyder, Charles Derleth, Jr., and C. B. Wing, "Report of Committee on Fire and Earthquake Damage to Buildings," *Transactions of the American Society of Civil Engineers* 59 (1907): 223-244, on 232; "Strength is Feature of Building Laws," *SFC*, May 18, 1906, p. 9. For the development of wind bracing of steel frame buildings, see Carl W. Condit, "The Wind Bracing of Buildings," *Scientific American* 230, no. 2 (1974): 92-105, esp. 98-99.

31. Carl W. Condit, *American Building: Materials and Techniques from the First Colonial Settlements to the Present,* 2nd ed. (Chicago: U of Chicago P, 1982), chaps. 14, 18; John W. Snyder, "Buildings and Bridges for the Twentieth Century," *California History* 63 (1984): 280–292, on 281–284; Richard Michael Levy, "The Professionalization of American Architects and Civil Engineers, 1865–1917" (Ph.D. diss., University of California, Berkeley, 1980), chap. 1; "Report by Prof. C. Derleth, Jr.," 503; "Effect of the San Francisco Earthquake on a Reinforced Concrete Building," *EN* 55 (1906): 520; W. P. Day, "Concrete To Be a Factor in Rebuilding," *A&EC* 5, no. 3 (1906): 75.

32. Charles W. Dickey, "Lessons of the San Francisco Earthquake and Fire," *A&EC* 5, no. 1 (1906): unpaginated.

33. Maurice C. Couchot, "Reinforced Concrete and Fireproof Construction in the San Francisco Disaster," *EN* 55 (1906): 622–623, on 622 (quote); "Injuries by Earthquake to the Buildings of Stanford University," *EN* 55 (1906): 509–510; Wm. B. Gester, "Reinforced Concrete in the Earthquake," *A&EC* 5, no. 1 (1906): unpaginated.

34. John B. Leonard, "How Reinforced Concrete Stood Earthquake and Fire," *A&EC* 5, no. 1 (1906): unpaginated.

35. Wm. Ham Hall, "Some Lessons of the Earthquake and Fire: VI. Stone and Concrete Construction," *SFC*, June 1, 1906, p. 15. For Hall's work on behalf of reinforced concrete, see W. H. Hall, undated and untitled typescript [written ca. 1907], William Hammond Hall papers, Bancroft Library, Berkeley, CA, carton 10, folder 24.

36. "Brickmaker's Opposition to Re-enforced Concrete," *A&EC* 5, no. 2 (1906): 68; "Defends Use of Concrete," *SFC*, June 12, 1906, p. 9.

37. "Reinforced Concrete in the New San Francisco Building Law," *EN* 56 (1906): 96–97; Wm. Ham. Hall, "The Rebuilding of San Francisco-Reinforced Concrete Buildings," *A&EC* 9, no. 3 (1907): 61–67.

38. Himmelwright, *The San Francisco Earthquake and Fire,* 261–262.

39. "Brick In San Francisco," *A&EC* 5, no. 3 (1906): 60; W. F. Barnes, "Concrete From a Brick Man's Standpoint," ibid., 6, no. 2 (1906): 61–62; "Discussion," *Transactions of the American Society of Civil Engineers* 59 (1907): 264–329, on 264, 266, 282–283, 285, 286.

40. "Building Ordinance Before Supervisors," *SFC*, June 2, 1906, p. 16; "Blunders Delay the Building Ordinance," ibid., June 19, 1906, p. 8.

41. "Partial Collapse of the Bixby Hotel at Long Beach," *A&EC* 7, no. 1 (1906): 45; Joseph Simmons, "The Bixby Hotel Disaster as Viewed by a Brick Man," ibid., no. 2 (1906): 51–52.

42. Charles Derleth, Jr., "The Destructive Extent of the California Earthquake of 1906: Its Effect Upon Structures and Structural Materials Within the Earthquake Belt," in David Starr Jordan, ed., *The California Earthquake of 1906* (San Francisco: A. M. Robertson, 1907), 79–212, on 201 (quote); Charles Derleth, Jr., "Structural Lessons of the Earthquake Disturbance," *A&EC* 5, no. 1 (1906), unpaginated; idem, "Reinforced Concrete Construction: Its Proper Application in Earthquake Countries," *Journal of the Association of Engineering Societies* 38 (1907): 72–82, on 79.

43. Derleth, "Destructive Extent" (in *EN* 55 [1906]), 708.

44. "Report by Prof. C. Derleth Jr."; Derleth, "Supplementary Report"; idem, "Some Effects of the San Francisco Earthquake on Water-Works, Streets, Sewers, Car Tracks and Buildings," *EN* 55 (1906): 548–554; idem, "Structural Lessons of the Earthquake Disturbance"; idem, "Destructive Extent" (in *EN* 55 [1906]); idem, "Destructive Extent of the San Francisco Earthquake," *ESABR* 1, no. 15 (1906): 5–11; idem, "Reinforced Concrete Construction"; idem, "Destructive Extent" (in Jordan, *The California Earthquake*); [Derleth], "Concrete-Block Housing Construction," *ESABR* 1, no. 2 (1906), unpaginated; "Meeting of August 9, 1906," ibid., no. 9 (1906): 13–14.

45. Robert W. Gardner, "Reinforced Brick Work," *A&EC* 5, no. 2 (1906): 43; F. C. Davis, "Anchoring of Masonry," *ESABR* 1, no. 13 (1906): 6–9; Lewis A. Hicks, "Masonry Bond and Anchorage," ibid.: 5–6; W. J. Cuthbertson, "The New Building Law of San Francisco," *A&EC* 6, no. 1 (1906): 27–30, on 30.

46. W. W. Breite, "More First-Class Designs and Good Steel Construction Wanted," *A&EC* 5, no. 3 (1906): 63–64, on 64.

47. Edwin Duryea, Jr., "A Report by Edwin Duryea, Jr., on Earthquake Damages at San Francisco and Palo Alto," *EN* 55 (1906): 510–511, on 511.

48. H. L. Haehl, Otto von Geldern, Luther Wagoner, and C. E. Gilman, "Report of Committee on the Geology of the Earthquake," *Transactions of the American Society of Civil Engineers* 59 (1907): 216–222, on 221.

49. "Meeting of Architects," *SFC,* May 4, 1906, p. 12.

50. Morgan, "A Los Angeles Architect's Impressions"; Wm. Ham Hall, "Some Lessons of the Earthquake and Fire: V. Brick Construction," *SFC,* May 28, 1906, p. 3; Charles Peter Weeks, "Who Is to Blame for San Francisco's Plight," *A&EC* 5, no. 2 (1906): 28–29, quotes on 28.

51. "Meeting of Architects"; A. McBean, "Discussion on Terra Cotta," *ESABR* 2, no. 4 (1906): 5–6.

52. Lawson to Pardee, undated telegram, Pardee Papers, box 77 (quote); Pardee, "To Whom It May Concern," Apr. 21, 1906, ibid., box 30; J. C. Branner to R. S. Woodward, Apr. 24, 1906, and Branner to Lawson, Apr. 24, 1906, both in JBP, box 6, vol. 22, p. 462.

53. G. K. Gilbert to C. W. Hayes, May 7, 1906, Records of the USGS (RG 57), National Archives II, College Park, MD, entry 81, box 29, file 549-33. In later years, seismologists would claim that California's boosters had sought to stymie the commission's work by refusing to give it financial support; see, for example, Stephen Taber, "The Earthquake Problem in Southern California," *BSSA* 10 (1920): 276–289, on 287; J. C. Branner, "Earthquakes and Structural Engineering," ibid., 3 (1913): 1–5, on 3. I have not found, in the contemporary letters and memoranda of various commission members that I have examined, any evidence for such an attempt at suppression. On the contrary, as Gilbert's letter shows, commission members for very good reasons did not even seek support from the local business community.

54. Branner to Woodward, Apr. 24, 1906, JBP, box 6, vol. 22, p. 462; Lawson to Branner, Apr. 26, Apr. 28, and May 2, 1906, all in ibid., box 38, folder 176; Woodward to Branner, June 4, 1906, ibid., box 39, folder 183; Gilbert to Hayes, May 7, 1906,

Records of the USGS, entry 81, box 29, file 549-33; Lawson to James Phelan, June 9, 1906, ALP, box 15, folder "1905-1906"; Lawson to Woodward, Nov. 4, 1907, ibid., box 15, folder "July-Dec. 1907." For the establishment of the Carnegie Institution and its impact on the funding of science, see Howard S. Miller, *Dollars for Research: Science and Its Patrons in Nineteenth-Century America* (Seattle: U of Washington P, 1970), chap. 9, and Robert E. Kohler, *Partners in Science: Foundations and Natural Scientists, 1900-1945* (Chicago: U of Chicago P, 1991), chap. 2.

55. Lawson to Branner, Apr. 28, 1906, JBP, box 38, folder 176.

56. Andrew C. Lawson, "The Post-Pliocene Diastrophism of the Coast of Southern California," *Bulletin of the Department of Geology, University of California* 1 (1893): 115-160, on 149-151; F. M. Anderson, "The Geology of the Point Reyes Peninsula," ibid., 2 (1899): 119-153, on 143-145, 153; Vance C. Osmont, "A Geological Section of the Coast Ranges North of the Bay of San Francisco," ibid., 4 (1905): 39-87; Andrew C. Lawson, "Sketch of the Geology of the San Francisco Peninsula," in *Fifteenth Annual Report of the United States Geological Survey to the Secretary of the Interior 1893-94* (Washington, DC: GPO, 1895), 399-476, on 435-442, 460-463, and 465-468.

57. Branner to Lawson, Apr. 30, 1906, JBP, box 6, vol. 22, p. 464; Lawson to Branner, May 2, 1906, ibid., box 38, folder 176.

58. For early statements of this conclusion, see Branner to Professor Winchell, May 2, 1906, JBP, box 6, vol. 22, p. 467; Branner to R. D. Salisbury, May 5, 1906, ibid., box 6, vol. 22, pp. 479-480; Gilbert to Hayes, May 7, 1906, Records of the USGS, entry 81, box 29, file 549-33.

59. Minutes of the State Earthquake Investigation Commission, May 22, 1906, George Davidson Papers, Bancroft Library, Berkeley, CA, carton 17, folder "Earthquakes and Volcanoes 1"; Lawson to Davidson, June 7, 1906, ibid., box 15, folder "Lawson"; Andrew C. Lawson and A. O. Leuschner, *Preliminary Report of the State Earthquake Investigation Commission* (n.p., 1906).

60. Lawson to The Governor of California, Mar. 6, 1907, ALP, box 15, folder "Jan.-Mar. 1907"; Minutes of the State Earthquake Investigation Commission, June 20, 1906, Davidson Papers, carton 17, folder "Earthquakes and Volcanoes 1"; Andrew C. Lawson and A. O. Leuschner, "The California Earthquake," *Science* 23 (1906): 961-967; "Report of the State Earthquake Commission," in *After Earthquake and Fire*, 148-159.

61. "Split in Mountain Caused Earthquake," *SFE*, May 8, 1906, p. 16 (quote); A. S. Cooper, "Science Tells the Story of Shock," *SFC*, May 1, 1906, p. 12; "Studies of the Big Fault," *SFC*, May 27, 1906, p. 14; W. H. Storms, "Earthquake Lines," in *After Earthquake and Fire*, 86-90; G. K. Gilbert, "The Investigation of the San Francisco Earthquake," *Popular Science Monthly* 69 (1906): 97-115; David Starr Jordan, "The Earthquake Rift of 1906," ibid.: 289-309; J. C. Branner, "Earthquake of the 18th," *Sierra Educational News and Book Review* 3, no. 3 (1907): 12; Stephen Taber, "Some Local Effects of the San Francisco Earthquake," *Journal of Geology* 14 (1906): 303-315; Derleth, "Some Effects of the San Francisco Earthquake"; idem, "Destructive Extent" (in *EN* 55 [1906]); H. L. Haehl et al., "Report of Committee on the Geology of the Earthquake."

62. Gilbert, "Investigation of the San Francisco Earthquake," 114-115.

63. Lawson to Gilbert, June 21, 1906, ALP, box 15, folder "1905-1906"; Gilbert to Lawson, Sept. 8, 1906, and Jan. 15, 1907, both ibid., box 5, folder "Gilbert."

64. "To Study Big Earthquake," *SFC*, May 16, 1906, p. 16; "Japanese Expert Studies Quake," ibid., May 23, 1906, p. 8; F. Omori, "Preliminary Note on the Cause of the California Earthquake of 1906," in Jordan, *The California Earthquake*, 283; Minutes of the State Earthquake Investigation Commission, May 22, 1906, Davidson Papers, carton 17, folder "Earthquakes and Volcanoes 1." For biographical information on Omori, see Charles Davison, "Fusakichi Omori and His Work on Earthquakes," *BSSA* 14 (1924): 240-255.

65. "[Omori] Declares There Is No Danger of Severe Shock in Near Future," *SFC*, June 6, 1906, p. 8.

66. F. Omori, "Observations of Distant Earthquakes," in *After Earthquake and Fire*, 140-147; "Gives Explanation of Earth Tremors," *SFC*, June 11, 1906, p. 2; "As to Earthquakes," ibid., June 11, 1906, p. 6 (quote); see also F. Omori, "One Big Shake Insures Us There'll Not Be Another," *SFE*, June 14, 1906, p. 16.

67. Quotes from "No More Big Quakes for at Least Fifty Years," *SFC*, June 19, 1906, p. 14, and "No More Shocks for California," ibid., Aug. 2, 1906, p. 6; see also "Zone of Greatest Shock Located," ibid., July 23, 1906, p. 2; F. Omori, "Omori Says Pacific Coast Has No Reason to Expect More Earthquakes," *Sunset Magazine* 17 (1906): 240.

68. Alexander G. McAdie, "The Scientific Side of It," *Sunset Magazine* 17 (1906): 42-44, on 44 (first quote); "Talks of Cause of Earthquakes," *SFC*, Nov. 16, 1906, p. 9 (second quote).

69. Geo. H. Ashley, "The Geological Prelude to the San Francisco Earthquake," *Popular Science Monthly* 69 (1906): 69-75, on 75 (quote). For biographical information on Ashley, see S. H. Catchcart, "George Hall Ashley (1866-1951)," *Bulletin of the American Association of Petroleum Geologists* 36 (1952): 536-538.

70. Lawson to F. M. Clarke, Aug. 20, 1907, ALP, box 15, folder "July-Dec. 1907"; Lawson to Wm. H. Hobbs, Oct. 19, 1908, ibid., box 15, folder "1908."

71. Lawson, *The California Earthquake*, I:5-151. For a good assessment of this report within the context of the history of seismological theory, see Dennis R. Dean, "The San Francisco Earthquake of 1906," *Annals of Science* 50 (1993): 501-521, on 517-521.

72. Lawson, *The California Earthquake*, I:152-366, 434-451. The figure of 300 observers is mentioned in Lawson to The President and the Board of Trustees of the Carnegie Institution of Washington, May 1, 1907, ALP, box 15, folder "Apr.-June 1907," and Lawson to Chas. S. Fee, Nov. 8, 1907, ibid., box 15, folder "July-Dec. 1907."

73. Lawson, *The California Earthquake*, vol. II. For the elastic rebound theory, see also Harry Fielding Reid, "The Elastic-Rebound Theory of Earthquakes," *Bulletin of the Department of Geology, University of California*, 6 (1911): 413-443.

74. John F. Hayford and A. L. Baldwin, "The Earth Movements in the California Earthquake of 1906," in *Report of the Superintendent of the Coast and Geodetic Survey Showing the Progress of the Work From July 1, 1906, to June 30, 1907* (Washington, DC: GPO, 1907), 67-104, esp. 94-95. Hayford and Baldwin's report was reprinted in Lawson, *The California Earthquake*, I:114-145.

75. Lawson, *The California Earthquake,* II:16–28, 31–32.

76. Ibid., I:19.

77. Minutes of the State Earthquake Investigation Commission, July 11, 1906, Davidson Papers, carton 17, folder "Earthquakes and Volcanoes 1."

78. Harry Fielding Reid, A. O. Leuschner, and C. F. Marvin, "Report of the Committee on a State Seismological Survey, and the Type of Seismographs to be Used," Aug. 7, 1906; C. F. Marvin, "Notes on Seismological Instruments and the Organization of a State Seismological Survey," undated; and Minutes of the State Earthquake Investigation Commission, Aug. 7, 1906, all in Davidson Papers, carton 17, folder "Earthquakes and Volcanoes I."

79. McAdie to Branner, Aug. 22, 1906, JBP, box 39, folder 178; McAdie to Davidson, Sept. 4, 1906, Davidson Papers, box 16, folder "McAdie"; Perry Byerly, "History of the Seismological Society of America," *BSSA* 54 (1964): 1723-1741, on 1723-1724, 1731–1734; "Earthquake Society Organizes for Work," *Daily Californian,* Nov. 22, 1906, p. 3; Louderback to Branner, Dec. 31, 1906, JBP, box 38, folder 177; Louderback to Lawson, June 23, 1907, ALP, box 12, folder "Seismological Society"; McAdie to Marvin, May 27, 1907, Records of the U.S. Weather Bureau (RG 27), National Archives II, College Park, MD, entry 47 (NC3), box 2064, file 3222.

80. On the social makeup of California environmental reform groups, see Susan R. Schrepfer, *The Fight to Save the Redwoods: A History of Environmental Reform 1917-1978* (Madison: The U of Wisconsin P, 1983), 10, 13–17.

81. For Muir and Olney, see Holway R. Jones, *John Muir and the Sierra Club: The Battle for Yosemite* (San Francisco: Sierra Club, 1965), chap. 1; on the involvement of McAdie and Davidson in the Sierra Club, see Michael L. Smith, *Pacific Visions: California Scientists and the Environment, 1850-1915* (New Haven: Yale UP, 1987), 144–148.

82. Jones, *John Muir,* fig. 44; Joseph N. LeConte to S. D. Townley, Jan. 14, 1910 [*sic,* 1911], SSAC, carton 1, folder "Correspondence 1911–1912."

83. Stephen J. Pyne, *Grove Karl Gilbert: A Great Engine of Research* (Austin: U of Texas P, 1980), esp. chap. 6.

84. G. K. Gilbert, "Earthquake Forecasts," *Science* 29 (1909): 121–138, quotes on 133.

85. Ibid., 134.

86. Ibid., 136–138, quotes on 138.

Chapter 2

1. Stephen J. Pyne, *Grove Karl Gilbert: A Great Engine of Research* (Austin: U of Texas P, 1980), 214–215.

2. A. O. Leuschner to Andrew C. Lawson, Feb. 28, 1907, ALP, box 9, folder "Leuschner"; Lawson to Leuschner, Mar. 12, 1907, ibid., box 15, folder "Jan.-Mar. 1907"; Lawson to William H. Hobbs, Dec. 10, 1907, ibid., box 15, folder "July-Dec. 1907"; George D. Louderback to John C. Branner, Mar. 16, 1907, JBP, box 38, folder 177; Alexander McAdie to Branner, Aug. 26, 1908, ibid., box 41, folder 193; Louderback to Branner, Oct. 2, and Dec. 5, 1910, ibid., box 42, folder 204; Hobbs to Branner, Feb. 8, 1911

(quote), ibid., box 44, folder 212; Harry Fielding Reid to Branner, Sept. 14, 1911, ibid., box 44, folder 215.

3. Sidney D. Townley, "John Casper Branner," *BSSA* 12 (1922): 1–11; R. A. F. Penrose, Jr., "Memorial to John Casper Branner," *Bulletin of the Geological Society of America* 36 (1925): 15–44; [Branner], "Chronology of John C. Branner 1850–1920," J. C. Branner Collection, Huntington Library, San Marino, CA, box 2; George H. Nash, *Herbert Hoover and Stanford University* (Stanford: Hoover Institution Press, 1988), chap. 2; Branner to T. C. Chamberlin, Aug. 15, 1910, JBP, box 8, vol. 29, p. 345; Michael L. Smith, *Pacific Visions: California Scientists and the Environment, 1850–1915* (New Haven: Yale UP, 1987), 145.

4. J. C. Branner, "Impressions Regarding the Relations of Surface Geology to Intensity in the Mendoza, Valparaiso, Kingston and San Francisco Earthquakes," *BSSA* 1 (1911): 38–43; Branner to Hobbs, May 11, 1908, JBP, box 7, vol. 24, p. 20.

5. Branner to George D. Louderback, Jan. 2, 1907, JBP, box 6, vol. 23, p. 323; Branner to Lawson, Apr. 5, 1907, ibid., p. 476; Branner to Louderback, June 5, 1908, ibid., box 7, vol. 24, p. 80; Branner to McAdie, Dec. 24, 1908, ibid., vol. 25, p. 267; Branner to McAdie, Jan. 16, 1909, ibid., vol. 26, pp. 45–46.

6. J. C. Branner, "Suggested Organization for Seismologic Work on the Pacific Coast," *BSSA* 1 (1911): 5–8, quotes on 5, 6, and 7.

7. "Control of Quakes Is Discussed," *SFE*, Dec. 14, 1913, p. 37.

8. John Casper Branner, "One of the Scientific Problems at Our Doors," *BSSA* 7 (1917): 45–48, quote on 47.

9. John C. Branner, "Earthquakes and Structural Engineering," *BSSA* 3 (1913): 1–5, quotes on 2, 3.

10. The earliest available list of Seismological Society members is "Members of the Seismological Society of America, Dec. 29, 1911," *BSSA* 1 (1911): 191–199, which lists 362 members. By subtracting the names of those added as members during various meetings of the society's board of directors in 1911 [see ibid., 1 (1911): 31, 92–94, 135, 190; 2 (1912): 103], one can obtain a list of those who were members before Branner's extensive membership campaign.

11. Louderback to Branner, Sept. 19, 1910, JBP, box 42, folder 204; Branner to Solon Shedd, Oct. 15, 1910, ibid., box 9, vol. 30, p. 109; Branner to George Davidson, Sept. 18, 1911, ibid., vol. 31, p. 285; Branner to Clarence B. Osborne, Sept. 19, 1911, ibid., p. 300; Branner to Gordon Surr, Oct. 2, 1911, ibid., p. 340; Branner to J. S. Rossiter, Jan. 8, 1912, ibid., box 10, vol. 32, p. 22; Townley to Reid, Apr. 29, 1912, SSAC, carton 1, folder "Correspondence 1912–1914"; S. D. Townley, "Secretary's Report From Dec. 1910 to Apr. 6, 1912," *BSSA* 2 (1912): 148. On Moffitt, Keesling, and Grant, see William Issel and Robert W. Cherny, *San Francisco, 1865–1932: Politics, Power, and Urban Development* (Berkeley: U of California P, 1986), 36, 46–47, 163; Susan R. Schrepfer, *The Fight to Save the Redwoods: A History of Environmental Reform, 1917–1978* (Madison: U of Wisconsin P, 1983), 13, 21.

12. S. D. Townley, "Minutes of the Meeting of the Board of Directors of the Seismological Society of America, Dec. 15, 1910," *BSSA* 1 (1911): 31; D. L. Webster, J. V. Uspen-

sky, and James W. McBain, "Memorial Resolution: Sidney Dean Townley, 1867–1946," School of Engineering Records (SC 165), Department of Special Collections, Stanford University Libraries, ser. 1, box 9, folder 26; Branner to J. B. Woodworth, Dec. 27, 1910, JBP, box 9, vol. 30, pp. 266–267; Branner to Louderback, Oct. 25, 1910, ibid., p. 131; Branner to Reid, Sept. 19, 1911, ibid., vol. 31, p. 298; Woodworth to Branner, Mar. 22, Sept. 20, Nov. 27, and Thanksgiving Day, 1911, all in ibid., box 44, folder 216; "Mr. Sayles' Gift," *BSSA* 1 (1911): 173.

13. S. D. Townley, "Minutes of the Meeting of the Board of Directors of the Seismological Society of America, Apr. 6, 1912," *BSSA* 2 (1912): 150–151, on 150; Perry Byerly, "History of the Seismological Society of America," ibid., 54 (1964): 1723–1741, on 1738; Townley to N. H. Heck, Nov. 23, 1927, SSAC, carton 1, folder "Correspondence July–Dec. 1927"; Reid to Townley, May 14, 1913, ibid., carton 1, folder "Correspondence 1912–1914"; McAdie to Townley, May 30, 1915, ibid., carton 1, folder "Correspondence 1914–1916."

14. R. S. Holway, "Report of the Board of Election of the Seismological Society of America," *BSSA* 8 (1918): 100; Townley to Lawson, May 10, 1917, and Louderback to Townley, Jan. 23, 1918, both in SSAC, carton 1, folder "Correspondence 1916–1918"; Lawson to Townley, Apr. 9, 1918, Townley to W. W. Campbell, Jan. 21, 1919, and Townley to Louderback, Jan. 30, 1919, all in ibid., carton 1, folder "Correspondence 1918–1920." The lack of a proper quorum at these meetings would later come to haunt the society; see Bailey Willis to Charles P. Berkey, June 6, 1928, BWP, box 24, folder 59.

15. H. O. Wood, "The Observation of Earthquakes: A Guide for the General Observer," *BSSA* 1 (1911): 48–82; "Reports of Earthquakes," ibid.: 29; "Seismological Notes," ibid.: 131–133; S. D. Townley, "Minutes of the Meeting of the Seismological Society of America, Apr. 6, 1912," ibid., 2 (1912): 147.

16. Branner to Stephen Taber, Sept. 28, 1911, JBP, box 9, vol. 31, p. 324; Branner to Campbell, Aug. 21, 1911, ibid., p. 156; Branner to Harry O. Wood, Sept. 13, 1911, ibid., p. 275; Branner to Reid, Nov. 22, 1911, ibid., p. 554; "Notes on the California Earthquake of July 1, 1911," *BSSA* 1 (1911): 110–121; E. C. Templeton, "The Central California Earthquake of July 1, 1911," ibid.: 167–169; H. O. Wood, "On the Region of Origin of the Central Californian Earthquakes of July, August and September, 1911," ibid., 2 (1912): 31–39, on 38–39.

17. Branner to Robert W. Sayles, Jan. 6, 1913, JBP, box 10, vol. 34, p. 10; Branner to Lawrence Martin, Jan. 8, 1913, ibid., p. 22; Branner, "Earthquakes and Structural Engineering," 4; "Control of Quakes Is Discussed." For more on the Weather Bureau's efforts to become involved in seismology, see Carl-Henry Geschwind, "Embracing Science and Research: Early Twentieth-Century Jesuits and Seismology in the United States," *Isis* 89 (1998): 27–49, on 38–41.

18. C. F. Marvin to Branner, May 7, 1914, Records of the U.S. Weather Bureau (RG 27), National Archives II, College Park, MD, entry 50, box 2456, file "Branner"; C. F. Marvin, "Resumption of Seismological Work," July 1, 1914, ibid., entry 50, box 2458, file "Washington"; "Earthquakes and Instructions for their Noninstrumental Observation," *Nashville Tennessean and American,* Jan. 24, 1915, clipping in ibid., entry 50, box

2456, file "Nashville"; Marvin to Branner, Mar. 11, 1915, ibid., entry 50, box 2457, file "Stanford"; Branner to Marvin, May 13, 1914, JBP, box 11, vol. 36, p. 310; Andrew H. Palmer to Branner, Jan. 11, 1915, ibid., box 53, folder 260; Andrew H. Palmer, "California Earthquakes During 1915," *BSSA* 6 (1916): 8-25; idem, "California Earthquakes During 1916," ibid., 7 (1917): 1-17, on 7-8.

19. Carl H. Beal, "The Earthquake in the Santa Cruz Mountains, California, Nov. 8, 1914," *BSSA* 4 (1914): 215-219. On Beal's status as Branner's student, see Branner to Beal, Sept. 15, 1914, JBP, box 11, vol. 37, pp. 245-246; Branner to Roderic Crandall, Jan. 11, 1915, ibid., box 12, vol. 38, p. 16.

20. Carl H. Beal, "The Earthquake at Los Alamos, Santa Barbara County, California, Jan. 11, 1915," *BSSA* 5 (1915): 14-25; Branner to Woodworth, Mar. 3, 1915 (quote), JBP, box 12, vol. 38, p. 132.

21. Carl H. Beal, "The Earthquake in the Imperial Valley, California, June 22, 1915," *BSSA* 5 (1915): 130-149; S. D. Townley, "Minutes of the Meeting of the Board of Directors of the Seismological Society of America, February 16, 1916," ibid., 6 (1916): 52.

22. Homer Hamlin to Branner, Nov. 9, Nov. 16, Dec. 19, and Dec. 29, 1916, all in JBP, box 54, folder 267; Hamlin to Branner, May 15, and June 18, 1917, both in ibid., box 56, folder 277; John Casper Branner, "The Tejon Pass Earthquake of Oct. 22, 1916," *BSSA* 7 (1917): 51-59; A. C. Mattei, "Two Santa Barbara Channel Earthquakes," ibid.: 61-66; Homer Hamlin, "Miscellaneous Earthquakes in Southern and Eastern California," ibid.: 113-118. On Hamlin, see "Hamlin, Homer," *The National Cyclopaedia of American Biography* 19 (1926): 307-308.

23. Ralph Arnold to Branner, Mar. 20, Mar. 30, and Aug. 23, 1917, JBP, box 55, folder 274. On Arnold, see Homer J. Steiny, "Ralph Arnold (1875-1961)," *Bulletin of the American Association of Petroleum Geologists* 45 (1961): 1897-1900.

24. Branner to Arnold, Apr. 23, 1918, JBP, box 13, vol. 44, p. 209; Townley to Rollin T. Chamberlin, Apr. 23, 1918, SSAC, carton 1, folder "Correspondence 1918-1920"; Sidney D. Townley, "The San Jacinto Earthquake of Apr. 21, 1918," *BSSA* 8 (1918): 45-62; idem, "Report of Secretary-Treasurer From April 6, 1918, to April 5, 1919," ibid., 9 (1919): 52-53.

25. Branner to Lawson, May 6, 1918, JBP, box 13, vol. 44, p. 233; Branner to Arnold, May 6, 1918, ibid., p. 234; Arnold to Branner, June 3, 1918, ibid., box 57, folder 282; Frank Rolfe and A. M. Strong, "The Earthquake of April 21, 1918, in the San Jacinto Mountains," *BSSA* 8 (1918): 63-67; Ralph Arnold, "Topography and Fault System of the Region of the San Jacinto Earthquake," ibid.: 68-73.

26. Stephen Taber, "The Inglewood Earthquake in Southern California, June 21, 1920," *BSSA* 10 (1920): 129-145, quote on 145. On Taber, see Branner to Ray Lyman Wilbur, July 2, 1920, Branner Collection (Huntington Library), box 1, letterbook 1920-1921; Stephen Taber, "The South Carolina Earthquake of January 1, 1913," *BSSA* 3 (1913): 6-13; idem, "Seismic Activity in the Atlantic Coastal Plain Near Charleston, South Carolina," ibid., 4 (1914): 108-160; idem, "Earthquakes in South Carolina During 1914," ibid., 5 (1915): 96-99; idem, "The Earthquake in the Southern Appalachians

Feb. 21, 1916," ibid., 6 (1916): 218–226; Harry Fielding Reid and Stephen Taber, "The Porto Rico Earthquakes of October–November, 1918," ibid., 9 (1919): 95–127.

27. Stephen Taber, "The Los Angeles Earthquakes of July, 1920," *BSSA* 11 (1921): 63–79, quote on 77–78.

28. Branner to Woodworth, Sept. 21, 1917 (first quote), JBP, box 13, vol. 43, p. 21; Branner to Woodworth, May 2, 1918, J. B. Woodworth Papers, General Correspondence (HUG 1880.305), Harvard University Archives, Cambridge, MA, box 5, folder "1918"; Branner to Robert T. Hill, July 31, 1920, Branner to S. B. Morris, Aug. 2, 1920, Branner to Spaulding, Aug. 4, 1920 (second quote), and Branner to Homer Laughlin, Jr., Aug. 6, 1920, all in Branner Collection (Huntington Library), box 1, letterbook 1920–1921; Branner to Arnold, Aug. 24, 1920, and Hill to Arnold, Aug. 29, 1920, both in RAP, box 234, folder 1.

29. Arnold to Branner, Sept. 8, 1920, RAP, box 234, folder 1; Robert T. Hill, "The Rifts of Southern California," *BSSA* 10 (1920): 146–149; Stephen Taber, "The Earthquake Problem in Southern California," ibid.: 276–289; Ralph Arnold, "The Earthquake Problem in Southern California," ibid.: 297–299; Laughlin to Townley, Jan. 17, 1921, SSAC, carton 1, folder "Correspondence 1920–1921."

30. Byerly, "History of the Seismological Society of America," 1740; Branner to Taber, Jan. 2, 1913, JBP, box 10, vol. 34, p. 1; Branner to A. H. Philips, Nov. 25, 1914, ibid., box 11, vol. 37, p. 414; Branner to McAdie, July 29, 1916, ibid., box 12, vol. 41, p. 102; Townley to Branner, Jan. 25, 1916, SSAC, carton 1, folder "Correspondence 1914–1916"; Branner to H. T. Clifton, Apr. 3, 1920, Branner Collection (Huntington Library), box 1, letterbook 1920–1921; S. D. Townley, "Report of Secretary-Treasurer of the Seismological Society of America From April 3, 1920, to April 2, 1921," *BSSA* 11 (1921): 149–151, on 149.

31. For an exception to this statement, see "An Earthquake-Proof Concrete Tower," *EN* 74 (1915): 308–309.

32. "Educational Summary—Harry O. Wood," undated typescript [ca. 1919], HWP, box 7, folder 12; H. O. Wood, "Distribution of Apparent Intensity in San Francisco," in Andrew C. Lawson, ed., *The California Earthquake of April 18, 1906: Report of the State Earthquake Investigation Commission,* 2 vols. (Washington, DC: Carnegie Institution of Washington, 1908–1910), I:220–241; W. M. Davis to Wood, Nov. 18, 1921, Wood to Davis, Nov. 28, 1921, and May 18, 1926, all in HWP, box 2, folder 14; Pyne, *Grove Karl Gilbert,* 215.

33. Lawson to Benjamin Ide Wheeler, Mar. 2, 1909, ALP, box 15, folder "1909"; Wood, "On the Region of Origin"; idem, "The Observation of Earthquakes"; idem, "California Earthquakes: A Synthetic Study of Recorded Shocks," *BSSA* 6 (1916): 55–180, quotes on 57.

34. Wood to Branner, Apr. 10, and June 5, 1912, JBP, box 47, folder 228; Lawson to Branner, Apr. 12, 1912, ibid., box 46, folder 223; Jaggar to Wood, Aug. 20, 1917, HWP, box 23; Harry O. Wood, "The Hawaiian Volcano Observatory," *BSSA* 3 (1913): 14–19; T. A. Jaggar, "Seismometric Investigation of the Hawaiian Lava Column," ibid.,

10 (1920): 155–275, on 172–174; Russell A. Apple, "Thomas A. Jaggar, Jr., and the Hawaiian Volcano Observatory," in Robert W. Decker, Thomas L. Wright, and Peter H. Stauffer, eds., *Volcanism in Hawaii,* USGS Professional Paper 1350 (Washington, DC: GPO, 1987), 1619–1644, on 1637–1642.

35. T. H. Peck to Wood, Aug. 1, 1917, Jaggar to Wood, Aug. 20, 1917, Wood to W. F. Meyer, Dec. 31, 1917, and Arthur L. Day to Wood, Oct. 17, 1916, all in HWP, box 23; Wood to Arthur Hannon, Apr. 9, 1922, ibid., box 4, folder 2; Wood to Branner, Aug. 4, 1917, JBP, box 57, folder 281; Wood to Louderback, Jan. 17, 1918, George D. Louderback Papers, Bancroft Library, Berkeley, CA, box 28, folder "Wood."

36. Wood to R. A. Millikan, Nov. 26, 1917, HWP, box 23; "Data Concerning Military Service" and "Educational Summary—Harry O. Wood," ibid., box 7, folder 12; Wood to Woodworth, June 20, 1921, ibid., box 10, folder 6; Wood to Hannon, Apr. 9, 1922, ibid., box 4, folder 2; Wood to Townley, Jan. 16, 1918, SSAC, carton 1, folder "Correspondence 1916–1918"; J. B. Woodworth to Townley, Apr. 18, 1917, ibid., carton 1, folder "Correspondence 1916–1918"; Woodworth to Otto Klotz, Aug. 4, 1917, Woodworth Papers, box 5, folder "1917"; Harry O. Wood, "Earth Vibrations: Outline Report on the Investigation of Vibrations in the Earth," May 1918, Records of the National Research Council, National Academy of Sciences Archives, Washington, DC, folder "PS: Projects: Sound Ranging: Seismological Investigations: Reports 1918"; "Report on Conference of Physics and Engineering Divisions of the National Research Council, Washington, July 18, 1918," ibid., folder "PS: Projects: Submarine Detection: Crystal Detectors: Reports 1918," on p. 8.

37. Wood to Millikan, Nov. 26, 1917, HWP, box 23; "Directory of National Research Council," Apr. 1, 1921, ibid., box 7, file 2; Wood to Louderback, Jan. 17, 1918, Louderback Papers, box 28, folder "Wood"; Wood to Woodworth, Jan. 17, 1918, Woodworth Papers, box 5, folder "1918"; Wood to W. F. G. Swann, July 11, 1919, National Research Council Records, folder "G & G: General 1919–23." For background on the National Research Council, see Ronald C. Tobey, *The American Ideology of National Science, 1919–1930* (Pittsburgh: U of Pittsburgh P, 1971), chap. 2; Daniel J. Kevles, *The Physicists: The History of a Scientific Community in Modern America* (New York: Knopf, 1977), chaps. 8 and 9; Nathan Reingold and Ida H. Reingold, *Science in America: A Documentary History, 1900–1939* (Chicago: U of Chicago P, 1981), chap. 10.

38. Harry O. Wood, "The Earthquake Problem in the Western United States," *BSSA* 6 (1916): 197–217, first quote on 203, second quote on 201.

39. Harry O. Wood, "Seismology in the United States—Its Problems and Needs," spring 1919, John C. Merriam Papers, Manuscript Division, Library of Congress, box 224, folder "Wood," p. 11; Ernest A. Hodgson to Millikan, July 16, 1921, RMP, box 33, folder 1.

40. Wood, "Earthquake Problem," 210–213; Harry O. Wood, "The Study of Earthquakes in Southern California," *BSSA* 8 (1918): 28–33.

41. "Minutes of the Second Meeting of the Division of Geology and Geography," June 5 and 6, 1919, National Research Council Records, folder "G & G: Meetings 1919"; Wood to G. E. Hale, Mar. 10, 1919, Wood to Day, Apr. 9, 1920, Wood to Swann,

Apr. 26, 1920, Wood to Reid, Apr. 26, 1920, and Vernon Kellogg to C. Derleth, Jr., Sept. 15, 1920, all in ibid., folder "G & G: Com on Seismology 1919-22"; "Minutes of the Meeting of the Division of Geology and Geography of the National Research Council, held on April 30, 1920," ibid., folder "G & G: Meetings Annual 1920"; Wood, "Southern California Province: Detailed Seismological Program," winter 1919-20, and untitled, undated [ca. spring 1920] memorandum, both in Merriam Papers, box 224, folder "Wood"; Woodworth to Harry Fielding Reid, Apr. 19, 1920, and Woodworth to Otto Klotz, July 31, 1920, both in J. B. Woodworth Papers, Correspondence Regarding American Geophysical Union (HUG 1880.308), Harvard University Archives, Cambridge, MA; Wood to Louderback, June 7, 1920, Louderback Papers, box 28, folder "Wood."

42. Chester Stock, "John Campbell Merriam as Scientist and Philosopher," in *Cooperation in Research* (Washington, DC: Carnegie Institution of Washington, 1938), 765–778; Frank F. Bunker, "Cooperative Research, Its Conduct and Interpretation," ibid., 713-752.

43. "Meeting of the Executive Committee Thursday, March 10, 1921: Order of Business," CIWR, folder "Trustees: Root, Elihu 1921-1930," item 7; see also Minutes of National Research Council Division of Geology and Geography, April 28, 1921, National Research Council Records, folder "G & G: Meetings Annual 1921," pp. 4-5; "Minutes of the Meeting of the Executive Committee, Thursday, Mar. 10, 1921," copy in CIWR, p. 26.

44. Merriam to Anderson, May 16, 1921, CIWR, folder "Seismology: Name, Plan, Scope"; Millikan to Hale, July 28, 1920, in Reingold and Reingold, *Science in America,* 360-363, on 361; Merriam to Hale, May 9, 1921, George Ellery Hale Papers, Mount Wilson Observatory Archives, Pasadena, CA, reel 24, frame 842.

45. Harry O. Wood, "On a Piezo-Electrical Accelerograph," *BSSA* 11 (1921): 15-57; Wood to Day, June 1, 1921, HWP, box 12, folder 7; J. A. Anderson to Day, Dec. 16, 1921, CIWR, folder "Seismology: Equipment"; Wood to Merriam, Mar. 21, and Sept. 1, 1922, copies of both in RMP, box 33, folder 1.

46. Day to Wood, Nov. 25, 1922, and Wood to Day, Aug. 2, Sept. 11, Oct. 6, Nov. 7, and Nov. 21, 1922, all in HWP, box 12, folder 6; Wood to Wenner, Sept. 21, and Nov. 15, 1922, ibid., box 9, folder 21; Wood to Merriam, Sept. 25, 1922, ibid., box 13, folder 9; Wood to Day, Jan. 8, Jan. 22, May 7, and Sept. 19, 1923, and Day to Wood, Mar. 10, June 2, Sept. 27, and Dec. 1, 1923, all in ibid., box 12, folder 5; Wood to J. P. Buwalda, Jan. 23, 1923, ibid., box 1, folder 23; Willis to Wood, Nov. 12, 1923, ibid., box 10, folder 3; William Bowie to Wood, Mar. 6, 1924, ibid., box 16, folder 2. The theory of the torsion seismograph is treated in great mathematical detail in J. A. Anderson and H. O. Wood, "Description and Theory of the Torsion Seismometer," *BSSA* 15 (1925): 1-72. On the development of the torsion seismograph, see also Judith R. Goodstein, "Waves in the Earth: Seismology Comes to Southern California," *Historical Studies in the Physical Sciences* 14 (1984): 201-230, on 208-211.

47. Wood to Merriam, Sept. 1, 1923, HWP, box 13, folder 9, pp. 7-8; Wood to Day, Sept. 19, 1923, and Day to Wood, Oct. 24, 1923, both in ibid., box 12, folder 5; Day

to Merriam, Sept. 6, 1923, and Merriam to Day, Dec. 31, 1923, both in CIWR, folder "Seismology: Finance-Allotments"; Day to Members of the Advisory Committee in Seismology, Nov. 28, 1923, RMP, box 33, folder 1.

48. Wood to Day, Jan. 4, 1924, Wood Papers, box 12, folder 4; Wood to Day, Nov. 25, 1927, ibid., box 12, folder 1; "Building Against Shocks," undated clipping from unidentified newspaper, Arnold to Day, Mar. 12, 1929, and Arnold to Herbert C. Hoover, Mar. 12, 1929, all in RAP, box 234; Arnold to Townley, Jan. 4, 1924, Arnold to Day, Feb. 4, 1924, S. C. Evans to Arnold, Sept. 4, and Dec. 11, 1924, Arnold to William Wrigley, Jr., Feb. 11, 1924, and Arnold to Willis, Mar. 29, 1924, all in ibid., box 234, folder "Seismological Society of America 1924"; Evans to Arnold, June 16, 1925, ibid., box 234, folder "Earthquakes 1925"; Arnold to Wood, Jan. 28, 1925, copy in Merriam Papers, box 13, folder "Arnold, Ralph"; Day to Merriam, Mar. 14, 1927, CIWR, folder "Seismology: Buildings, Quarters, Grounds."

49. Wood to T. W. Vaughan, Feb. 15, 1921, HWP, box 8, folder 11; Day to Wood, Jan. 13, 1927, ibid., box 12, folder 1; Wood to Day, Sept. 23, 1925, ibid., box 12, folder 3; Day to Merriam, Aug. 1, 1924, CIWR, folder "Seismology: Name, Plan, Scope," p. 3; Wood to Merriam, Aug. 8, 1925, ibid., folder "Seismology: Reports 1921-1935"; Wood to Merriam, Mar. 15, and May 11, 1926, Merriam to Wood, Mar. 20, 1926, and Day to Merriam, June 10, 1926, all in ibid., folder "Seismology: California Institute of Technology." On Pritchett, see Abraham Flexner, *Henry S. Pritchett: A Biography* (New York: Columbia UP, 1943), esp. 176-188.

50. Day to Hale, May 6, 1924, Day to Merriam, Aug. 1, 1924, Merriam to Day, Aug. 11, 1924, and Merriam to Hale, Oct. 11, 1924, all in CIWR, folder "Seismology: Name, Plan, Scope"; Hale to Merriam, June 8, 1924, and Hale to Arthur H. Fleming, Mar. 18, 1925, copy in RMP, box 33, folder 2; Day to Wood, Nov. 7, 1923, HWP, box 12, folder 5; Wood to Day, Dec. 21, 1925, ibid., box 12, folder 3. On Hale's role in revitalizing Caltech, see Robert H. Kargon, "Temple to Science: Cooperative Research and the Birth of the California Institute of Technology," *Historical Studies in the Physical Sciences* 8 (1977): 3-31. For more on the negotiations between the Carnegie Institution and the California Institute of Technology, see Goodstein, "Waves in the Earth," 212-214.

51. Merriam to Wood, Dec. 4, 1926, CIWR, folder "Seismology: Buildings, Quarters, Grounds"; Wood to Millikan, Dec. 8, 1926, RMP, box 33, folder 3; Wood to Merriam, June 22, 1927, and July 2, 1928, both in HWP, box 13, folder 8.

52. Lawson, ed., *The California Earthquake*, I:114-145, and II:16-28.

53. For Lawson's attempts to claim credit at the expense of Reid, see Day to Wm. Gilbert, Dec. 6, 1923, CIWR, folder "Seismology: Reports 1921-1935."

54. Lawson to Campbell, Sept. 11, 1920, Lawson to R. H. Tucker, Sept. 21, and Oct. 13, 1920, all in ALP, box 17, folder "1920"; Campbell to Lawson, Sept. 18, 1920, ibid., box 3, folder "Campbell"; Andrew C. Lawson, "Mobility of the Coast Ranges of California," *Bulletin of the Geological Society of America* 32 (1921): 45; idem, "The Mobility of the Coast Ranges of California: An Exploitation of the Elastic Rebound Theory," *Bulletin of the Department of Geology, University of California* 7 (1921): 431-

473. On the Ukiah station, see W. W. Campbell, "The Variation of Terrestrial Latitudes," *Transactions of the Commonwealth Club of California* 16 (1921): 350–357.

55. John F. Hayford and A. L. Baldwin, "The Earth Movements in the California Earthquake of 1906," in *Report of the Superintendent of the Coast and Geodetic Survey Showing the Progress of the Work From July 1, 1906, to June 30, 1907* (Washington, DC: GPO, 1907), 67–104.

56. Wood to Bowie, Feb. 17, 1921, and Mar. 18, 1925, and Bowie to Wood, July 1, 1921, all in HWP, box 16, folder 2; Bowie to Lawson, May 27, 1921, ALP, box 13, folder "U.S. Coast and Geodetic Survey"; Day to Merriam, May 28, 1921, CIWR, folder "Geophysical Laboratory: Miscellaneous 1908–1938 / 3"; Day to Reid, Sept. 7, 1921, ibid., folder "Seismology: Reports 1921–1935"; Day to Millikan, Anderson, and Arnold, Aug. 19, 1921, and Day, "Memorandum for Members of the Advisory Committee in Seismology," Mar. 10, 1922, both in RAP, box 234, folder 1; *Hearing Before Subcommittee of House Committee on Appropriations . . . in Charge of Departments of Commerce and Labor Appropriation Bill for 1923* (Washington, DC: GPO, 1922), 355, 734–737.

57. E. Lester Jones to Francis A. Tondorf, Oct. 13, 1923, Francis A. Tondorf Papers, Georgetown University Archives, Washington, DC, box 1, folder "Correspondence October '23–December '23"; Wood to Bowie, Oct. 13, 1923, HWP, box 16, folder 2; William Bowie, *Earth Movements in California*, U.S. Coast and Geodetic Survey Special Publication no. 106 (Washington, DC: GPO, 1924).

58. S. D. Townley, "Minutes of the Meeting of the Board of Directors of the Seismological Society of America, May 25, 1921," *BSSA* 11 (1921): 146–148, on 146; Eliot Blackwelder, "Bailey Willis," *Biographical Memoirs of the National Academy of Sciences* 35 (1961): 333–350; and Aaron C. Waters, "Memorial to Bailey Willis (1857–1949)," *Geological Society of America Bulletin* 73 (1962): P55–P72.

59. Willis to Arnold, July 26, and Sept. 17, 1921, and Wayne Loel to Sumner P. Hunt, Oct. 21, 1921, all in RAP, box 234, folder 1; Sumner Hunt, "Building for Earthquake Resistance," *BSSA* 12 (1922): 12–18. For biographical information on Hunt, see "Sumner P. Hunt," *Pacific Coast Architect* 30, no. 1 (1926): 39.

60. Wood to Merriam, Mar. 21, and Sept. 1, 1922, copies of both in RMP, box 33, folder 1; "Earthquakes Are To Be Predicted," *Pasadena Star-News,* July 21, 1921, p. 13; "Plan Forecast of Earthquake," *LAT,* July 21, 1921, II:1; Arnold to Willis, July 25, 1921, RAP, box 234, folder 1; Wood to Willis, Nov. 14, 1921, HWP, box 10, folder 3

61. Harry O. Wood, "Minutes of Meeting of Sub-Committee on Fault-Zone Geology," June 28, 1922, Willis to Wood, July 6, and Sept. 13, 1922, all in HWP, box 10, folder 3; Willis to David White, June 30, 1922, and White to Willis, July 13, 1922, both in Records of the USGS (RG 57), National Archives II, College Park, MD, entry 82, file 663.2; Willis to Merriam, July 30, 1922, Merriam Papers, box 183, folder "Willis 1921–1931"; S. D. Townley, "Report of the Secretary-Treasurer of the Seismological Society of America From April 1, 1922, to April 7, 1923," *BSSA* 13 (1923): 86–88, on 87.

62. "Members of the Seismological Society of America, December 20, 1920," *BSSA* 10 (1920): 330–341; "Members of the Seismological Society of America, February 20,

1925," ibid., 14 (1924): 279-298; Issel and Cherny, *San Francisco, 1865-1932*, first quote on 37; J. C. Branner, W. W. Campbell, A. C. Lawson, S. D. Townley, and Bailey Willis, "The Seismological Society of America: Its Work and Needs," undated [fall 1921] (second quote), copy in Stanford School of Engineering Records, ser. 1, box 3, folder 33; Willis to Arnold, Sept. 17, 1921, RAP, box 234, folder 1; Willis to Lawson, Oct. 14, 1921, ALP, box 12, folder "Seismological Society"; Willis to Merriam, July 30, and Aug. 20, 1922, Merriam Papers, box 183, folder "Willis 1921-1931"; S. D. Townley, "Minutes of the Meeting of the Board of Directors of the Seismological Society of America October 3, 1922," *BSSA* 12 (1922): 254-255, on 255; Willis to Wood, May 25, 1922, HWP, box 10, folder 3.

63. Arnold to Day, Nov. 28, 1921, Arnold to Miller, Nov. 30, 1921, and Arnold to Willis, Apr. 24, 1922, all in RAP, box 234, folder 1; Wood to Laughlin, May 9, 1922, HWP, box 5, folder 9; Day to Willis, Aug. 9, 1922, BWP, box 12, folder 56; Willis to F. P. Keppel, undated (early 1924), SSAC, carton 1, folder "Correspondence 1923-1924"; Merriam to Day, June 19, 1924, CIWR, folder "Seismology: Finance-Allotments"; Arthur L. Day and Bailey Willis, "Cooperation in Seismology," *Science* 60 (1924): 217-218.

64. Willis to Wood, Sept. 13, 1922, HWP, box 10, folder 3; "Carbon From Air Baffles Scientists," *NYT*, Dec. 29, 1922, p. 3; Townley to Arnold, June 23, 1923, SSAC, carton 1, folder "Correspondence 1923-1924"; "California Quake Map Is Published," *SFC*, Oct. 11, 1923, p. 6; Bailey Willis, "A Fault Map of California," *BSSA* 13 (1923): 1-12, quote on 11. This issue of the *Bulletin* is dated March 1923, but at that time the journal was about half a year behind in its publication schedule.

65. "Earthquake Map Given to Library," undated clipping from *Palo Alto Times* in BWP, box 48, folder 2; "Forum Speaker Urges Ball-Bearing Houses," *San Jose Mercury Herald*, Oct. 26, 1923, clipping in ibid.; "California Quake Map"; "Science Sees New Key to Earthquakes," *SFE*, Oct. 11, 1923, p. 17; Willis to Wood, Nov. 12, 1923, HWP, box 10, folder 3; "Fault Map of the State of California" (advertisement), *Science Advertising Supplement*, Jan. 4, 1924, p. ix; S. D. Townley, "Report of the Secretary-Treasurer of the Seismological Society of America From April 7, 1923, to April 5, 1924," *BSSA* 14 (1924): 77-79, on 77; idem, "Report of the Secretary-Treasurer of the Seismological Society of America From April 5, 1924 to April 4, 1925," ibid., 15 (1925): 156-158, on 157; Willis to Arnold, Jan. 19, 1924, RAP, box 234, folder "Seismological Society of America 1924."

Chapter 3

1. Eliot Blackwelder, "Bailey Willis," *Biographical Memoirs of the National Academy of Sciences* 35 (1961): 333-350; Aaron C. Waters, "Memorial to Bailey Willis (1857-1949)," *Geological Society of America Bulletin* 73 (1962): P55-P72; Bailey Willis, "Water Circulation and Its Control," undated [ca. 1908], BWP, box 7, folder 7; idem, "National Watersheds" and "Engineering Control of Surface Flow," both undated [late 1908], and "National Watersheds as a Step Towards Regulation of Navigable Streams," Dec. 1908, all in ibid., box 48, folder 5; Willis to Gifford Pinchot, Nov. 27, 1908, and July 23, 1909, both in ibid., box 30, folder 41; George Bird Grinnell to Willis, Aug. 11, 1908,

.ibid., box 14, folder 29; Willis to Margaret D. Willis, May 23, and Aug. 23, 1925, both in ibid., box 39, folder 2; Bailey Willis, *A Yanqui in Patagonia* (Stanford: Stanford UP, 1947).

2. Bailey Willis, "Earthquake Risk in California [Part 1 of 5]," *BSSA* 13 (1923): 89–99, on 89, emphasis in original. Although this issue of the *Bulletin* is dated September 1923, it did not appear until the spring of 1924 due to the continuing fiscal difficulties of the Seismological Society; see Bailey Willis, "Cooperation," *BSSA* 13 (1923): 119-123, on 122.

3. Willis, "Earthquake Risk [Part 1]," 90.

4. Ibid., 90, 92.

5. Ibid., 92-96. On p. 96, Willis advertised the availability of the fault map.

6. Ibid. [Part 2], *BSSA* 13 (1923): 147-154, quotes on 153.

7. Ibid. [Part 3], *BSSA* 14 (1924): 9-25, quotes on 15, 21, 23.

8. Ibid. [Part 4], *BSSA,* 14 (1924): 150-164.

9. Ibid. [Part 5], *BSSA,* 14 (1924): 256-264, on 257-258.

10. Ibid., 258-263.

11. "Science Sees New Key to Earthquakes," *SFE,* Oct. 11, 1923, p. 17; "Shock-Proof Houses Urged," ibid., Oct. 20, 1923, p. 7; "Forum Speaker Urges Ball-Bearing Houses," *San Jose Mercury Herald,* Oct. 26, 1923, and "Earthquakes Certain, Says Willis," *Palo Alto Times,* Oct. 11, 1923, both clippings in BWP, box 48, folder 2; Willis to Wood, Nov. 12, 1923, HWP, box 10, folder 3; Willis, "Application for a Grant Addressed by the Seismological Society of America to the Carnegie Corporation," undated (spring 1924), SSAC, carton 1, folder "Correspondence 1923-1924," p. 2; "New Quake Coming in About 15 Years, Rotary Club Told," *Palo Alto Times,* Dec. 18, 1923, clipping in BWP, box 48, folder 2; San Francisco Chapter, American Institute of Architects, "Monthly Bulletin," *Pacific Coast Architect* 25, no. 3 (1924): 45.

12. "More Substantial Building Urged," *Carmel Pine Cone,* Mar. 15, 1924, p. 1, clipping in BWP, box 48, folder 1.

13. John R. Kibbey to Townley, May 17, 1924, BWP, box 16, folder 58.

14. S. D. Townley, "Meeting of the Committee on Building for Safety Against Earthquakes of the Seismological Society of America," July 26, 1924, SSAC, carton 1, folder "Correspondence July 1924-March 1925"; Willis to A. Emory Wishon, Aug. 5, 1924 (quote), RAP, box 234, folder "Seismological Society of America 1924."

15. Townley, "Meeting of the Committee on Building for Safety"; Bailey Willis, "Committee on Building for Safety against Earthquakes, Seismological Society of America: Minutes of Meeting, February 13 and 14, 1925," RAP, box 234, folder "Earthquakes 1925."

16. Henry D. Dewell, "Report of Committee on Building for Safety Against Earthquakes: Preliminary Report of Subcommittee on Framed Structures: Wood, Steel, and Ferro-Concrete," *BSSA* 15 (1925): 175-195; Robert E. Andrews, "Report of Committee on Building for Safety Against Earthquakes: Preliminary Report on Fire Protection," ibid.: 196-212; C. T. Manwaring, "Report of Committee on Building for Safety Against Earthquakes: Preliminary Report on Guarding Against Panic," ibid.: 213-221; Sumner

Hunt, "Committee on Building for Safety-Preliminary Report on Construction of Walls and Floors," ibid., 16 (1926): 266-271.

17. Bailey Willis, "A Study of the Santa Barbara Earthquake of June 29, 1925," *BSSA* 15 (1925): 255-278; Henry D. Dewell and Bailey Willis, "Earthquake Damage to Buildings," ibid.: 282-301.

18. Robert M. Fogelson, *The Fragmented Metropolis: Los Angeles, 1850-1930* (Cambridge: Harvard UP, 1967), chaps. 4 and 5; Kevin Starr, *Material Dreams: Southern California through the 1920s* (New York: Oxford UP, 1990), chap. 5; Tom Zimmerman, "Paradise Promoted: Boosterism and the Los Angeles Chamber of Commerce," *California History* 64 (1985): 22-33.

19. B. M. Rastall to Mr. Rainey, Aug. 11, 1925, and B. M. Rastall, "Special Report to Members on Santa Barbara Earthquake Activities" (quotes), copies of both in F. F. Peabody to Henry S. Pritchett, Nov. 23, 1925, Henry S. Pritchett Papers, Manuscript Division, Library of Congress, box 9, folder "Santa Barbara Earthquake, June 29, 1925 / Santa Barbara Relief Fund Committee Nov. 1925-Aug. 1926."

20. Los Angeles Chamber of Commerce to Chamber of Commerce, Baltimore, Maryland, July 1, 1925, copy in Peabody to Pritchett, Nov. 23, 1925, Pritchett Papers, box 9, folder "Santa Barbara Earthquake, June 29, 1925 / Santa Barbara Relief Fund Committee Nov. 1925-Aug. 1926."

21. "Pictures Tell True Damage," *LAE,* July 2, 1925, I:2; "East Assured by L. A. This City Not Damaged," ibid., July 1, 1925, I:3; George I. Cochran, "Facts Show California Safe Place," ibid., June 30, 1925, I:9 (quotes); idem, "Rally to Fight Harmful Tales," *LAT,* July 1, 1925, II:2; "Quake Deaths First In More Than 100 Years," *Santa Barbara Daily News,* July 2, 1925, p. 1.

22. "Santa Barbara Earthquake," *California Journal of Development* 15, no. 7 (1925): 14-15; "Survey to Reveal Actual Damage," *LAE,* July 2, 1925, I:2; "Hale Issues Quake Survey," *SFC,* July 3, 1925, p. 5; "State to Back Santa Barbara," *LAT,* July 3, 1925, I:2; Arthur Brisbane, "Today" (column), *LAE,* July 2, 1925, I:1.

23. "Splendid, Santa Barbara," *LAT,* July 2, 1925, II:4; " 'Business As Usual' Her Slogan," *LAE,* July 3, 1925, I:5; "Calming Fears of Lightning Dodgers," *SFC,* July 2, 1925, p. 24; "Man's Fortitude Is Secret Of His Life," ibid., July 1, 1925, p. 30; "And He's Afraid of Earthquakes!" ibid., July 2, 1925, p. 24; "Check Exaggeration of Earthquake," *Pasadena Star-News,* July 2, 1925, p. 4; "Rebuilding Loans Are Pledged," ibid., July 3, 1925, p. 4; C. G. Milham, "Earthquake Victims 800 In All Time," *Santa Barbara Morning Press,* July 7, 1925, I:1; "Quake Not Worst Disaster," *Santa Barbara Daily News,* July 10, 1925, I:12.

24. "Santa Barbara Earthquake," quote on 15.

25. "City to Gather 'Quake Data," *Santa Barbara Morning Press,* Aug. 18, 1925, II:2 (quote); Pritchett to Santa Barbara Relief Fund Committee, Sept. 24, 1925, and Pritchett to E. F. MacDonough, Nov. 7, 1925, both in Pritchett Papers, box 9, folder "Santa Barbara Earthquake June 29, 1925 / Santa Barbara Relief Fund Committee July-Oct., 1925"; Peabody to Pritchett, Nov. 23, 1925, ibid., box 9, folder "Santa Barbara Earthquake June 29, 1925 / Santa Barbara Relief Fund Committee Nov. 1925-Aug. 1926"; "An

Unfortunate Condition," *Santa Barbara Daily News,* Sept. 11, 1925, p. 16; "Just One of the Common People," ibid., Sept. 26, 1925, p. 12.

26. Dewell and Willis, "Earthquake Damage"; see also "Report of Engineering Committee on the Santa Barbara Earthquake," *BSSA* 15 (1925): 302-304; Augustine Hobrecht, "Santa Barbara Mission and the Earthquake of June 29, 1925," ibid.: 251-254; Herbert Nunn, "Municipal Problems of Santa Barbara," ibid.: 308-319; Perry Byerly, "Notes on the Intensity of the Santa Barbara Earthquake Between Santa Barbara and San Luis Obispo," ibid.: 279-281.

27. Willis, "Study of the Santa Barbara Earthquake," quotes on 256, 259, 60. On the Seismological Society's membership at the end of 1925, see "Members of the Seismological Society of America," *BSSA* 15 (1925): 338-361.

28. "Quake Expert in Midst of Upheaval," *Palo Alto Times,* June 29, 1925, p. 1; "Expert Who Predicted Quake Is In Temblor," *Washington Star,* June 29, 1925, clipping in BWP, box 48, folder 2; "Phenomenon is Studied," *LAT,* June 30, 1925, I:3.

29. "Faulty Buildings Blamed by Bailey Willis for Most of Santa Barbara's Loss," *Palo Alto Times,* July 1, 1925, pp. 1, 6; "Quake Peril Removed," *LAT,* July 2, 1925, I:1, 2; "Stanford Scientist Tells Quake Cause," *LAE,* July 2, 1925, I:3; "Another Hard Earth Tremor Improbable," *Santa Barbara Morning Press,* July 2, 1925, I:2; "No More Serious Shocks Says Scientist Who Was Here," *Santa Barbara Daily News,* July 7, 1925, p. 2; "Blame Builders More Than Quake," *New York Sun,* July 2, 1925, clipping in BWP, box 48, folder 2; Bailey Willis, "We Live As We Build," *Home Builder [of Los Angeles]* 3, no. 8 (1925): 6; "City of Santa Barbara Will Be Rebuilt," *Southwest Builder and Contractor,* July 3, 1925, p. 46; Bailey Willis, "Designing Against Earthquakes," *Pacific Coast Architect* 28, no. 5 (1925): 6, 47-53; Rastall, "Special Report."

30. "Santa Barbara Loss Explained," *SFC,* July 3, 1925, p. 5; Willis to Wood, July 3, 1925, HWP, box 10, folder 2; Arthur L. Day, "Earthquakes and Their Effect on Buildings," *American Concrete Institute, Proceedings of the Annual Convention* 22 (1926): 72-78, quote on 75.

31. Bailey Willis, "The Santa Barbara Earthquake and Local Earthquake Prospects," *The Commonwealth: Transactions of the Commonwealth Club of California* 20 (1925): 215-225, quotes on 216 (emphasis in original), 219; Henry D. Dewell, "Building Against Earthquake Shock," ibid.: 226-233.

32. J. M. Cumming to Willis, July 25, 1925, BWP, box 12, folder 31; "Expert Warns of Temblors," *SFC,* Sept. 30, 1925, p. 6; "Earthquake Protection," *Berkeley Daily Gazette,* Nov. 21, 1925, clipping in BWP, box 48, folder 2; "Is Working Out Standards," *National Underwriter,* July 23, 1925, p. 31; "Willis Discusses Earthquake Hazard," ibid., July 30, 1925, p. 30.

33. "Santa Barbara Earthquake," *Proceedings of the San Francisco Section, American Society of Civil Engineers* 21 (1925), unnumbered special issue; "Engineers Discuss Earthquake Experiences and Cautions," *ENR* 95 (1925): 271-272; Henry D. Dewell, "Earthquake Damage to Santa Barbara Buildings," ibid.: 68-72; idem, "Building Damage at Santa Barbara-Second Report," ibid.: 149-153; "Earthquake Effects at Santa Barbara," ibid.: 46-47; "Earthquake Hazards," ibid.: 126-127; "Notes and Comments," *Southwest*

Builder and Contractor, July 17, 1925, pp. 41–42; Henry D. Dewell, "Building Against Earthquakes," *Pacific Coast Architect* 28, no. 4 (1925): 33–37.

34. Sumner Hunt to Wood, July 18, 1925, HWP, box 4, folder 16. For Hunt's efforts to spread knowledge about the architectural effects of the Santa Barbara earthquake, see also "Santa Barbara Earthquake Discussed by Architects," *Southwest Builder and Contractor,* July 17, 1925, p. 44.

35. Bailey Willis, "Faults (Natural and Unnatural)," *Bulletin Allied Architects Association of Los Angeles* 1, no. 10 (1925), unpaginated. A copy of this special issue can be found in John R. Freeman Papers (MC 51), Institute Archives, Massachusetts Institute of Technology, Cambridge, MA, box 34, file "Misc. Data on Various Earthquakes 1627 to 1931."

36. Harry O. Wood, "The Practical Lesson of the Santa Barbara Earthquake," *Bulletin Allied Architects Association of Los Angeles* 1, no. 10 (1925); Roy C. Mitchell, "Life Saving vs. Dollar Saving," ibid.; in the same issue, see also Loyall F. Watson, "Suggested Measures for Protection Against Earthquakes"; T. Beverly Keim, Jr., "The Santa Barbara Earthquake"; Wm. Richards, "Santa Barbara Earthquake"; Sumner P. Hunt, "As to Earthquakes"; and Henry D. Dewell, "Building for Safety Against Earthquakes As Evidenced by the Santa Barbara Earthquake."

37. "Lessons from the Santa Barbara Earthquake," *Pacific Coast Architect* 64 (1925): 689–699; "Observations on the Santa Barbara Earthquake," *Architecture and Building* 57 (1925): 93–94; "Experts Tell How Buildings May Be Made More Secure Against Earthquake Shock," *Southwest Builder and Contractor,* Sept. 1, 1925, pp. 41–42; "Data on Quake Effects Ready," *LAT,* Aug. 24, 1925, II:18.

38. Willis to Margaret D. Willis, June 12, 1925, BWP, box 39, folder 2; Willis to Wood, Aug. 10, 1926, HWP, box 10, folder 1. For the text of the amendment, see "Protection Against Earthquakes in P. A. Proposed in Report," *Palo Alto Times,* July 25, 1925, pp. 1, 8; Bailey Willis, "Construction Lessons From Santa Barbara," *Pacific Coast Architect* 28, no. 3 (1925): 5, 53.

39. "News Briefs," *Palo Alto Times,* July 22, 1925, p. 3; "Protection Against Earthquakes in P. A. Proposed in Report," ibid., July 25, 1925, pp. 1, 8; "C. of C. Backs New 'Quake Ordinance'," ibid., July 31, 1925, p. 1; "Building Laws to Stop Temblor Loss Urged," ibid., July 14, 1925, p. 1; "Earthquake-Proof Buildings," ibid., June 30, 1925, p. 4; "Should Be Written Into Law," ibid., July 2, 1925, p. 4; "Back the Building Code," ibid., July 10, 1925, p. 4; "Now Is the Time to Act," ibid., July 15, 1925, p. 4; "Plans for Proposed New Police Building Will Be Submitted; Other Points," ibid., Sept. 4, 1925, p. 1. On the civic involvement of Marx and Wing, see "C. D. Marx Dies," *Palo Alto Times,* Jan. 1, 1940, pp. 1, 9; "Death Takes C. B. Wing, Stanford Engineer and Ex-Mayor of City," ibid., Aug. 23, 1945, pp. 1, 2.

40. Willis to Wood, Aug. 10, 1926 (quote), HWP, box 10, folder 1; "Earthquake Ordinance Is Drafted," *Palo Alto Times,* Apr. 6, 1926, pp. 1, 6; "Council Votes Ordinance on Bond Election," ibid., Apr. 13, 1926, p. 1; "Council Acts on Plans to Widen Streets," ibid., Apr. 20, 1926, p. 1.

41. "Changes Offered in Construction Laws of the City," *Santa Barbara Morning*

Press, Sept. 1, 1925, II:2; "Revised Building Code Now Ready," *Santa Barbara Daily News,* Sept. 1, 1925, p. 9. For the full text of the new building code, as finally passed after considerable revision, see "Ordinance No. 1278," *Santa Barbara Morning Press,* Dec. 22, 1925, sec. III.

42. Starr, *Material Dreams,* chap. 10, quotes on 278, 280. For opposition to the new building code, see, for example, "Those Emergency Ordinances," *Santa Barbara Daily News,* Sept. 11, 1925, p. 16; "That Building Ordinance," ibid., Sept. 21, 1925, p. 12; "That Remarkable Building Ordinance," ibid., Sept. 22, 1925, p. 12; "New Building Code Illegal, Extravagant, Menacing," ibid., Sept. 25, 1925, pp. 9, 11; H. L. Sweeney, "Let Us Have The Truth About The Building Ordinance" (advertisement), ibid., Sept. 28, 1925, p. 5; idem, "Building Codes and Building Codes" (advertisement), ibid., Dec. 12, 1925, p. 3; "That Building Ordinance," ibid., Dec. 15, p. 20.

43. "New Council Is Divided on Building Code," *Santa Barbara Daily News,* Dec. 18, 1925, p. 13.

44. Bernhard Hoffmann, "The Rebuilding of Santa Barbara," *BSSA* 15 (1925): 323–328; Vern Hedden, "Building Department Problems After the Santa Barbara Earthquake," ibid.: 320–322; "Quake Student Flays Faulty Building," *Santa Barbara Daily News,* Oct. 6, 1925, p. 9; Nunn, "Municipal Problems"; idem, "Lessons to be Drawn From an Earthquake," *Pacific Municipalities,* 39 (1925): 373–381, quote on 373; "Experiences Worth While Out of the Earthquake," *Santa Barbara Morning Press,* Sept. 24, 1925, I:4; "Building Code Gets Laudation," ibid., Dec. 22, 1925, I:4.

45. "Man's Fortitude Is Secret Of His Life"; "Calming Fears of Lightning Dodgers"; "And He's Afraid of Earthquakes!"

46. "Safety A Matter of Good Design and Workmanship," *SFC,* July 4, 1925, p. 18; "Lax Building Only Earthquake Peril," ibid., July 7, 1925, p. 24; "Why Every City Needs Strict Building Code," ibid., undated clipping in BWP, box 48, folder 2.

47. "Santa Barbara Earthquake," *Proceedings of the San Francisco Section, American Society of Civil Engineers* 21 (1925): unnumbered issue; "Proceedings of the One Hundred and Twenty-Second Regular Meeting," ibid., vol. 21, no. 5 (1925); "Proceedings of the One Hundred and Twenty-Third Regular Meeting," ibid., no. 6 (1925); "Proceedings of the One Hundred and Twenty-Fourth Regular Meeting," ibid., vol. 22, no. 1 (1926); "Proceedings of the One Hundred and Twenty-Ninth Regular Meeting," ibid., no. 6 (1926); "Proceedings of the One Hundred and Thirty-Sixth Regular Meeting," ibid., vol. 23, no. 6 (1927). Copies of this publication can be found in the Engineering Societies Library collection, now housed at the Linda Hall Library, Kansas City, MO.

48. Willis to J. C. Merriam, Jan. 22, 1925, CIWR, file "Seismology: Name, Plan, Scope." See also Commonwealth Club of California, "Minutes of the Meeting of the Section on Scientific Research," Mar. 5, 1925, HWP, box 10, folder 2.

49. "Seismographs Urged For Bay," *SFC,* July 1, 1925, p. 4; "Expert Warns of Temblors," ibid., Sept. 30, 1925, p. 6; Cumming to Willis, July 25, 1925, BWP box 12, folder 31; "Federated C. of C. Committees Are Appointed," *Palo Alto Times,* July 17, 1925, p. 6; Willis to Wood, Oct. 17, 1925, HWP, box 10, folder 2; Willis to Merriam, Jan. 12, 1926, John C. Merriam Papers, Manuscript Division, Library of Congress, box 183, folder

"Willis 1921–1931"; Willis to Wood, Feb. 11, and Feb. 26, 1926, and J. K. Moffitt to S. D. Townley, May 18, 1926, all in HWP, box 10, folder 1; Townley to G. D. Louderback, Oct. 6, 1926, SSAC, carton 1, folder "Correspondence June–Dec. 1926."

50. S. D. Townley, "Report of the Secretary-Treasurer of the Seismological Society of America From April 2, 1927, to April 7, 1928," *BSSA* 18 (1928): 138–140, on 138; idem, "Seismometer Stations in the San Francisco Bay Region," ibid., 19 (1929): 117; Wood to Willis, Feb. 19, 1929, HWP, box 10, folder 1.

51. [Willis], "Experimental Studies on Earthquake-Resistant Building," Aug. 9, 1934, BWP, box 3, folder 15; Willis to Theodore J. Hoover, Dec. 16, 1925, and Jan. 11, 1926 (first quote), E. P. Lesley to Hoover, Jan. 11, and Jan. 25, 1926, and Hoover to Charles Moser, Mar. 3, 1926 (second quote), all in School of Engineering Records, ser. 1, box 9, folder 39.

52. Bailey Willis, "Buildings and Lateral Thrust," Apr. 2, 1926 (quote), Willis to Hoover, July 9, 1926, and Willis to Ray Lyman Wilbur, Aug. 28, 1926, all in School of Engineering Records, ser. 1, box 9, folder 39; Willis to Margaret D. Willis, Apr. 13, 1926, BWP, box 39, folder 3; Willis to J. A. McCarthy, Sept. 11, 1926, ibid., box 29, folder 39.

53. Hoover to Henry Anderson, Feb. 4, and Nov. 11, 1927, and "Rough Draft of Shorthand Notes on Vibration Research Committee, Nov. 10, 1927," all in School of Engineering Records, ser. 1, box 9, folder 39; Lydik S. Jacobsen, "Earthquake Engineering Research at Stanford in the Period of 1906–1960," *EERIN* 10, no. 3 (1976): 1–10.

54. [Harry O. Wood], "Memorandum Concerning Seismologic Studies," June 1920, and [Wood], "Memorandum Concerning Seismology in the Western United States," Sept. 1920, both in Merriam Papers, box 224, folder "Wood"; Bailey Willis, "Earthquake Risk in California [Part 3]," 21–23; William Bowie, *Earth Movements in California*, U.S. Coast and Geodetic Survey Special Publication no. 106 (Washington, DC: GPO, 1924).

55. Willis to Wood, July 3, 1925, HWP, box 10, folder 2; Arthur L. Day to Wood, Aug. 12, 1925, ibid, box 12, folder 3; "California Earthquake Predicted by University Geology Professor," *Daily Californian,* Sept. 22, 1925, p. 3; Wood, "The Practical Lesson of the Santa Barbara Earthquake."

56. "Dr. Bailey Willis Clears Statement Regarding Quakes," *Daily Palo Alto,* Nov. 4, 1925, p. 1, clipping in BWP, box 48, folder 2; "Prof. Willis Predicts Los Angeles Tremors," *NYT,* Nov. 4, 1925, p. 9; "Faux Pas," *Time,* Nov. 16, 1925, p. 26; "Earthquake Prediction," *Adelaide Mail,* Feb. 27, 1926, clipping in BWP, box 48, folder 2.

57. Bailey Willis, *A Rational Basis of Earthquake Insurance* (New York: n.p., 1926), quotes on 4–5, 15.

58. Willis to Joseph Jensen, Oct. 19, 1927, Robert T. Hill Papers, DeGolyer Library, Southern Methodist University, University Park, TX, box 6, folder 166; see also Willis to P. R. Kent, Dec. 11, 1927, Ray Lyman Wilbur Presidential Papers (SC 64A), Department of Special Collections, Stanford University Libraries, box 68, folder "Geology."

59. John R. Freeman to Irving B. Crosby, Aug. 20, 1926, Crosby Family Papers (MC 81), Institute Archives, Massachusetts Institute of Technology, Cambridge, MA, box 1, folder "Scientific + Professional, 1923–1928: F–G." For the data on which this judgment

was based, see entries for June 14 to July 31, 1926, in Freeman's diary, Freeman Papers, box 3.

60. Arthur M. Brown, "Insuring the Earthquake Hazard," *The Commonwealth: Transactions of the Commonwealth Club of California* 20 (1925): 234-238, on 235; "Santa Barbara Loss Placed at $2,000,000," *National Underwriter*, July 9, 1925, p. 5; "Largest Quake Risk," ibid., p. 12; "Large Quake Lines Still Being Written," ibid., July 16, 1925, p. 5; "Company Officials Alarmed," ibid., July 23, 1925, p. 24; "Earthquake Situation Serious for Companies," ibid., p. 17; "Is Working Out Standards" ibid., p. 31; "Willis Discusses Earthquake Hazard," ibid., July 30, 1925, p. 30; P. A. Pflueger et al., "Bankers Study Insurance Problems," *Bulletin, California Bankers Association* 8 (1927): 556, 582, 589-590, on 582.

61. "F. U. A. P. Set New High Mark," *National Underwriter*, Feb. 17, 1927, p. 26; "Scientist Sees Little Relief in Earthquake," *LAT*, Jan. 11, 1927, II:14; E. W. Bannister to Andrew Lawson, Jan. 12, 1927, ALP, box 2, folder "Board of Fire Underwriters"; Robert E. Andrews to Wood, Jan. 13, 1927, HWP, box 7, folder 1. For the figure of 85%, see "Figures on San Francisco Business," *National Underwriter*, Apr. 21, 1927, p. 36.

62. Lawson to Bannister, Feb. 23, 1927, "Digest of Mr. Bannister's interviews with men best fitted to express authoritative views on the subject of Earthquakes in Southern California," Mar. 5, 1927 (quote), and Wood to H. F. Badger, Feb. 26, 1927, copies of all in RMP, box 33, folder 4.

63. Edward W. Bannister, "Rating Buildings for Liability to Earthquake Damage," *ENR* 98 (1927): 1052-1054; "Pacific Board Adopts New Earthquake Rates," *National Underwriter*, May 12, 1927, p. 7; Pflueger et al., "Bankers Study Insurance Problems," 582, 589.

64. Pflueger et al., "Bankers Study Insurance Problems," quote on 556; see also "Quake Insurance Not Essential to Safe Financing of Buildings," *ENR* 100 (1928): 580; "Protection Against Earthquakes as Viewed by a Banker," *Western Construction News*, Apr. 25, 1928, pp. 266-267.

65. Mark C. Cohn, "State Development Association Begins Building Code Work," *Pacific Coast Architect* 33, no. 5 (1928): 48-49; "Proceedings of the One Hundred and Thirty-Sixth Meeting"; "Proceedings of the One Hundred and Thirty-Eighth Regular Meeting," *Proceedings of the San Francisco Section, American Society of Civil Engineers* 24, no. 2 (1928); P. A. Pflueger, "Supplementary Report by the Insurance Committee," *Bulletin, California Bankers Association* 9 (1928): 280-281, on 280; Henry D. Dewell, "The Earthquake Resistance of Buildings from the Standpoint of the Building Code," *BSSA* 19 (1929): 96-100.

66. A. L. Lathrop, "The Earthquake Insurance Situation in Los Angeles," undated [Dec. 1927], copy in Freeman Papers, box 34, file "Earthquake Data, Reprints and Correspondence 1925-30," folder 1, pp. 1-2.

67. Henry M. Robinson to Robert A. Millikan, Apr. 8, 1927 (quote), RMP, box 33, folder 4; Wood to Watson Davis, Aug. 18, 1927, HWP, box 8, folder 10. For the proposed public statement, see R. R. Martel to Members of the Southern California Council on Earthquake Protection, Mar. 10, and Mar. 12, 1927, both in ibid., box 2, folder 4

68. Nancy Alexander, *Father of Texas Geology: Robert T. Hill* (Dallas: Southern Methodist UP, 1976), esp. 219, 223, 240–241; Robert T. Hill, "The Rifts of Southern California," *BSSA* 10 (1920): 146–149; idem, untitled draft of legal deposition prepared in late Aug. 1928, Hill Papers, box 22, folder "Legal Papers—evidence against the Secretary California," pp. 3–5, 9.

69. Mr. Marion to Mr. Harman, July 16, 1927, copy in Hill Papers, box 22, folder "Earthquake Book—Correspondence 1929–30." See also Mr. Jost to Mr. Marion, July 14, 1927, copy in ibid.; "Geologist Discounts Quake Fear," *LAT*, July 14, 1927, II:11.

70. Edward W. Pew to Hill, July 24, 1927, and Hill to Ford J. Carpenter, July 29, 1927 (quoted), both in Hill Papers, box 22, folder "Earthquake Book—Correspondence 1929–30"; Hill, draft of legal deposition, 6–9, 13–18.

71. Bowie, *Earth Movements in California,* esp. 6, 7; "Report of the Advisory Committee in Seismology," *BSSA* 17 (1927): 25–48, on 40–42; "Report of the Advisory Committee in Seismology," ibid.: 249–254, on 250.

72. Hill to W. Bowie, Oct. 26, 1927, and R. L. Faris to Hill, Oct. 27, 1927, both in Robert T. Hill, *Southern California Geology and Los Angeles Earthquakes* (Los Angeles: Southern California Academy of Sciences, 1928), 56; Faris to Hill, Oct. 29, 1927, in ibid., 58–61. Another copy of this letter is in HWP, box 4, folder 9. For a more detailed explanation of the survey's adjustment of its triangulation results for "Laplace azimuths," see William Bowie, *Comparison of Old and New Triangulation in California,* U.S. Coast and Geodetic Survey Special Publication no. 151 (Washington, DC: GPO, 1928), 13–19.

73. Day to Wood, Dec. 13, 1927, and Wood to Day, Dec. 23, 1927, both in HWP, box 12, folder 1; Henry D. Dewell to Townley, Jan. 27, 1928, SSAC, carton 1, folder "Correspondence Jan.–June 1928." For celebrations of the Coast and Geodetic Survey's precision, see, for example, Day to Wood, Mar. 31, 1922, HWP, box 10, folder 3; Wood to Townley, Jan. 9, 1928, SSAC, carton 1, folder "Correspondence Jan.–June 1928"; Day to Lawson, Jan. 18, 1928, ALP, box 3, folder "Carnegie Institution."

74. Lathrop, "The Earthquake Insurance Situation" (first four quotes); "City Found Safe From Temblors," *LAT,* Dec. 2, 1927, II:11 (last two quotes); Otheman Stevens, "Earthquake Predictions Laid to Error," *LAE,* Dec. 2, 1927, I:6; C. A. Copper to Freeman, Jan. 3, 1928, Freeman Papers, box 34, file "Earthquake Data, Reprints & Correspondence 1925–30," folder 1; Hill, draft of legal deposition, 11–12.

75. Robert T. Hill, "Earthquake Conditions in Southern California," *Bulletin of the Geological Society of America* 39 (1928): 188–189; Hill, draft of legal deposition, 20–21; "Temblor Talk Tumbled Over," *LAT,* Jan. 19, 1928, II:1, 12; C. A. Copper, "Scientific World Backs Dr. R. T. Hill's Findings on L. A. Quake Situation," *[Los Angeles] Commercial and Financial Digest,* Feb. 1, 1928, p. 8, clipping in Hill Papers, box 22, folder "Earthquake Book—Correspondence 1929–1930"; "Science's Business," *Time,* Feb. 27, 1928, p. 19.

76. Hill to W. M. Davis, July 24, 1928, Hill Papers, box 22, folder "Earthquake Book—Correspondence 1929–30"; Hill to John Comstock, July 26, 1928, ibid., box 22, folder "John A. Comstock 1928"; Hill, *Southern California Geology,* quote on 211; idem, draft of legal deposition, 27–35.

77. This dust cover has not been preserved on the library copies of Hill's book that I have examined, but it is quoted in W. M. Davis, "The Earthquake Problem in Southern California" (book review), galley proof in Hill Papers, box 22, folder "1928: book review," and is described in Hill to Davis, July 12, 1928, ibid., box 2, folder 133.

78. "Book Refutes Prediction of Big Quake Here," *LAE,* Apr. 16, 1928, clipping in Regional History Center Clipping Collection, University of Southern California, Los Angeles, file "Bailey Willis"; "Expert Shoos Quake 'Bogey,' " *LAT,* Apr. 16, 1928, II:1; "Quake Insurance Arouses L. A. Building Owners," *Los Angeles Daily News,* Apr. 16, 1928, clipping in Hill Papers, box 22, folder "Earthquake Book—Correspondence 1929-30"; "Quake Rate Cut to be Extended," *LAT,* May 9, 1928, II:1; "Insurance Cut on Temblors," ibid., May 18, 1928, II:1, 2; "Our Debt to Geologist," *LAE,* May 10, 1928, clipping in Regional History Center Clipping Collection, file "Robert T. Hill."

79. Willis to Townley, Jan. 2, and Jan. 7, 1926, and Townley to Willis, Feb. 20, 1926, all in SSAC, carton 1, folder "Correspondence Jan.-May 1926"; Willis to Margaret D. Willis, May 12, 1926, BWP, box 39, folder 4. For the article in question, see Charles H. Lee, "The Future Development of the Metropolitan Area Surrounding San Francisco Bay," *BSSA* 16 (1926): 81-132.

80. Townley to Louis M. Blankenhorn, May 14, 1926, and Townley to Day, June 28, 1926 (quoted), both in SSAC, carton 1, folder "Correspondence Jan.-May 1926"; Townley to Guardian Fire Assurance Corporation of New York, Dec. 7, 1926, ibid., carton 1, folder "Correspondence June-Dec. 1926"; Townley to Shepard E. Barry, Jan. 10, 1927, ibid., carton 1, folder "Correspondence Jan.-June 1927."

81. Townley to Day, June 28, 1926, SSAC, carton 1, folder "Correspondence Jan.-May 1926"; Day to Townley, July 2, 1926, and Townley to Day, July 7, 1926, HWP, box 12, folder 2; Willis to Wood, Sept. 9, and Sept. 22, 1926, both in ibid., box 10, folder 1.

82. Townley to Charles H. Lee, Jan. 22, 1929, SSAC, carton 1, folder "Correspondence Jan.-May 1929"; N. H. Heck to Freeman, May 20, 1930, School of Engineering Records, ser. 1, box 3, folder 24; Bailey Willis, "Man Probes the Secret of Earthquakes: A Famous Geologist Explains Stanford's Shaking Table," *Baltimore Sun,* Mar. 8, 1931, clipping in BWP, box 48, folder 1; Petrie Mondell to Willis, Feb. 12, 1934, ibid., box 18, folder 42. For Willis's efforts to refute the continental drift hypothesis in the late 1920s and 1930s, see, for example, Naomi Oreskes, "The Rejection of Continental Drift," *Historical Studies in the Physical and Biological Sciences* 18 (1988): 311-348, on 334-337; idem, *The Rejection of Continental Drift: Theory and Method in American Earth Science* (New York: Oxford UP, 1999), 208-218; Robert P. Newman, "American Intransigence: The Rejection of Continental Drift in the Great Debates of the 1920's," *Earth Sciences History* 14 (1995): 62-83.

Chapter 4

1. Harry O. Wood to Arthur L. Day, Feb. 25, Oct. 13, and Nov. 25, 1927, HWP, box 12, folder 1; Wood to John C. Merriam, July 1, 1929, ibid., box 13, folder 7, p. 4; Wood to Merriam, June 22, 1927, ibid., box 13, folder 8, p. 5; Charles F. Richter, "Report for

the Period Oct. 1, 1927 to June 11, 1928," ibid.; Richter, "Report to June, 1929," ibid., box 13, folder 7; Wood to Merriam, July 1, 1930, ibid., box 13, folder 7, pp. 5–9; Wood to Merriam, July 1, 1931, ibid., box 13, folder 7, pp. 5–6; Judith R. Goodstein, "Waves in the Earth: Seismology Comes to Southern California," *Historical Studies in the Physical Sciences* 14 (1984): 201–230, on 215–217.

2. Wood to Merriam, July 1, 1931, HWP, box 13, folder 7, p. 7; Perry Byerly and Neil R. Sparks, "Earthquakes in Northern California and the Registration of Earthquakes at Berkeley-Mount Hamilton-Palo Alto from October 1, 1931, to March 31, 1932," *Bulletin of the [University of California] Seismographic Stations* 3 (1933): 53–96; Perry Byerly, "Northern California Earthquakes, April 1, 1932, to April 1, 1933," *BSSA* 22 (1934): 115–117; Wood to N. H. Heck, Oct. 4, 1928, HWP, box 17, folder 3; Wood to R. S. Patton, Aug. 28, 1933, ibid., box 18, folder 6; Thomas J. Maher, "The United States Coast and Geodetic Survey—Its Work in Collecting Earthquake Reports in California," *BSSA* 19 (1929): 77–79.

3. Harry O. Wood and Frank Neumann, "Modified Mercalli Intensity Scale of 1931," *BSSA* 21 (1931): 277–283. For the Rossi-Forel scale and other intensity scales developed around the turn of the century, see Charles Davison, "On Scales of Seismic Intensity and on the Construction and Use of Isoseismal Lines," ibid., 11 (1921): 95–129.

4. Goodstein, "Waves in the Earth," 222–226.

5. Wood to T. Z. Franklin, Nov. 14, 1929, HWP, box 3, folder 14; R. W. Stewart to Wood, Dec. 13, 1930, ibid., box 5, folder 14; Patton to Wood, Feb. 18, and Feb. 27, 1933, and Wood to Patton, Feb. 23, 1933, all in ibid., box 18, folder 6; Heck to Wood, May 22, 1934, ibid., box 17, folder 2; U.S. Coast and Geodetic Survey, *Destructive and Near-Destructive Earthquakes in California and Western Nevada, 1769-1933*, U.S. Coast and Geodetic Survey Special Publication 191 (Washington, DC: GPO, 1934).

6. Wood to Alfred D. Flinn, July 5, 1928, HWP, box 3, folder 7, and published in "Engineering Foundation Committee on Arch Dam Investigation: Discussion," *American Society of Civil Engineers Proceedings* 54 (1928): 2607–2610, on 2609; S. B. Morris and C. E. Pearce, "Design of Gravity Dam in San Gabriel Canyon to Resist Earthquakes," *BSSA* 19 (1929): 143–155; idem, "Earthquake Forces on Dams," ibid., 21 (1931): 204–215; H. M. Westergaard, "Water Pressures on Dams During Earthquakes," *American Society of Civil Engineers Transactions,* 98 (1933): 418–472, esp. 443–452.

7. "Henry Dievendorf Dewell, M. ASCE," *American Society of Civil Engineers Transactions* 112 (1947): 1432–1434; Perry Byerly, "Report of the Executive Committee of the Seismological Society of America for the Calendar Year 1932," *BSSA* 23 (1933): 41; Henry D. Dewell, "Earthquake Expectancy in California," *ENR* 96 (1926): 114–116; idem, "Building Against Earthquake Shock," *The Commonwealth: Transactions of the Commonwealth Club of California* 20 (1925): 226–233, first quote on 232; idem, "Building Against Earthquakes," *Pacific Coast Architect* 28, no. 4 (1925), 33–37, second quote on 33; idem, "Earthquake-Resistant Construction: I-Data of Design," *ENR* 100 (1928): 650–655; "The New Zealand Earthquake of February 3, 1931," *BSSA* 21 (1931): 251–260; Tachu Naito, "Earthquake-Proof Construction," ibid., 17 (1927): 57–94; Henry D. Dewell, "Earthquake-Resistant Construction: II-Principles of Design," *ENR* 100 (1928):

699–702; idem, "A Survey of the 1923 Earthquake Damage to Buildings in Tokyo, Japan," ibid., 105 (1930): 51–54.

8. H. M. Engle, "The Earthquake Resistance of Buildings from the Underwriter's Point of View," *BSSA* 19 (1929): 86–95; Wood to S. D. Townley, Sept. 14, 1929, and Henry D. Dewell to Townley, Sept. 17, 1929, both in SSAC, carton 1, folder "Correspondence June–Dec. 1929."

9. "Romeo Raoul Martel, F. ASCE," *American Society of Civil Engineers Transactions* 131 (1966): 904; "Proceedings of Special Meeting Held on March 18, 1930," *Proceedings of the San Francisco Section, American Society of Civil Engineers* 27, unnumbered issue (1930); "For Uniform Building Rules," *Pasadena Star-News*, Aug. 13, 1927, p. 11; A. H. Brandon to Bailey Willis, Oct. 26, 1927, School of Engineering Records (SC 165), Department of Special Collections, Stanford University Libraries, ser. 1, box 9, folder 39; L. H. Nishkian, "Design of Tall Buildings for Resistance to Earthquake Stresses," *A&E* 88, no. 3 (1927): 73–83; C. R. Harding, "Location and Design of Southern Pacific Company's Suisun Bay Bridge as Affected by Consideration of Earthquakes," *BSSA* 19 (1929): 162–166; Norman Green, "Concrete Walls for Earthquake Bracing of Tall Buildings," *ENR* 107 (1931): 364–366; H. H. Tracy, "Welded Joints for Seismic Stresses in a Tall Building," ibid., 109 (1932): 312–313.

10. Lydik S. Jacobsen, "Vibration Research at Stanford University," *BSSA* 19 (1929): 1–27; idem, "Earthquake Engineering Research at Stanford in the Period of 1906–1960," *EERIN* 10, no. 3 (May 1976): 1–10; idem, "Experimental Study of the Dynamic Behavior of Models of Timber Walls Subjected to an Impulsive, Horizontal, Ground Vibration," *BSSA* 20 (1930): 115–146; idem, "Motion of a Soil Subjected to a Simple Harmonic Ground Vibration," ibid.: 160–195; Henry D. Dewell, "Some Remarks on the Shaking Table Investigations," ibid.: 231–236; Dewell to Theodore Hoover, Dec. 29, 1927, John B. Leonard to Hoover, Apr. 2, 1928, Jacobsen, "Minutes of Vibration Research Meeting of April 10, 1928," Jacobsen to Hoover, Apr. 25, 1928, and L. S. Jacobsen, "Report on Tests of 'Fyr Kwake' Valve Characteristics Made at Stanford University," Nov. 6, 1928, all in School of Engineering Records, ser. 1, box 9, folder 39; Henry D. Dewell, Lydik S. Jacobsen, and A. W. Pioda, "Report of Sub-Committee on Dynamic Lateral Forces," Feb. 1931, BWP, box 6, folder 22; Leander M. Hoskins and Lydik S. Jacobsen, "Water Pressure in a Tank Caused by a Simulated Earthquake," *BSSA* 24 (1934): 1–32.

11. Walter E. Spear, "John Ripley Freeman, Past-President and Hon. M. Am. Soc. C. E.," *American Society of Civil Engineers Transactions* 98 (1933): 1471–1476; John R. Freeman, "First Draft for Circular to Members, Earthquake Insurance—Wind Insurance," July 20, 1925, and Manufacturers Mutual Fire Insurance Company and Associated Companies, "Earthquake Insurance," undated [ca. Sept. 1925], both in John R. Freeman Papers (MC 51), Institute Archives, Massachusetts Institute of Technology, Cambridge, MA, box 34, file "Earthquake Data," folder 1; Freeman to Kyoji Suyehiro, Dec. 17, 1930, and Freeman to Dewell and others, June 27, 1931, both in Freeman Papers, box 51, file "Letters Regarding Suyehiro"; Freeman to Louderback, Aug. 15, 1931, George D. Louderback Papers, Bancroft Library, Berkeley, CA, box 15, folder "Freeman"; Kyoji

Suyehiro, "Engineering Seismology: Notes on American Lectures," *Proceedings of the American Society of Civil Engineers* 58, no. 4 (1932), part 2; John Ripley Freeman, *Earthquake Damage and Earthquake Insurance* (New York: McGraw-Hill, 1932).

12. John R. Freeman, "Earthquake Data for the Structural Engineer Through Improved Seismology" (letter to editor), *ENR* 99 (1927): 74; Freeman to Patton, Mar. 17, 1930, copy in Louderback Papers, box 15, folder "Freeman"; Freeman to Wood, Apr. 10, 1930, HWP, box 3, folder 15; R. P. Lamont to Freeman, May 22, 1930, copy in School of Engineering Records, ser. 1, box 3, folder 24; N. H. Heck, "Accurate Records of Strong Earthquake Motions," *BSSA* 21 (1931): 285–288; *Earthquake Investigations in California, 1934–1935,* U.S. Coast and Geodetic Survey Special Publication 201 (Washington, DC: GPO, 1936); Dean S. Carder, ed., *Earthquake Investigations in the Western United States, 1931–1964,* U.S. Coast and Geodetic Survey Publication 41-2 (Washington, DC: GPO, 1965), 1–4.

13. Townley to Wood, Oct. 23, 1928, SSAC, carton 1, folder "Correspondence July–Dec. 1928"; Robert E. Andrews to Townley, Mar. 2, 1929, ibid., folder "Correspondence Jan.-May 1929"; Andrews to Byerly, Feb. 8, 1933, ibid., folder "Correspondence 1933 M–Z"; Andrews to Wood, May 27, 1929, HWP, box 7, folder 1.

14. Stanley Burne, "Engineer Explains Quake Safety Building Needs" (letter to editor), *San Diego Union,* Mar. 23, 1933, I:8.

15. Mark M. Falk, minutes of organizational meeting of Structural Engineers Association of Los Angeles County, office files, Structural Engineers Association of Southern California, Whittier, historical file; H. J. Brunnier to Keith Carlin, Apr. 10, 1930, Harold B. Hammill, "Structural Engineers' Association," [Feb. 1930], idem, "Minutes of Meeting of Structural Engineers Association of Northern California, November 10, 1930," idem, "Minutes of Meeting of the Structural Engineers Association of Northern California, December 8, 1930," and "Minutes of Meeting-Structural Engineers Association of Northern California," Jan. 12, 1932, all in office files, Structural Engineers Association of Northern California, San Francisco, drawer "Board, Misc., 1931–1948," folder "Minutes 1930–1936"; A. V. Saph, Jr., "List of Members of the Structural Engineers Association of Northern California," Nov. 18, 1932, ibid., folder "Correspondence [1930s]"; "Structural Engineers Association of Southern California Roster as of March 1, 1933," ibid., unlabeled folder; "Old Timers Luncheon," Mar. 24, 1958, ibid., folder "S.E.O.N.C. History."

16. C. H. Snyder, "Minimum Requirements for Lateral Bracings," Jan. 27, 1931, Jesse Rosenwald to Structural Engineers Association, Apr. 15, 1931, [Hammill], "Lateral Forces," Apr. 17, 1931, L. H. Nishkian, "Recommendations for Lateral Forces," Apr. 22, 1931, C. N. Bley et al., "Minimum Requirements for the Lateral Forces to be Resisted by Structures," May 12, 1931, Earle Russell et al., "Report of Committee on Lateral Forces," May 12, 1931, all in Structural Engineers Association of Northern California Office Files, drawer "Seismology I," folder "Earthquake Provisions 1931–1933"; Hammill, "Minutes of Meeting of the Structural Engineers Association of Northern California, Feb. 10, 1931," idem, "Minutes of Special Meeting of Structural Engineers Association of Northern California, Feb. 24, 1931," idem, "Minutes of Meeting of Structural Engineers Asso-

ciation of Northern California, March 10, 1931," idem, "Minutes of Meeting of Structural Engineers Association of Northern California, March 31, 1931," idem, "Minutes of Meeting of Structural Engineers Association of Northern California, May 12, 1931," idem, "Minutes of Meeting of Structural Engineers Association of Northern California, July 14, 1931," idem, "Minutes of Meeting of Structural Engineers Association of Northern California, Oct. 13, 1931," Saph, "Minutes of Meeting-March 29, 1932," idem, "Minutes of Meeting-May 10, 1932," all in ibid., drawer "Board, Misc., 1931-1948," folder "Minutes 1930-1936." There are no comparable records surviving for the early years of the Structural Engineers Association of Southern California; thus, I could not determine whether earthquake provisions were also discussed there.

17. Henry D. Dewell, "The Earthquake Resistance of Buildings from the Standpoint of the Building Code," *BSSA* 19 (1929): 96-100; David J. Witmer, "Uniform Building Code, California Edition," *California Journal of Development* 23, no. 4 (1933): 7-8, 25; "Proceedings of the One Hundred and Thirty-Sixth Regular Meeting," *Proceedings of the San Francisco Section, American Society of Civil Engineers* 23, no. 6 (1927); "State to be Made Safest for Lives," *LBPT,* Mar. 16, 1933, p. 9; "Building Code Offered Cities," *LAT,* Apr. 15, 1933, II:1. For the earthquake provisions of the Uniform Building Code, as adopted in 1933 by the California Division of Architecture, see Division of Architecture, *Appendix A: Temporary Regulation No. 5 Relating to the Safety of Design and Construction of Public School Buildings in California* (Sacramento: California State Printing Office, 1933).

18. Harry O. Wood, "Preliminary Report on the Long Beach Earthquake," *BSSA* 23 (1933): 43-56; Charles F. Richter, "An Instrumental Earthquake Magnitude Scale," ibid., 25 (1935): 1-32, on 22-23; National Board of Fire Underwriters, *Report on the Southern California Earthquake* (n.p.: n.d.), copy in Freeman Papers, box 34, file "Earthquake Data," folder 3.

19. Wood to Willis, Mar. 13, 1933, HWP, box 10, folder 1; Wood to Arthur Hannon, Mar. 24, 1933, ibid., box 4, folder 2; Richter to Robert Millikan, Mar. 20, 1933, RMP, box 24, folder 9; "Biggest Quake Yet to Come, Say Scientists After Studying Data," *San Diego Union,* Mar. 17, 1933, I:6; Walter B. Clausen, "Science Supplies Means to Avert Earthquake Loss," *LBPT,* Mar. 22, 1933, A:2.

20. "Fatality Rates in Quakes Low," *LAT,* Mar. 13, 1933, I:6; "The Bad Man From the West!" ibid., Mar. 18, 1933, I:1; "The Country Should Know The Truth," ibid., Mar. 15, 1933, I:9; "Tell World of New Long Beach," *LBPT,* Mar. 18, 1933, A:12; "Chamber Gives Warning," *Pasadena Star-News,* Mar. 28, 1933, p. 13; Charley Paddock, "Follow Facts," ibid., Apr. 5, 1933, p. 9.

21. "Aerial View Shows Sturdiness of Modern Buildings," *LAT,* Mar. 14, 1933, I:3; "Long Beach From the Air Yesterday-Big Buildings Undamaged," *LAE,* Mar. 14, 1933, I:3; "National Paper to Show Skyline of Long Beach," *LBPT,* Mar. 27, 1933, B:1, B:2; "Optimism Found in Nation," *LAT,* May 19, 1933, II:2 (quotes).

22. Compare "Los Angeles City Building for March Classified," *Southwest Builder and Contractor,* Apr. 7, 1933, p. 13, with "Los Angeles Building Permits for June," ibid., July 3, 1925, p. 46.

23. "Bills Seek Quake Aid," *LAT*, Mar. 14, 1933, I:1, 2; "Legislature Votes $50,000 To Aid Earthquake Sufferers," *SFC*, Mar. 14, 1933, p. 6; "$5,000,000 Relief Voted In Congress," *LAE*, Mar. 15, 1933, II:1, 8; "Quake-Damage Aid Authorized," *LAT*, Mar. 24, 1933, II:1, 2. For opposition to advertising southern California's devastation in this manner, see the editorial "Our Indomitable Spirit," ibid., Mar. 24, 1933, II:4.

24. "L. A. County Acts For Protection Against Quakes," *San Diego Union*, Mar. 21, 1933, I:1, 2; "School Damage Surveyed," *LAT*, Mar. 15, 1933, I:6; "City Closes Schools," ibid., Mar. 13, 1933, I:1, 5; "Class Work Resumed at 335 Schools," *LAE*, Mar. 21, 1933, II:1; "Quake Loans Are Sought," *Pasadena Star-News*, Mar. 13, 1933, p. 1; "Quake Exposes Public Schools as Sepulchres, Pretty Outside," *San Diego Union*, Mar. 17, 1933, I:1, 2; "Faulty School Construction Attacked," *LBPT*, Mar. 18, 1933, A:1.

25. "Survey of Quake-Damaged Schools Begun to Prevent Future Disaster," *LAE*, Mar. 15, 1933, I:3; "Fitts Will Probe Construction of Damaged Schools," ibid., Mar. 16, I:1, 2; "Statewide Drive Begun to Force Quakeproof School Construction," ibid., Mar. 17, I:3; "Fitts Promises Thorough Grand Jury Inquiry Into School Damage by Quake," ibid., Mar. 20, 1933, I:1, 2; "Quake Inquest Set Today," ibid., Mar. 21, I:1, 2; "Reconstruction Problems," *LAT*, Mar. 14, 1933, II:4; "The Earthquake Fault We Must Find!" ibid., Mar. 16, 1933, I:1.

26. "Good Buildings Immune to Quakes, Savants Say," *LAT*, Mar. 14, 1933, I:2 (quote); "Make All Public Buildings Safe," *LBPT*, Mar. 15, 1933, p. 10.

27. "Poor Building Methods Blamed for Quake Loss," *LAE*, Mar. 15, 1933, II:1.

28. R. R. Martel, "Caltech Engineering Head Blames Politics," *LAE*, Mar. 18, 1933, I:2; Samuel Harwick, "Class A Construction Logical School Type," ibid.; A. E. Holt, "Buildings Must Meet Condition," ibid.; Sumner P. Hunt, "Damage Could Have Been Cut," ibid.; C. J. Derrick, "New Legislation Urgent Need," ibid., I:2, 5; Charles C. Kinne, "Proper Design, Adequate Supervision Needed," ibid., I:2; Raymond G. Osborne, "Specification Codes Should Be Rewritten," ibid.; Paul W. Penland, "There Must Be Rigid Inspection On All Jobs," ibid.; John C. Austin, "Austin Backs 'Class A' Work," ibid.; Edward W. Cunningham, "Cunningham Cites Lack of Construction Detail," ibid.; Albert C. Martin, "Less Politics, Better Building Work Urged," ibid.; "Examiner Experts Analyze Cause of School Damage in Quake; Offer Safety Advice," ibid., I:1, 2; "ALL School Buildings Must Be SAFELY Built," ibid., Mar. 20, 1933, I:10 (quote).

29. "Coroner to Name Building Experts on Inquest Jury," *LAE*, Mar. 14, 1933, I:5 (quote); "Nine Experts to Serve on Quake Inquest Jury," *LAT*, Mar. 14, 1933, I:3; "Nine Experts Put On Quake Inquest Jury," *LAE*, Mar. 19, 1933, I:9; "Most of City's Schools Will Open Tomorrow," *LAT*, Mar. 19, 1933, I:1, 10; "Quake Facts Related By Pasadenan," *Pasadena Star-News*, Mar. 21, 1933, p. 9; "Quake Inquest to Open Today," *LAT*, Mar. 21, 1933, II:1, 2; "Huntington Park Ruin Laid to False Economy," *LAE*, Mar. 22, 1933, I:2.

30. "Quake Facts Related by Pasadenan"; "Huntington Park Ruin"; "Building Faults Cited At Inquest Over Quake Dead," *LAE*, Mar. 22, 1933, I:1, 2; "Safer Schools Demanded at Quake Quiz," ibid., Mar. 23, 1933, I:1, 2; "Compton City Inspector On Stand At Quiz," ibid., I:2; "Inquiry Reveals Building Faults in Recent Quake," *San Diego*

Union, Mar. 22, 1933, I:1, 2 (quotes); "Point to Future Earthquake Safety in Buildings of State," ibid., Mar. 23, 1933, I:1; "Quake Inquest Hears Bouelle," *LAT,* Mar. 22, 1933, II:1, 2; "Witness Lauds Newer Schools," *LAT,* Mar. 23, 1933, II:1, 7; "Jury Verdict Unhurried," *Pasadena Star-News,* Mar. 23, 1933, p. 10; "Wrecked Schools Hold Attention at Earthquake Inquest," *LBPT,* Mar. 22, 1933, B:1. See also "Best Types of School Buildings," ibid., Mar. 22, 1933, B:8; "Build Strongly, As Safeguard," *Pasadena Star-News,* Mar. 22, 1933, p. 4; "Experts Show Quake Lessons," *LAT,* Mar. 29, 1933, II:1, 2; "Lessons of the Quake," ibid., Mar. 30, 1933, II:4; "Quake Inquest Jury Demands School Safety," *LAE,* Mar. 29, 1933, I:1; "Safer Building Seen by Jury as Lesson of Quake," *LBPT,* Mar. 29, 1933, A:5; "Coroner's Jury Urges One-Story School Buildings for L. A.," *SFC,* Mar. 29, 1933, p. 5; "Rigid Building Code Urged to Guard Against Earthquake Damage," *Southwest Builder and Contractor,* Mar. 31, 1933, pp. 10, 11; "Better Construction in Earthquake Zone Recommended by Jury," *ENR* 110 (1933): 480–481; Ralph E. Homann, "Why They Failed," *Progressive Contractor,* May 1933, pp. 6–8. For the full text of the coroner's jury's verdict, see Victor L. Martins, "A Study of Public Schools in Southern California Damaged by the Earthquake of March 10, 1933" (master's thesis, University of Southern California, 1933), 129–144.

31. Wood, "Preliminary Report;"; "Report on the Southern California Earthquake," esp. 13.

32. Lynn Atkinson to Millikan, Mar. 23, 1933, Ormond A. Stone to Millikan, Mar. 24, 1933, Hoffmann to Henry S. Pritchett, Mar. 27, 1933, R. L. Daugherty and A. F. Barnard to Millikan, Mar. 27, 1933, Henry H. Baskerville to Millikan, Mar. 27, 1933, Palmer Sabin to Millikan, Mar. 29, 1933, Millikan to Hoffmann, Apr. 3, 1933, and Millikan to William Simpson, Apr. 12, 1933, all in RMP, box 24, folder 9; "R. F. C. Appoints Aid Fund Group," *LAT,* Mar. 28, 1933, II:8; "Schools Given Grand Jury Aid," ibid., Apr. 6, 1933, II:3; "Quake Building Parleys Start," ibid., Apr. 16, 1933, I:18; Joint Technical Committee on Earthquake Protection, *Earthquake Hazard and Earthquake Protection* (Los Angeles: n.p., 1933), preface. On Millikan's stature as a "sage," see Robert H. Kargon, *The Rise of Robert Millikan: Portrait of a Life in American Science* (Ithaca: Cornell UP, 1982), chap. 5.

33. Joint Technical Committee, *Earthquake Hazard,* quotes on 3–4, 6, 8, 13.

34. Wm. A. Simpson, undated cover letter attached to copy of Millikan Report in the Long Beach Earthquake File of the Long Beach Public Library.

35. "The Temblor Report," *LAT,* June 18, 1933, I:16; see also "Notes and Comment," *Southwest Builder and Contractor,* June 30, 1933, p. 19.

36. "Quake Exposes Public Schools as Sepulchres, Pretty Outside," *San Diego Union,* Mar. 17, 1933, I:1, 2; Richard S. Requa, "Architect Urges Quake-Resisting Buildings Here," ibid., Mar. 19, 1933, II:4; idem, "Temblor Severe on Apartments; Lessons Shown," ibid., Mar. 26, 1933, II:8; John S. Siebert, "Long Beach Building Methods Open Invitation to Earthquake Disaster" (letter to the editor), ibid., Apr. 17, 1933, I:2.

37. "When Building Codes Fail to Protect!" *SFC,* Mar. 20, 1933, p. 8; Albert J. Evers, "Safe Schools Cost But Little More" (letter to editor), ibid., Mar. 21, 1933; Anne de Guichy Treadwell to Willis, Mar. 25, 1933, BWP, box 22, folder 78; P. F. Gardiner to

Willis, Apr. 7, 1933, ibid., box 13, folder 82; Willis to A. V. Bowyer, Apr. 14, 1933, ibid., box 25, folder 9.

38. S. D. Townley, "Minutes of the Meeting of the Board of Directors of the Seismological Society of America, October 17, 1930," *BSSA* 20 (1930): 291–296, on 293; C. D. Wailes, Jr., and A. C. Horner, "Earthquake Damage Analyzed by Long Beach Officials," *ENR* 110 (1933): 684–686; "Quake-Proof Method of Building Drafted," *LBPT,* Mar. 20, 1933, A:7; "Council Adopts Safety Measures for Construction," ibid., Mar. 24, 1933, B:1; "Long Beach Adopts Special Code for Masonry in Buildings," *Southwest Builder and Contractor,* Mar. 31, 1933, pp. 12–13; [J. E. Shield], "Daily Report—Earthquake," Mar. 20–24, 1933, Structural Engineers Association of Southern California Office Files, Long Beach Earthquake File, vol. 1, item 25; R. W. Binder, "Steel Tie for Masonry Buildings to Resist Earthquake Shocks," *Southwest Builder and Contractor,* Apr. 14, 1933, p. 12.

39. "County Quake Law Passed," *LAT,* Mar. 21, 1933, II:1, 2; "New Building Ordinance In Effect At Once," *LAE,* Mar. 21, 1933, I:2; "Los Angeles County Adopts Uniform Building Code," *Southwest Builder and Contractor,* Apr. 14, 1933, p. 10.

40. "Quake-Damage Aid Authorized," *LAT,* Mar. 24, 1933, II:1, 2; "Quake Safety Moves Pushed," ibid., Mar. 30, 1933, II:1; "Board Studies Quake-Proof Building Plans," *LAE,* Mar. 30, 1933, II:10; "Revision of Los Angeles Code to Committee," *Southwest Builder and Contractor,* Mar. 31, 1933, p. 11; "Questionnaire by Mr. Blaine Noice," undated, Structural Engineers Association of Southern California Office Files, Long Beach Earthquake File, vol. 3, item 39; "Quake Code Up To Council," *LAT,* June 12, 1933, II:1, 2; "Earthquake Provisions of Los Angeles City Building Code Adopted," *Southwest Builder and Contractor,* Aug. 25, 1933, pp. 16–17; "Earthquake Regulations for Masonry Construction," ibid., Sept. 22, 1933, p. 14; N. A. Bowers, "California Makes Progress Against Earthquake Hazard," *ENR* 113 (1934): 14–17, on 15.

41. "Demand Action to Make Schools All Quakeproof," *San Diego Union,* Mar. 17, 1933, I:1, 3; "Statewide Drive Begun to Force Quakeproof School Construction," *LAE,* Mar. 17, 1933, I:3; "Safety Bill Up In Legislature," ibid., Mar. 19, 1933, I:6; "Quake Inquest Set Today; State Acts for School Safety," ibid., Mar. 21, 1933, I:1, 2; A. M. Rochlen, "State Will Act Today to Forbid Flimsy Schools," ibid., Mar. 22, 1933, I:1, 2; "Safer Schools Demanded at Quake Quiz; Legislature Gets Sharp 'Death Trap' Bill," ibid., Mar. 23, 1933, I:1, 2; "Safe School Legislation Demand Gains," ibid., Mar. 27, 1933, II:1, 10; A. M. Rochlen, "Builders Join In Appeal For Safe Schools," ibid., Mar. 30, 1933, II:1, 10; "Building of Schools to be Watched," *LBPT,* Mar. 29, 1933, A:1.

42. "Schools Can and Must Be Made Safe for Children," *SFC,* Mar. 24, 1933, p. 10; "Make Schools Safe Before Earthquake Makes Its Test," ibid., Mar. 25, 1933, p. 10; "Safety First for ALL School Buildings," *LAE,* Mar. 28, 1933, I:8; Chester Rowell, "Observations," *Pasadena Star-News,* Mar. 23, 1933, p. 4; "Teachers Pledge Code Legislation," ibid., p. 7; "School Safety First Requisite," *LBPT,* Mar. 28, 1933, B:8; "Women Score Poor Building," *LAT,* Apr. 5, 1933, II:1, 2; Thos. F. Chace, "A Contingency Parents Ought to Think About" (letter to editor), *SFC,* Mar. 31, 1933, p. 10; "Safe School Legislation Demand Gains," *LAE,* Mar. 27, 1933, II:1, 10; "Speed Urged In Protection For Students," ibid., Mar. 28, 1933, I:5; A. M. Rochlen, "Senators O. K. Bill for Quake-

proof School Buildings," ibid., Mar. 31, 1933, I:1; "Safe School Bill Up for Action Today," ibid., Apr. 3, 1933, II:1; "School Safety Beats Economy in Vote," ibid., Apr. 5, 1933, II:1; "Legislative Votes School Safety Laws," ibid., Apr. 6, 1933, I:1; "Rolph Signs Bill To Make Schools Safe," ibid., Apr. 11, 1933, I:1; J. H. M., "Northern California Chapter," *A&E* 113, no. 1 (1933): 57–58. For the full text of the Field Act as finally passed, see "School Designs to be Checked by State Architect," ibid., no. 2 (1933): 59–60; "State Architect Will Pass On Plans for All School Buildings in California," *Southwest Builder and Contractor*, Apr. 14, 1933, p. 8.

43. "Proceedings of the One Hundred and Thirty-Eighth Regular Meeting," *Proceedings of the San Francisco Section, American Society of Civil Engineers* 24, no. 2 (1928); "Building Code Offered Cities," *LAT*, Apr. 15, 1933, II:1; Division of Architecture, *Rules and Regulations Relating to the Safety of Design and Construction of Public School Buildings in California* (Sacramento: California State Printing Office, 1934); "Structural Engineers Wanted By State To Inspect Public Buildings," *Southwest Builder and Contractor*, July 28, 1933, p. 17; "Opportunity for Engineers," *A&E* 114, no. 2 (1933): 66–67; "Old Timers Luncheon"; A. V. Saph, Jr., "Minutes of Meeting of Board of Directors," Aug. 8, 1933, Structural Engineers Association of Northern California Office Files, drawer "Board, Misc., 1931–1948," folder "Minutes-Board of Directors, 1930–1936"; "Earthquake Research Committee Will Be Set Up By Architects," *Southwest Builder and Contractor*, Sept. 22, 1933, p. 8; Erle L. Cope, "The Field and Riley Acts Defined," *A&E* 117, no. 3 (1934): 53–57, on 57.

44. "State Building Code Supported As Emergency," *SFC*, Mar. 29, 1933, p. 5; "Quake-Proof Building Plans Made by Experts," ibid., Mar. 30, 1933, p. 20; "Building Code To Be Offered," ibid., Apr. 25, 1933, p. 8; Witmer, "Uniform Building Code, California Edition," 8; "Bills On Quake Relief Signed," *LAT*, May 28, 1933, I:6; "A Law for Earthquake Safety," *ENR*, 110 (1933): 721; N. H. Heck and Frank Neumann, "Destructive Earthquake Motions Measured for First Time," ibid.: 804–807; "Minutes of Regular Meeting," Mar. 13, 1934, and A. V. Saph, Jr., "Minutes of Regular Meeting— July 10, 1934," both in Structural Engineers Association of Northern California Office Files, drawer "Board, Misc., 1931–1948," folder "Minutes 1930–1936"; A. L. Bolton to Saph, May 3, 1934, ibid., folder "Correspondence [1930s]"; "Safety in Frankness," *SFC*, June 1, 1933, p. 10. For the full text of the Riley Act as finally passed, see "Earthquake Resistant Construction Is Provided For In Bill," *Southwest Builder and Contractor*, May 26, 1933, p. 8.

45. "Easing of Quake Law Advocated," *LAT*, Jan. 11, 1934, I:2; "Earthquake Law Flayed," ibid., Feb. 15, 1934, I:9; C. H. Kromer, "Earthquake-Resistant Construction Applied to California Schools," *ENR* 115 (1935): 856–860, on 857; Herbert J. Powell, "Santa Monica Plan As Developed in New School Building at Beach City," *Southwest Builder and Contractor*, Dec. 28, 1934, pp. 17–18; Frederick W. Jones, "Schools," *A&E* 120, no. 2 (1935): 17–21; Ralph C. Flewelling, "Schools, Earthquakes, and Progress," *California Art and Architecture*, Sept. 1935, pp. 20–21, 29–30, and Oct. 1935, pp. 18–19, 36; see also Louise H. Ivers, "The Evolution of Modern Architecture in Long Beach," *Southern California Quarterly* 68 (1986): 257–291.

46. "Schools Name Building Group," *LAT*, July 28, 1933, II:16; "Experts to Redesign School Buildings To Resist Earthquakes," *Southwest Builder and Contractor*, Aug. 4, 1933, p. 17; "Redesign School Buildings," *A&E* 114, no. 3 (1933): 75; "School Board Must Wreck 100 Buildings," *LAE*, Nov. 18, 1933, and "School Quake-Proofing Cost Set At $31,287,497," ibid., Dec. 27, 1933, both clippings in Regional History Center Clipping Collection, University of Southern California, Los Angeles, file "Allan E. Sedgwick"; "Architects Board Endorses School Bond Issue," *Southwest Builder and Contractor*, Mar. 9, 1934, p. 17; "Rigid Tests Revealed for New Schools," *LAE*, Mar. 14, 1934, II:1; "School Bonds Defeated," ibid., Mar. 21, 1934, I:1, 2; "School Board Going Ahead With Reconstruction," *Southwest Builder and Contractor*, Apr. 6, 1934, pp. 14–15; "P. W. A. and W. P. A. Approval of City School Reconstruction Given," ibid., Sept. 27, 1935, p. 14; "Institute Architects Vote Support Of School Bond Issue," ibid., Oct. 18, 1935, p. 15; H. C. Chambers, "The School Bond Issue—Safe School Construction Most Essential," ibid., Nov. 8, 1935, p. 17; "Los Angeles School Bonds Carry Three to One," ibid., Nov. 22, 1935, p. 16.

47. J. E. Byers, "Repairing Earthquake Damage," *ENR* 118 (1937): 362–366; A. S. Nibecker, Jr., "Safeguarding Schools Against Earthquakes," ibid.: 359–362.

48. Pasadena Board of Education, *81 Years of Public Education in Pasadena* (n.p., 1955), 35–36; "School Safety Plans Speeded," *LAT*, Jan. 22, 1934, I:6; "Quake-Proofing of City Schools Said Completed," *San Diego Union*, Mar. 16, 1937, I:16; "San Francisco School Program," *A&E* 116, no. 1 (1934): 59; "Busy On School Work," ibid., 118, no. 2 (1934): 62; "Berkeley School Work," ibid., 120, no. 2 (1935): 58.

49. Schools Committee to Los Angeles County Grand Jury, Dec. 27, 1933, HWP, box 5, folder 15, pp. 6–8; "California Earthquake Law Closes Five Berkeley Schools," *ENR* 111 (1933): 727; "Quake Proofing of City Schools Puzzles Board," *SEB*, Jan. 9, 1934, p. 4; "Quakes Shaded by Auto Deaths," *LAT*, Jan. 23, 1934, II:5; C. H. Kromer, "State Asked to Inspect More Than One Thousand School Plants," *Southwest Builder and Contractor*, Aug. 3, 1934, pp. 12–13, 16–17; idem, "Earthquake-Resistant Construction."

50. C. D. Wailes, Jr., "Reconstruction in Long Beach Following the Earthquake," *ENR* 112 (1934): 263–267; Edward L. Mayberry, "Large Los Angeles Store Given New Strength and Appearance," ibid., 114 (1935): 415–417; "Rehabilitated Long Beach City Hall In Modernistic Style," *Southwest Builder and Contractor*, Dec. 14, 1934, p. 10; "Trinity Telephone Building Notable Example of Modernization," ibid., Sept. 21, 1934, p. 19; H. M. Engle, "Elevated Tanks Strengthened After Earthquake Hazard Survey," *ENR* 116 (1936): 807–809; "Power Plants and Earthquakes," ibid., 117 (1936): 122–123; W. G. B., "Some California Engineering," ibid., 120 (1938): 689–690, quote on 689.

51. Robert H. Orr, "Many Bills Affecting Architects and Builders Before Legislature," *Southwest Builder and Contractor*, Mar. 1, 1935, pp. 11, 14; "Disposition of Bills in Legislature in Which Architects of State Were Interested," ibid., Aug. 9, 1935, pp. 17–18; Chas. Bursch, "Financing," *A&E* 121, no. 1 (1935): 39–43; "Minutes of Sacramento Meeting—March 22, 1935," Structural Engineers Association of Northern California Office Files, drawer "Board, Misc., 1931–1948," folder "Minutes 1930–1936"; Clarence H. Kromer to Byerly, Apr. 16, 1935, Byerly to Engle, Apr. 29, 1935, and Byerly

to Dewell, May 3, 1935, all in SSAC, carton 2, folder "Correspondence 1935 Field + Riley Acts"; "Notes of Meeting of Senate Investigating Committee, Safety of Design and Construction of Public School Buildings," Jan. 6, 1936, and "Hearing Before Special Senate Committee on School Buildings," Nov. 30, 1936, both in Legislative Papers, California State Archives, Sacramento, entry "Hearings on Field Act, 1936"; "Report of Special Senate Committee on Investigation of School Buildings," Apr. 30, 1937, ibid., entry "Hearing Before Assembly Standing Committee on Education on Field Act, 1952"; Thomas C. Lynch, "Opinion No. 65/324," May 4, 1966, and Fred W. Cheesebrough, "Statement to Assembly Interim Committee on Education," July 20, 1966, both in ibid., entry "Committee Hearing on Field Act July 20-21, 1966."

Chapter 5

1. Charles Louderback to H. R. Aldrich, July 28, 1945, and Perry Byerly to Harold Jeffreys, July 7, 1945, both in SSAC, carton 2, folder "Corres. 1945 A–M"; John P. Buwalda to Byerly, July 10, 1946, ibid., carton 3, folder "Corres. 1946 A–R"; Stanley Scott (interviewer), *John A. Blume,* Connections: The EERI Oral History Series, vol. 2 (Oakland: Earthquake Engineering Research Institute, 1994), 25; Charles F. Richter to A. Blake, Aug. 23, 1943, CRP, box 25, folder 3; C. F. Richter, "Memorandum on the Southern California Seismological Network Program," Jan. 28, 1963, Caltech Historical Files, Institute Archives, California Institute of Technology, Pasadena, folder A10.2; P. Byerly, "Minutes of the Meeting of the Board of Directors of the Seismological Society of America, Mar. 20, 1943," BSSA 33 (1943): 125-126.

2. Daniel J. Kevles, *The Physicists: The History of a Scientific Community in Modern America* (New York: Alfred A. Knopf, 1977), 287-392; Brian Balogh, "Reorganizing the Organizational Synthesis: Federal-Professional Relations in Modern America," *Studies in American Political Development* 5 (1991): 119-172.

3. Statistics from T. H. Watkins, *California: An Illustrated History* (New York: Weathervane Books, 1973), 435-436, 450, 464-465. On Stanford, see also Stuart W. Leslie, *The Cold War and American Science: The Military-Industrial-Academic Complex at MIT and Stanford* (New York: Columbia UP, 1993).

4. William H. Chafe, *The Unfinished Journey: America since World War II,* 2nd ed. (New York: Oxford UP, 1991), chaps. 6, 11; Brian Balogh, "Introduction," *Journal of Policy History* 8 (1996): 1-33; Hugh Heclo, "The Sixties' False Dawn: Awakenings, Movements, and Postmodern Policy-making," ibid.: 34-63; Robert Cameron Mitchell, "From Conservation to Environmental Movement: The Development of the Modern Environmental Lobbies," in Michael J. Lacey, ed., *Government and Environmental Politics: Essays on Historical Developments since World War Two* (Washington, DC: Woodrow Wilson Center Press, 1989), 81-113; John McPhee, *Encounters with the Archdruid* (New York: Farrar, Straus and Giroux, 1971).

5. Richter, "Memorandum"; Perry Byerly, "The Beginnings of Seismology in America," in O. H. Hildebrand et al., *Symposium on the Physical and Earth Sciences* (Berkeley: U of California P, 1958), 42-52; C. F. Richter, *Elementary Seismology* (San Francisco: W. H. Freeman, 1958), 333; idem, "The Seismological Laboratory and Earthquake

Study," *Earthquake and Science* [Caltech alumni magazine], Feb. 1948, pp. 11-12; Byerly, "Location of Epicenters," undated [ca. 1948] draft, SSAC, carton 3, folder "Corres. 1948 A-F"; C. R. Allen, P. St. Amand, C. F. Richter, and J. M. Nordquist, "Relationship Between Seismicity and Geologic Structure in Southern California," *BSSA* 55 (1965): 753-797.

6. Byerly to Buwalda, July 16, 1946, and George D. Louderback to Buwalda, Oct. 30, 1946, both in SSAC, carton 3, folder "Corres. 1946 A-R"; Byerly to Frederick S. Duhring, May 23, 1947, and Byerly to Buwalda, July 21, 1947, both in ibid., carton 3, folder "Corres. 1947 A-G."

7. C. A. Whitten, "Horizontal Earth Movement, Vicinity of San Francisco, California," *Transactions, American Geophysical Union* 29 (1948): 318-323.

8. Perry Byerly, "Report of the Secretary, February 12, 1944, to February 17, 1945," *BSSA* 35 (1945): 85; Karl V. Steinbrugge, "Report of Secretary for the Period May 6, 1960, to March 27, 1961," ibid., 51 (1961): 470.

9. Byerly to San Francisco Chamber of Commerce, Aug. 20, 1948, SSAC, carton 3, folder "Corres. 1948 A-F"; Byerly to R. C. Strange, Nov. 16, 1945, ibid., carton 2, folder "Corres. 1945 N-Z"; Byerly to James B. Macelwane, Jan. 19, 1946, ibid., carton 3, folder "Corres. 1946 A-R"; Perry Byerly, Alexis I. Mei, and Carl F. Romney, "Dependence on Azimuth of the Amplitudes of P and PP," *BSSA* 39 (1949): 269-284; Beno Gutenberg, "On the Layer of Relatively Low Wave Velocity at a Depth of About 80 Kilometers," ibid., 38 (1948): 121-148.

10. Karl V. Steinbrugge and Donald F. Moran, "An Engineering Study of the Southern California Earthquake of July 21, 1952 and Its Aftershocks," *BSSA* 44 (1954): 199-462, quote on 250; idem, "Damage Caused by the Earthquakes of July 6 and August 23, 1954," ibid., 46 (1956): 15-33; idem, "An Engineering Study of the Eureka, California, Earthquake of December 21, 1954," ibid., 47 (1957): 129-153; idem, "Engineering Aspects of the Dixie Valley-Fairview Peak Earthquakes," ibid.: 335-348; see also George G. Shor, Jr., and Ellis Roberts, "San Miguel, Baja California Norte, Earthquakes of February 1956: A Field Report," ibid., 48 (1958): 101-116. For biographical details on Steinbrugge, see Stanley Scott (interviewer), *Henry J. Degenkolb*, Connections: The EERI Oral History Series, vol. 1 (Oakland: Earthquake Engineering Research Institute, 1994), 31, 172.

11. Richter, *Elementary Seismology*, 228.

12. Scott, *John A. Blume*, 49-51; idem, *George W. Housner*, Connections: The EERI Oral History Series, vol. 4 (Oakland: Earthquake Engineering Research Institute, 1997), 131-135; Advisory Committee on Engineering Seismology, "Minutes of Meeting Held 18 Sept. 1948" (quote), Lydik S. Jacobsen Papers (SC 235), Stanford University Archives, box 1, folder "Earthquake Research Institute, 1948"; John A. Blume, "Plans for the Earthquake Engineering Research Institute," Apr. 12, 1949, and Earthquake Engineering Research Institute, "Minutes of Meeting of Board of Directors, November 19, 1949," both in ibid., box 1, folder "Earthquake Research Institute, 1949"; "Minutes, Earthquake Engineering Research Institute Meeting of 1 April 1950," ibid., folder "Earthquake Research Institute, 1950"; Franklin P. Ulrich, "Minutes of the EERI Board of Di-

rectors, December 2, 1950," ibid., folder "Earthquake Engineering Research Institute"; John A. Blume, "Minutes, Earthquake Engineering Research Institute," Feb. 24, 1951, ibid., folder "Seismology."

13. Byerly, "Beginnings of Seismology"; Richter to G. W. Quade, Apr. 14, 1955, and Richter to P. S. Barker, Aug. 23, 1955, both in CRP, box 30, folder 15; Richter to J. A. O'Gorman, Oct. 17, 1955, ibid., box 29, folder 12; Hugo Benioff and Richter to Philip S. Rowan, Feb. 24, 1956, John H. Beatty to Richter, Oct. 10, 1957, and Richter to Beatty, Oct. 14, 1957, all in ibid., box 27, folder 3; W. A. Ackerman to Richter, Oct. 26, 1956, ibid., box 24, folder 10; M. R. Stokesbury to Richter, Jan. 23, 1958, ibid., box 27, folder 2.

14. *Crustal Strain and Fault Movement Investigation Progress Report*, Department of Water Resources Bulletin No. 116-1 (Sacramento: Department of Water Resources, 1963), vi-vii, 1, 15; *Geodimeter Fault Movement Investigations in California*, Department of Water Resources Bulletin No. 116-6 (Sacramento: Department of Water Resources, 1968), 1-3; Laurence James, "The California Water Plan," *Geotimes*, Nov. 1964, pp. 9–11; Scott, *George W. Housner*, 152, 159, 167–169.

15. "Calls Parapet Walls an Earthquake Hazard," *ENR*, May 6, 1948, p. 2; "Well Built," *LAT*, July 25, 1952, II:4.

16. Frederic A. Chase to Herbert R. Klocksiem, Dec. 27, 1951, and J. A. Russell to Francis Dunn, Dec. 27, 1951, both in California State Archives, Sacramento, entry "Hearing before Assembly Standing Committee on Education on Field Act, 1952."

17. "Minutes, Sub-Committee on the 'Field Act,'" Mar. 4, 1952, Legislative Papers, California State Archives, entry "Hearing before Assembly Standing Committee on Education on Field Act, 1952."

18. "Highlights of Municipal Developments," *ENR*, June 26, 1947, p. 58; "San Francisco Building Code is Revised" (quote), ibid., Aug. 28, 1947, p. 3; *Building Codes: City and County of San Francisco, 1948* (San Francisco: Stark-Roth Printing & Publishing, 1948); for the debates discussed in this paragraph, see also Scott, *John A. Blume*, 36–41; idem, *Henry J. Degenkolb*, 134–135; idem, *Michael V. Pregnoff, John E. Rinne*, Connections: The EERI Oral History Series, vol. 3 (Oakland: Earthquake Engineering Research Institute, 1996), 49–50, 103–125; George W. Housner, "Current Practice in the Design of Structures to Resist Earthquakes," *BSSA* 39 (1949): 169–179; H. M. Engle, "Earthquake Provisions in Building Codes," ibid., 43 (1953): 233–237.

19. Arthur W. Anderson et al., "Lateral Forces of Earthquake and Wind," *American Society of Civil Engineers Transactions* 117 (1952): 716–780.

20. "Codes for Tall Buildings," *Proceedings, 26th Annual Convention, Structural Engineers Association of California, 1957*, pp. 13–45; T. P. Tung and N. M. Newmark, "Numerical Analysis of Earthquake Response of a Tall Building," *BSSA* 45 (1955): 269–278; Clarence J. Derrick, "The Damage Potential of Earthshocks," *Proceedings, Structural Engineers Association of California Twenty-Third Annual Convention, 1954*, individually paginated.

21. Steinbrugge and Moran, "Engineering Study of the Southern California Earthquake," esp. 220.

22. "Schools Safe at Tehachapi, Survey Shows," *LAT,* July 23, 1952, I:7; "County Building Code Proves Merit in Quakes," ibid., July 24, 1952, I:21; "Well Built."

23. "Quake Causes City to Review Disaster Setup," *SFC,* Mar. 26, 1957, pp. 1, 4; "Lessons of Quake Reviewed by City," ibid., Mar. 27, 1957, pp. 1, 6; Charles F. Richter, "Expert: The Next Great Earthquake Is Not Far Off," ibid., p. 2; Charles Raudebaugh, "No Cause to Fear A Quake Like '06," ibid., Apr. 1, 1957, pp. 1, 5; Richter to Merrill Windsor, Apr. 1, 1957, CRP, box 33, folder 1. For earlier expressions of smugness by earthquake engineers, see Derrick, "The Damage Potential of Earthshocks"; "S. F. Building Techniques 'Superior,'" *SFC,* June 15, 1956, p. 14.

24. C. Martin Duke and Morris Feigen, "Earthquake and Blast Effects on Structures: A Summary of the Symposium," *Proceedings, Structural Engineers Association of California Annual Convention, 1952,* individually paginated, quote on 25; see also Fred N. Severud and Anthony F. Merrill, *The Bomb: Survival and You* (New York: Reinhold, 1954), 236–239.

25. Duke and Feigen, "Earthquake and Blast Effects"; *Proceedings of the Symposium on Earthquake and Blast Effects on Structures: Los Angeles, California, June 1952* (n.p., 1952); Earthquake Engineering Research Institute, "Minutes of Meeting of Board of Directors, November 19, 1949," Jacobsen Papers, box 1, folder "Earthquake Research Institute, 1949"; Frank Neumann to Blume, Jan. 9, 1950, ibid, box 1, folder "Earthquake Research Institute, 1950"; Franklin P. Ulrich, "Minutes of the EERI Board of Directors, December 2, 1950," ibid., folder "Earthquake Engineering Research Institute . . ."; Scott, *George W. Housner,* 133–134.

26. J. Morley English, "Structures Research at the University of California, Los Angeles," *Proceedings, Structural Engineers Association of California Annual Convention, 1952,* individually paginated, 4–5; Jack R. Benjamin, "Structural Research at Stanford University," ibid., 2–3; N. M. Newmark to New Research Personnel in Structural Research, June 9, 1953, and "Minutes-Meeting of Structural Research Staff," Mar. 15, 1956, both in Nathan M. Newmark Papers, University of Illinois Archives, Champaign, box 3, folder "Minutes"; H. B. Zackrison to Newmark, Oct. 5, 1953, ibid., box 3, folder "Engineering Approach"; G. W. Housner, R. R. Martel, and J. L. Alford, "Spectrum Analysis of Strong-Motion Earthquakes," *BSSA* 43 (1953): 97–119; Scott, *George W. Housner,* 48.

27. Benjamin, "Structural Research at Stanford," 2–3; Lydik S. Jacobsen, "Earthquake Engineering Research at Stanford in the Period of 1906–1960," *EERIN* 10, no. 3 (1976): 1–10, on 8.

28. Nathan M. Newmark, "An Engineering Approach to Blast-Resistant Design," *Transactions of the American Society of Civil Engineers* 121 (1956): 45–64; Tung and Newmark, "Numerical Analysis of Earthquake Response."

29. Spencer R. Weart, *Nuclear Fear: A History of Images* (Cambridge: Harvard UP, 1988), 184–190, 199–214; Charles C. Bates, Thomas F. Gaskell, and Robert B. Rice, *Geophysics in the Affairs of Man: A Personalized History of Exploration Geophysics and Its Allied Sciences of Seismology and Oceanography* (New York: Pergamon Press, 1982), 118–122; Scott, *John A. Blume,* 55–61, 110–111, 119–121; Michael Drosnin, *Citizen Hughes* (New York: Holt, Rinehart and Winston, 1985), chaps. 6, 7; John A. Blume, "Response

of Highrise Buildings to Ground Motion From Underground Nuclear Detonations," *BSSA* 59 (1969): 2343-2370.

30. Bruce Bolt, *Nuclear Explosions and Earthquakes: The Parted Veil* (San Francisco: W. H. Freeman and Co., 1976), chap. 5; Bates et al., *Geophysics in the Affairs of Man*, 122-133; Kai-Henrik Barth, "Science and Politics in Early Nuclear Test Ban Negotiations," *Physics Today* 51 (Mar. 1998): 34-39. Barth is currently completing a dissertation on seismology and the nuclear weapons tests from 1945 to 1970.

31. Bolt, *Nuclear Explosions and Earthquakes*, 108-112; Bates et al., *Geophysics in the Affairs of Man*, 169-194.

32. For Lamont, see, for example, Ivan Tolstoy, Richard S. Edwards, and Maurice Ewing, "Seismic Refraction Measurements in the Atlantic Ocean (Part 3)," *BSSA* 43 (1953): 35-48, on 35; W. S. Jardetzky and Frank Press, "Crustal Structure and Surface-Wave Dispersion, Part III: Theoretical Dispersion Curves for Suboceanic Rayleigh Waves," ibid.: 137-144, on 137. For Caltech, see Richter to Markus Bath, July 14, 1958, CRP, box 24, folder 14; "Seismological Laboratory: Record of Growth," undated [ca. 1963] tabulation in FPP, box 4, folder 136.

33. National Research Council Committee on Seismology, *Seismology: Responsibilities and Requirements of a Growing Science, Part I: Summary and Recommendations* (Washington, DC: National Academy of Sciences, 1969), 34-35.

34. Karl V. Steinbrugge, "Report of the Secretary for the Period April 9, 1963 to March 25, 1964," *BSSA* 54 (1964): 1569; idem, "Minutes of the Meeting of the Board of Directors of the Seismological Society of America, April 8, 1963," ibid., 53 (1963): 859-862; idem, "Report of Secretary for the Period April 17, 1962 to April 8, 1963," ibid.: 863.

35. Based on count of acknowledgments in articles appearing in 1963 regular issues, excluding the two special issues on the 1960 Chilean earthquake. I assumed that authors listing affiliation with the Air Force Cambridge Research Laboratory were working on Vela Uniform project. Of the remaining 20 articles, 3 acknowledged non–Vela Uniform federal funding from the Office of Naval Research or the National Science Foundation, and 10 were written by authors at institutions outside the United States.

36. Richter, "Memorandum."

37. Paul Boyer, *By the Bomb's Early Light: American Thought and Culture at the Dawn of the Atomic Age* (New York: Pantheon Books, 1985), chap. 26; Guy Oakes, *The Imaginary War: Civil Defense and American Cold War Culture* (New York: Oxford UP, 1994); Anne Claire Fitzpatrick, "Civil Defense under the Truman Administration: The Impact of Politicians and Scientists" (master's thesis, Virginia Polytechnic Institute and State University, 1992).

38. Barbara Shaw Young, "Community Response to Earthquake Threat in Southern California, Part Three: The Organizational Response," final technical report on National Science Foundation grants NSF ENV76-24154 and NSF-PFR78-23887 (1980), copy in Natural Hazards Research Center Library, Boulder, CO, 5-10, 14-18; John W. Davis to Richter, Mar. 22, 1963, CRP, box 31, folder 18.

39. Richter to F. L. Campbell, Aug. 6, 1947, CRP, box 25, folder 12; Richter to Lucian

Bourdages, Feb. 27, 1956 (quote), ibid., box 31, folder 27; Richter, *Elementary Seismology*, chaps. 8, 24; C. F. Richter, "Seismic Regionalization," *BSSA* 49 (1959): 123–162.

40. Caroline M. Barrick to Richter, Sept. 24, 1962, and undated draft of Richter's talk, both in CRP, box 31, folder 14.

41. Charles F. Richter, "Earthquake Risk," *The Firemen's Grapevine*, Aug. 1963, pp. 4–7, copy in CRP, box 31, folder 18; Richter to Warren M. Dorn, Sept. 30, 1963, ibid., folder 12; "Governor's Conference on Disaster Preparedness," program, Nov. 1, 1963, and Richter, "Earthquake Disaster-Causes, Effects, Precautions," text of talk, Nov. 1, 1963, both in ibid., folder 22.

42. Evar P. Peterson to Richter, Aug. 5, 1963; U.S. Civil Defense Council and California Civil Defense and Disaster Association, "Joint Conference," program, May 20–22, 1964; Richter, "Notes to Program of CD Conference," May 21, 1964; and Richter, "Report on Earthquake Risk and Disaster Planning in California and the West," text of talk, May 21, 1964 (both quotes on 1), all in CRP, box 31, folder 19.

43. C. F. Richter, "Notes on California Earthquake Risk and Protection," Oct. 17, 1963, CRP, box 31, folder 31; C. F. Richter, "Our Earthquake Risk-Facts and Non-Facts," *California Institute of Technology Quarterly*, Jan. 1964, pp. 2–11.

44. Richter to William W. Ward, Oct. 22, 1963, CRP, box 31, folder 22; Richard F. Gordon to Richter, Mar. 11, 1964, ibid., box 34, folder 6; Richter to Evar P. Peterson, Mar. 17, and Mar. 19, 1964, both in ibid., box 31, folder 19; John W. Davis to Richter, June 3, 1964, ibid., box 25, folder 16; Richter to Price Berrien, Aug. 28, 1967, ibid., folder 2; Edward P. Joyce to Richter, Nov. 7, 1968, ibid., box 27, folder 14.

45. Kenneth Hahn to Richter, Aug. 20, 1963, Richter to Hahn, Aug. 27, 1963, and Hahn to Richter, Sept. 4, 1963, all in CRP, box 31, folder 12; Davis to Richter, May 12, 1964, ibid., box 25, folder 16.

46. Cleo Boschoff to Richter, Apr. 11, 1963, Melville I. Stark to Richter, Aug. 14, 1964, Richter to Stark, Aug. 17, 1964, and Stark to Richter, Oct. 5, 1964, all in CRP, box 27, folder 2; Emile L. Meine to Richter, Jan. 13, 1964, ibid., box 31, folder 28.

47. Brian Balogh, *Chain Reaction: Expert Debate and Public Participation in American Commercial Nuclear Power, 1945–1975* (Cambridge: Cambridge UP, 1991); George T. Mazuzan and J. Samuel Walker, *Controlling the Atom: The Beginnings of Nuclear Regulation, 1946–1962* (Berkeley: U of California P, 1985).

48. Thomas Raymond Wellock, *Critical Masses: Opposition to Nuclear Power in California, 1958–1978* (Madison: U of Wisconsin P, 1998), 18–30; J. Samuel Walker, *Containing the Atom: Nuclear Regulation in a Changing Environment, 1963–1971* (Berkeley: U of California P, 1992), 85, 87.

49. Wellock, *Critical Masses*, 33–51; Balogh, *Chain Reaction*, 241–244, 250–253; Ann Lage, "Oral History Interview with David E. Pesonen" (1992), copy of transcript in Bancroft Library, University of California, Berkeley, 42–47, esp. 45.

50. Wellock, *Critical Masses*, 54–55; Lage, "Interview with Pesonen," 58; Pierre Saint-Amand, *Geologic and Seismologic Study of Bodega Head* (n.p.: Northern California Association to Preserve Bodega Head and Harbor, 1963), copy in Bancroft Library. For Saint-Amand's involvement in cloud-seeding, which was used in a covert attempt to wash out

the Ho Chi Minh Path, see Senate Committee on Foreign Relations, *Prohibiting Military Weather Modification: Hearing*, 92nd Cong., 2nd sess., July 26 and 27, 1972; idem, *Weather Modification: Hearing*, 93rd Cong., 2nd sess., Jan. 25 and Mar. 20, 1974, pp. 32–50 (I thank Tanya Levin for pointing out these references to me).

51. Walker, *Containing the Atom*, 90–92; Wellock, *Critical Masses*, 51–56.

52. Julius Schlocker, Manuel G. Bonilla, Alfred Clebsch, Jr., and Jerry P. Eaton, "Geologic and Seismic Investigations of a Proposed Nuclear Power Plant Site on Bodega Head, Sonoma County, California," USGS Report TEI-837, Sept. 1963, copy in Elmer C. Marliave Papers, Water Resources Center Archives, Berkeley, CA, box 2, file 1.12. Quotes from Eaton on 50.

53. *Earthquakes, the Atom, and Bodega Head* (Berkeley: Northern California Association to Preserve Bodega Head and Harbor, 1964), copy in Bancroft Library; Harold L. Price to C. C. Whelchel, Oct. 2, 1963, Marliave Papers, box 2, file 1.12; "Bodega Is Unsafe, U.S. Expert Finds," *San Francisco News Call Bulletin*, Oct. 4, 1963, final edition, p. 1; "U.S. Report On Bodega-'It's Unsafe,'" *SFC*, Oct. 5, 1963, pp. 1, 9; see also "Bodega A—Plant Risks Quake Peril, Udall Says," ibid., May 22, 1963, p. 2. For a similarly incendiary article about possible earthquake damage at Bodega Bay issued by the St. Louis–based Committee for Nuclear Information, see Lindsay Mattison and Richard Daly, "A Quake at Bodega," *Nuclear Information*, Apr. 1964, pp. 1–12, clipping in Marliave Papers, box 1, file 1.1, folder 2.

54. James B. Koenig, "The Geologic Setting of Bodega Head," *Mineral Information Service* 16, no. 7 (July 1963): 1–10, on 10; see also Housner to Mautz, Jan. 29, 1964, Marliave Papers, box 1, file 1.1, folder 1.

55. "PG&E Comments on AEC Release of Oct. 4, 1963 Re Bodega Bay Plant," Marliave Papers, box 1, file 1.1, folder 2; Don Tocher to Ferd Mautz, Oct. 20, 1963, ibid., folder 1; Don Tocher and E. C. Marliave, "Geologic and Seismic Investigation of the Site for a Nuclear Electric Power Plant on Bodega Head, California," Jan. 1964, ibid., file 1.4.

56. Julius Schlocker and Manuel G. Bonilla, "Engineering Geology of the Proposed Nuclear Power Plant Site on Bodega Head, Sonoma County, California," USGS report TEI-844, Dec. 1963, Marliave Papers, box 2, file 1.13.

57. Mautz, "Areas of Disagreement with the Department of Interior Report on Bodega Site," Jan. 27, 1964, Marliave Papers, box 1, file 1.1, folder 2; John F. Bonner, "Docket No. 50-205 Amendment No. 7," Mar. 31, 1964, and "Docket No. 50-205 Amendment No. 8," July 20, 1964, both in ibid., box 2, file 1.9.

58. Walker, *Containing the Atom*, 96–99; David Okrent, *Nuclear Reactor Safety: On the History of the Regulatory Process* (Madison: U of Wisconsin P, 1981), 266–273; E. C. Marliave to Mautz, May 28, 1965, and Hugo Benioff to Mautz, June 7, 1965, both in Marliave Papers, box 1, file 1.1, folder 3.

59. Walker, *Containing the Atom*, 101–107; Wellock, *Controlling Masses*, 61–66; Okrent, *Nuclear Reactor Safety*, 273–277; for lawyerly fireworks, see especially "LA Consultant Resumes Stand," *Santa Monica Evening Outlook*, July 31, 1965, night final, p. 2.

60. Richard H. Jahns to John Benkart, Jan. 29, 1970 (quote), RJP, ser. 2, carton 18, file "B"; see also Jahns to Ian Campbell, Sept. 23, 1964, ibid., carton 25, file "Campbell, Ian"; Richter to A. E. Gaede, July 8, 1963, CRP, box 26, folder 11; Richter to Wayne N. Johnson, Oct. 13, 1965, ibid., box 24, folder 15; Clarence R. Allen to Frank C. DiLuzio, May 24, 1966, ibid., box 25, folder 13.

61. Harold Gilliam, "Coming: Another Shake," *SFC*, Nov. 10, 1963, "This World" section, pp. 2, 4; idem, "The Anatomy of a Quake," ibid., Nov. 17, 1963, "This World" section, pp. 2, 4, 22.

62. Jahns to Benkart, Jan. 29, 1970.

Chapter 6

1. National Research Council Committee on the Alaska Earthquake, *Toward Reduction of Losses from Earthquakes: Conclusions from the Great Alaska Earthquake of 1964* (Washington, DC: National Academy of Sciences, 1969), 27-31; Stanley Scott (interviewer), *George W. Housner,* Connections: The EERI Oral History Series, vol. 4 (Oakland: Earthquake Engineering Research Institute, 1997), first quote on 231; Gordon B. Oakeshott, "The Alaskan Earthquake," *Mineral Information Service* 17 (1964): 119-125, second quote on 124; see also Mark N. Christensen and Bruce A. Bolt, "Earth Movements: Alaskan Earthquake, 1964," *Science* 145 (1964): 1207-1216; *The Alaskan Earthquake,* Department of Water Resources Bulletin No. 116-5 (Sacramento: Department of Water Resources, 1965); Stanley Scott (interviewer), *Henry J. Degenkolb,* Connections: The EERI Oral History Series, vol. 1 (Oakland: Earthquake Engineering Research Institute, 1994), 68.

2. G. K. Gilbert, "Earthquake Forecasts," *Science,* new ser., 29 (1909): 121-138; H. Landsberg, "The Problem of Earthquake Prediction," ibid., 82 (1935): 37; Ernest A. Hodgson, "A Proposed Research into the Possibilities of Earthquake Prediction," *BSSA* 13 (1923): 100-104; T. A. Jaggar, "Predicting Earthquakes," *Scribner's Magazine* 76 (1924): 370-382; Andrew C. Lawson, "The Prediction of Earthquakes," *University of California Chronicle* 24 (1922): 315-336; "Earthquake Prediction by Geodetic Measurement," *ENR* 88 (1922): 393; Henry D. Dewell, "Earthquake Expectancy in California," ibid., 96 (1926): 114-116.

3. Harry O. Wood and B. Gutenberg, "Earthquake Prediction," *Science* 82 (1935): 219-220; J. B. Macelwane, "Forecasting Earthquakes," *BSSA* 36 (1946): 1-4.

4. "Quake Forecasts in 10 Years Seen," *NYT,* Apr. 23, 1961, p. 19; Chuji Tsuboi, Kiyoo Wadati, and Takahiro Hagiwara, *Prediction of Earthquakes: Progress to Date and Plans for Further Development* (Tokyo: Earthquake Research Institute, University of Tokyo, 1962); T. Hagiwara and T. Rikitake, "Japanese Program on Earthquake Prediction," *Science* 157 (1967): 761-768.

5. Jack Oliver, "Earthquake Prediction," *Science* 144 (1964): 1364-1365.

6. "Studying the San Andreas Fault," *Engineering and Science* [Caltech alumni magazine], Nov. 1964, pp. 12-14; Frank Press to Clarence Allen et al., Sept. 16, 1963, CRP, box 34, folder 6; Roy Hanson, "NSF Support of University Research Pertinent to Earthquake Prediction," undated [summer 1964], FPP, box 25, folder 770.

7. Biographical notes in inventory to FPP; see also Philip M. Boffey, "Frank Press, Long-Shot Candidate, May Become Science Advisor," *Science* 195 (1977): 763–766, on 766.

8. Press to Donald F. Hornig, Apr. 2, 1964, FPP, box 25, folder 771.

9. Robert G. Fleagle, "Memorandum for Ad Hoc Panel on Seismological Research," Apr. 9, 1964, and OST Ad Hoc Panel on Seismological Research, untitled memo, Apr. 19, 1964 (quotes), both in FPP, box 25, folder 771.

10. Press to Hornig, July 1, 1964, and Hornig to Press, July 8, 1964, both in FPP, box 25, folder 771.

11. Ad Hoc Panel on Earthquake Prediction, "Earthquake Prediction: A Proposal for a Ten Year Program of Research," FPP, box 25, folder 774 (bears publication date of Sept. 1965 but is essentially unchanged from a draft dated May 1965 that is located in box 25, folder 773).

12. Donald F. Hornig, "Memorandum for Members and Consultants-at-Large, President's Science Advisory Committee," June 25, 1965, FPP, box 25, folder 772; F. A. Long to Press, June 1, 1965, ibid., folder 776; Press, handwritten notes for PSAC presentation, June 8, 1965, ibid., box 27, folder 835; R. A. Frosch, "Federal Government Efforts Pertinent to Earthquake Prediction," Dec. 17, 1964 ibid., box 25, folder 775.

13. Donald F. Hornig, "Memorandum for the President," Sept. 16, 1965, Donald F. Hornig Papers, Lyndon B. Johnson Presidential Library, Austin, TX, box 2, folder "Chronological File July–September 1965."

14. State of California Resources Agency, *Earthquake and Geologic Hazards Conference* (n.p., 1965), 46; "New Probe of Quakes Proposed," *Pasadena Star-News,* Dec. 8, 1964, p. 32; David Perlman, "Delving Deep Into Our Jittery Earth," *SFC,* Dec. 8, 1964, I:16; "Live Compatibly With Earthquakes," *RCT,* Dec. 14, 1964, p. 22; "Earthquake-Hazards Conference," *Geotimes,* Feb. 1965, p. 9; untitled editorial, ibid., Feb. 1965, p. 7.

15. Press to Hornig, Aug. 4, 1965, FPP, box 25, folder 769; Hornig, "Memorandum for the President," Sept. 16, 1965.

16. Luther J. Carter, "Earthquake Prediction: ESSA and USGS Vie for Leadership," *Science* 151 (1966): 181–183.

17. Department of the Interior, "Earthquake Research Center Established by Interior's Geological Survey," news release for Oct. 8, 1965, and USGS, "United States Earthquake Research Program: Earthquake Prediction," Oct. 15, 1965, both in William W. Rubey Papers, Manuscript Division, Library of Congress, box 45, folder 12; William T. Pecora, "National Center for Earthquake Research, USGS," *Geotimes,* Dec. 1965, p. 13; Robert E. Wallace, interviewed by Stanley Scott, *Earthquakes, Minerals and Me: With the USGS, 1942–1995,* USGS Open-File Report 96-260 (Washington, DC: GPO, 1996), available online at http://quake.wr.usgs.gov/study/history/OFR96-260-wallace/wallace_contents.html, section VI; J. P. Eaton, *Microearthquake Seismology in USGS Volcano and Earthquake Hazards Studies: 1953–1995,* USGS Open-File Report 96-54 (Washington, DC: GPO, 1996), 5; Tom Nolan to Press, Oct. 23, 1964, FPP, box 25, folder 772; Donald F. Hornig, "Memorandum for the President," Dec. 18, 1964, White House Cen-

tral Files, Lyndon B. Johnson Presidential Library, Subject Files, box DI4, folder "Ex DI3".

18. Richard H. Jahns to W. H. Bradley, Nov. 6, 1965, Bill Pecora to Jahns, Nov. 9, 1965, Jahns to Hornig, Nov. 11, 1965, and Pecora to Jahns, Nov. 16, 1965, all in RJP, ser. 2, carton 22, file "U.S.G.S."; Ian Campbell to Hornig, Nov. 9, 1965, ibid., carton 25, file "Campbell, Ian"; Pecora to John C. Crowell, Nov. 9, 1965, and Crowell to Hornig, Nov. 29, 1965, both in Rubey Papers, box 45, folder 12; Charles C. Bates to Hornig, Nov. 15, 1965, FPP, box 32, folder 982; Allen F. Agnew to Winfield K. Denton, Dec. 1, 1965, [Johnson] White House Central Files, Subject Files, box SC2, folder "Gen SC1"; Weston E. Vivian to Hornig, Dec. 10, 1965, ibid., folder "Gen SC2"; Martin Van Couvering, "USGS Research" (letter to editor), *Geotimes,* Feb. 1966, p. 8.

19. Evert Clark, "Institute Set Up on Earth Studies," *NYT,* Nov. 21, 1965, p. 26. The Coast and Geodetic Survey was now part of the newly established Environmental Sciences Service Administration in the Department of Commerce.

20. Leroy R. Alldredge, "Preface," in *ESSA Symposium on Earthquake Prediction: Rockville, Maryland, February 7, 8, 9, 1966* (Washington, DC: GPO, 1966), iii (first quote); Robert M. White, "Welcome," ibid., 3 (second quote).

21. Donald F. Hornig, "Memorandum for Members, Federal Council on Science and Technology," Apr. 17, 1966; Federal Council for Science and Technology, "Minutes of Meeting, April 26, 1966"; and Hornig to Thomas F. Bates, June 2, 1966 (quote), all in Records of the Office of Science and Technology (OST) (RG 359), National Archives II, College Park, MD, entry 2, box 12, file "April 26, 1966"; see also J. Herbert Hollomon to Colin M. MacLeod, Apr. 29, 1966, and C. V. Kidd to Hornig and MacLeod, May 12, 1966, both in ibid., entry 3A, box 11, file "Earthquake Research."

22. Garrett Hardin, "Pop Research and the Seismic Market," *Per/Se* 2, no. 3 (1967): 19–25, quote on 23; reprinted with new preface as "Earthquake: Prediction Worse than Nature?" in Garrett Hardin, *Stalking the Wild Taboo,* 2nd ed. (Los Altos: William Kaufmann, 1978), 107–116. Hardin remains best known for his environmentalist analysis of the "tragedy of the commons."

23. George Getze, "Predictions Held More Dangerous Than Quakes," *LAT,* Oct. 3, 1967, II:6; Ralph Selph, "Earthquake!" *Los Angeles Herald-Examiner,* Oct. 15, 1967, F1; Allen to Press, Oct. 6, 1967 (quote), and L. C. Pakiser to Press, Oct. 12, 1967, both in FPP, box 9, folder 304; Richter to Garrett Hardin, Oct. 24, 1967, CRP, box 26, folder 13.

24. Frank Press, "A Strategy for an Earthquake Prediction Research Program," *Tectonophysics* 6 (1968): 11–15.

25. Keith Power, "Science and the Earthquake Hullabaloo," *SFC,* Apr. 1, 1969, pp. 1, 12.

26. Bruce A. Bolt to Press, Oct. 13, 1964, FPP, box 25, folder 772.

27. "Panel Discussion on the Earthquake Prediction Program," *ESSA Symposium on Earthquake Prediction,* 153–167, on 154, 157; see also Karl V. Steinbrugge to Press, Oct. 21, 1964, FPP, box 25, folder 770; *Earthquake and Geologic Hazards Conference,* 120; Karl V. Steinbrugge, *Earthquake Hazard in the San Francisco Bay Area: A Continuing Problem in Public Policy* (Berkeley: Institute of Governmental Studies, 1968), 60.

28. Undated Caltech news release quoting Richter issued shortly after July 21, 1965, CRP, box 30, folder 14. For examples of public inquiries and the frustration they produced for Richter, see Richter to Officer in Charge, California Highway Patrol, May 13, 1963, ibid., box 27, folder 2; Richter to Lois Abel, Nov. 19, 1964, and Eileen Gaines to Dear Sirs, Mar. 28, 1965, both in ibid., box 30, folder 2; Richter to Station KABC, Jan. 1, 1966, ibid., box 28, folder 1; Margaret Hedlund to Richter, Feb. 24, 1966, ibid., box 27, folder 2; David Branch to Richter, Sept. 3, 1966, ibid., folder 1.

29. "Prophecy of Quake Alerts Santa Barbara," *LAT*, July 21, 1965, I:3, 23; "Santa Barbara Unshaken by Quake Alert," ibid., July 22, 1965, I:3, 28; Richter to John W. Gaffney, July 23, 1965, and Richter, letter for general circulation, July 23, 1965 (quotes), both in CRP, box 30, folder 14.

30. Quotation recalled in Christopher Scholz, "What Ever Happened to Earthquake Prediction?" *Geotimes* 17 (1997): 16–19. The date of the quotation was provided by Scholz in an interview by the author, Palisades, NY, Mar. 19, 1998. For contemporary confirmation that such or a similar remark was made, see Richard W. Berry to Richter, Apr. 18, 1968, and Richter to Berry, Apr. 24, 1968, both in CRP, box 25, folder 2.

31. Charles F. Richter, "Earthquakes," *Natural History* 78, no. 10 (1969): 36–45, quote on 38; for similar sentiments, see also Richter to Branch, June 11, 1966, CRP, box 28, folder 1; and Richter to Peter A. Franken, Mar. 19, 1969, ibid., box 26, folder 6.

32. Richter to Harold L. Krivoy, Apr. 19, 1969, CRP, box 28, folder 1.

33. Frosch to Press, Apr. 8, 1965, FPP, box 25, folder 772 (quote); Pierre St. Amand to Press, Nov. 2, 1964, and Egon Orowan, "Earthquake Forecasting," Nov. 9, 1964, both in ibid., folder 771; Karl V. Steinbrugge, "Structural Damage, Soil Mechanics, and Foundation Engineering," in *ESSA Symposium on Earthquake Prediction*, 112–118, on 112, 117–118; H. J. Degenkolb, "An Engineer's Perspective on Geologic Hazards," in Robert A. Olson and Mildred M. Wallace, eds., *Geologic Hazards and Public Problems: Conference Proceedings* (Santa Rosa: Office of Emergency Preparedness, Region Seven, 1969), 183–195, on 192.

34. Hornig, "Memorandum for the President," Sept. 16, 1965; see also Hornig, "Memorandum for Members," June 25, 1965.

35. George Housner, "Engineering Seismology," handwritten memo for Apr. 19, 1964 meeting, FPP, box 25, folder 771; Don E. Hudson and George Housner, "The Relationship Between the Earthquake Prediction Problem and Earthquake Engineering," draft of Mar. 1965, ibid., folder 770; Housner to Press, Apr. 1, 1965, ibid., folder 773; Press, notes for PSAC presentation; Ad Hoc Panel, "Earthquake Prediction," 2 and Appendix 4; Don Tocher, handwritten notes on Oct. 3, 1966, meeting of Pecora panel, in EMLR, box 3, file "210.05 Ad Hoc Interagency Working Group for Earthquake Research"; Press, "Strategy for an Earthquake Prediction Research Program."

36. Ad Hoc Panel, "Earthquake Prediction," 38; marginal comment by Press on Hornig, "Memorandum for Members," June 25, 1965; Pakiser to Press, July 13, 1965, FPP, box 27, folder 835; John Walsh, "Earthquake Prediction: OST Panel Recommends 10-Year Program," *Science* 150 (1965): 321–323.

37. Scott, *George W. Housner*, 202–203, quotes on 202; Housner to A. B. Kinzel,

Aug. 19, 1965, National Academy of Sciences Archives, Washington, DC, Divisional Series, folder "DIV ES: Com on Seismology: General, 1966–1967"; "Earthquake Engineering," *Engineering and Science*, Jan. 1966, p. 30.

38. Wayne E. Hall, "Ad Hoc Interagency Working Group for Earthquake Prediction: Minutes of October 3, 1966 Meeting," EMLR, box 3, file "210.05 Ad Hoc Interagency Working Group for Earthquake Research"; Ad Hoc Interagency Working Group for Earthquake Research, "Proposal for a Ten-Year National Earthquake Hazards Program: A Partnership of Science and the Community," June 1967, in Records of the OST, entry 2, box 13, file "July 25, 1967," quote on p. 34; Pakiser to Jahns, Feb. 13, 1967, RJP, ser. 2, carton 22, file "USGS Crustal Studies Advisory Board."

39. Ad Hoc Interagency Working Group, "Proposal for a Ten-Year National Earthquake Hazards Program," quotes on 24, 39; Hall, "Minutes of Oct. 3, 1966 Meeting."

40. "Minutes, Earthquake Prediction Committee Meeting, June 21, 1967," EMLR, box 1, folder "210.10 ESSA Earthquake Prediction Committee"; National Academy of Engineering Committee on Earthquake Engineering Research, *Earthquake Engineering Research: A Report to the National Science Foundation* (Washington, DC: National Academy of Sciences, 1969), 12.

41. Carter, "Earthquake Prediction"; D. S. Greenberg, "LBJ's Budget: Lean Fare Set Forth for Research and Development," *Science* 155 (1967): 434–435; Philip M. Boffey, "LBJ's New Budget: Another Tight Year for Research and Development," ibid., 159 (1968): 509–511; "Minutes of the First Meeting of the Advisory Panel of the National Center for Earthquake Research," Sept. 30, 1966, RJP, ser. 2, carton 22, file "USGS Crustal Studies Advisory Board"; Press to Hugo Benioff, Nov. 18, 1966, FPP, box 4, folder 107.

42. Office of Science and Technology, "Proposal for a Ten-Year National Earthquake Hazards Program," news release for Dec. 31, 1968, Records of the OST, entry 3A, box 11, file "Earthquake Research."

43. Ad Hoc Interagency Working Group, "Proposal for a Ten-Year National Earthquake Hazards Program," 12a; "Earthquake Research Program," May 12, 1969, and Bill Joyner to John DeNoyer, undated [summer 1969], both in EMLR, box 4, file "200.15 Geological Survey"; Comptroller General of the United States, "Need for a National Earthquake Research Program," GAO Report B-176621, Sept. 11, 1972, p. 21.

44. Eaton, *Microearthquake Seismology,* 10–13, 24–33; J. P. Eaton, W. H. K. Lee, and L. C. Pakiser, "Use of Microearthquakes in the Study of the Mechanics of Earthquake Generation Along the San Andreas Fault in Central California," *Tectonophysics* 9 (1970): 259–282; Comptroller General, "Need for a National Earthquake Research Program," 35, 42, 44; Nicholas Wade, "Earthquake Research: A Consequence of the Pluralistic System," *Science* 178 (1972): 39–43.

45. "Earthquake Mechanism Laboratory Summary Report, FY 1967," July 12, 1967, EMLR, box 1, file "100 Reports"; Comptroller General, "Need for a National Earthquake Research Program," 34–35, 39, 42.

46. "Five to Ten San Andreas Quakes Daily," *SFC,* Feb. 23, 1967, p. 36; Robert Gillette, "$100,000 Earthquake Trap Set Up for a Big Shake," ibid., May 4, 1969, B7; Allen

to Roger W. Greensfelder, Dec. 21, 1970, EMLR, box 6, file "290 Universities and Colleges."

47. Comments by John Handin in Hall, "Minutes of Oct. 3, 1966 Meeting"; "Tectonophysics Center," *Geotimes* 12 (1967): 31; L. C. Pakiser, J. P. Eaton, J. H. Healy, and C. B. Raleigh, "Earthquake Prediction and Control," *Science* 166 (1969): 1467–1474, on 1469–1470; Frank Press and W. F. Brace, "Earthquake Prediction," ibid., 152 (1966): 1575–1584; W. F. Brace and A. S. Orange, "Electrical Resistivity Changes in Saturated Rock under Stress," ibid., 153 (1966): 1525–1526; W. F. Brace, "Current Laboratory Studies Pertaining to Earthquake Prediction," *Tectonophysics* 6 (1968): 75–87.

48. "Quake Warnings-New Clues," *SFC*, Aug. 23, 1963, pp. 1, 16; David Perlman, "Magnetic 'Signatures' May Help Predict Quakes," ibid., Oct. 7, 1967, p. 4; Sheldon Breiner and Robert L. Kovach, "Local Geomagnetic Events Associated with Displacements on the San Andreas Fault," *Science* 158 (1967): 116–118; idem, "Local Magnetic Events Associated With Displacement Along the San Andreas Fault (California)," *Tectonophysics* 6 (1968): 69–73; idem, "Search for Piezomagnetic Effects on the San Andreas Fault," *SUPGS* 11 (1968): 111–116.

49. Pradeep Talwani and Robert L. Kovach, "Geomagnetic Observations and Fault Creep in California," *Tectonophysics* 14 (1972): 245–256. Another sophisticated claim for a scientific method of earthquake prediction at this time came from Renner Hofmann, a researcher employed by the California Department of Water Resources, but his claim of predicting earthquakes on the basis of anomalies in geodimeter distance measurements across faults also did not pan out in the end. See *Geodimeter Fault Movement Investigations in California*, Department of Water Resources Bulletin No. 116-6 (Sacramento: Department of Water Resources, 1968), 71–97; Roger W. Greensfelder and Douglas Crice, "Geodimeter Fault Movement Investigations in California," *California Geology* 24 (1971): 105–109.

50. Karl V. Steinbrugge and Edwin G. Zacher, "Fault Creep and Property Damage," *BSSA* 50 (1960): 389–396; Don Tocher, "Creep Rate and Related Measurements at Vineyard, California," ibid.: 396–404; C. A. Whitten and C. N. Claire, "Analysis of Geodetic Measurements Along the San Andreas Fault," ibid.: 404–414.

51. Lloyd S. Cluff and Karl V. Steinbrugge, "Hayward Fault Slippage in the Irvington-Niles Districts of Fremont, California," *BSSA* 56 (1966): 257–279; M. G. Bonilla, "Deformation of Railroad Tracks by Slippage on the Hayward Fault in the Niles District of Fremont, California," ibid.: 281–289; F. B. Blanchard and G. L. Laverty, "Displacements in the Claremont Water Tunnel at the Intersection with the Hayward Fault," ibid.: 291–294; Dorothy H. Radbruch and Ben J. Lennert, "Damage to Culvert Under Memorial Stadium, University of California, Berkeley, Caused by Slippage in the Hayward Fault Zone," ibid.: 295–304; Bruce A. Bolt and Walter C. Marion, "Instrumental Measurement of Slippage on the Hayward Fault," ibid.: 305–316.

52. A. J. Pope, J. L. Stearn, and C. A. Whitten, "Surveys for Crustal Movement Along the Hayward Fault," *BSSA* 56 (1966): 317–323; Thomas H. Rogers and Robert D. Nason, "Active Faulting in the Hollister Area," *SUPGS* 11 (1968): 42–45; Manuel G. Bonilla, "Evidence for Right-Lateral Movement on the Owens Valley, California, Fault

Zone During the Earthquake of 1872, and Possible Subsequent Fault Creep," ibid.: 4–5; Dorothy H. Radbruch, "New Evidence of Historic Fault Activity in Alameda, Contra Costa, and Santa Clara Counties, California," ibid.: 46–54; Tom Juster, "Area Geologists Set Hayward Meet 'Where Action Is,'" *San Leandro Morning News*, Apr. 3, 1967, clipping, and "Minutes, Apr. 5, 1967 Meeting," both in EMLR, box 1, file "210.40 East Bay Council on Surveying and Mapping"; Joseph P. Carey to Tocher, Dec. 25, 1967, ibid., box 2, file "280 Studies and Projects."

53. Benioff to F. F. Mautz, Jan. 14, 1964, Elmer C. Marliave Papers, Water Resources Center Archives, Berkeley, CA, box 1, file 1.1, folder 1.

54. Clarence R. Allen, "The Tectonic Environments of Seismically Active and Inactive Areas Along the San Andreas Fault System," *SUPGS* 11 (1968): 70–82. For assent, see, for example, Pakiser et al., "Earthquake Prediction and Control," 1468. The conclusion was not universally accepted; see, for example, C. H. Scholz and Thomas J. Fitch, "Strain Accumulation Along the San Andreas Fault," *Journal of Geophysical Research* 74 (1969): 6649–6666; idem, "Strain and Creep in Central California," ibid., 75 (1970): 4447–4453; idem, "Reply to Comments by J. C. Savage and R. O. Burford," ibid., 76 (1971): 6480–6484.

55. Bob Lyhne, "The Loaded Charge in County's Hills," *RCT*, Apr. 15, 1964, "Peninsula Midweek" section, pp. 1–5, on 3.

56. Gordon B. Oakeshott, "Parkfield Earthquakes," *Mineral Information Service* 19 (1966): 150–151, 156; "Surface Break," *Geotimes*, Oct. 1966, p. 23; C. R. Allen et al., "The Borrego Mountain, California, Earthquake of 9 April 1968: A Preliminary Report," *BSSA* 58 (1968): 1183–1186; Eaton et al., "Use of Microearthquakes."

57. "New Studies Re-Emphasize Hazard Potential of San Andreas Fault," *Mineral Information Service* 19 (1966): 96; Robert E. Wallace, "Notes on Stream Channels Offset by the San Andreas Fault, Southern Coast Ranges, California," *SUPGS* 11 (1968): 6–21.

58. See, for example, Robert D. Brown, Jr., and Robert E. Wallace, "Current and Historic Fault Movement Along the San Andreas Fault Between Paicines and Camp Dix, California," *SUPGS* 11 (1968): 22–41.

59. Lloyd S. Cluff, "Urban Development Within the San Andreas Fault System," ibid.: 55–69.

60. California State Mining Board, "Minutes of Mining Board Meeting, Dec. 17, 1959," in bound copy of California State Mining and Geology Board minutes, Doe Library, University of California, Berkeley; *Earthquake and Geologic Hazards Conference*, 47–49; M. G. Bonilla, "City College Fault, San Francisco, California," in *Geological Survey Research 1961: Short Papers in the Geologic and Hydrologic Sciences*, USGS Professional Paper 424-C (Washington, DC: GPO, 1961), C190–C192.

61. "Earthquake Fault Scars Shown On New Map," *Mineral Information Service* 20 (1967): 150; "Surface Trace of Hayward Fault Zone Shown on Map," ibid., 21 (1968): 65; "San Andreas Fault Scars Shown on New USGS Map," ibid.: 80; "New Map Shows Earthquake Hazards Along San Andreas Fault Segment," USGS news release for July 13, 1970, EMLR, box 6, file "90.50 Publicity."

62. U.S. Department of the Interior and U.S. Department of Housing and Urban Development, "Program Design for San Francisco Bay Region Environment and Resources Planning Study," Oct. 1971, EMLR, box 9, folder "200.25 U.S. Geological Survey," quote on viii; see also Wallace, *Earthquakes, Minerals, and Me*, sec. 7.

63. R. D. Borcherdt, ed., *Studies for Seismic Zonation of the San Francisco Bay Region*, USGS Professional Paper 941-A (Washington, DC: GPO, 1975); Rachel Gulliver to Friends of the San Francisco Bay Region, Jan. 18, 1971, EMLR, box 7, folder "200.12 Geological Survey."

64. First quote from Ian Campbell, *Fifty-Eighth Report of the State Geologist, July 1960–June 1963* (San Francisco: California Division of Mines and Geology, 1963), 105; second quote from California State Mining and Geology Board, "Minutes of Meeting, June 2, 1968," in bound copy of board minutes, Doe Library. See also Gordon B. Oakeshott, "The California State Geological Surveys," *California Geology* 24 (1971): 23–25.

65. Campbell, *Fifty-Eighth Report*, 92, 98; California State Mining Board, "Minutes of Mining Board Meeting, Jan. 28, 1960," and idem, "Minutes of Meeting, August 22, 1961," both in bound copy of board minutes, Doe Library; California Senate Fact Finding Committee on Natural Resources, "The Economic Potential of the Geologic Resources of the State of California: Hearing," Jan. 23 and 24, 1964, copy in Doe Library, p. 67.

66. First quote from California State Mining Board, "Minutes of Meeting, April 9, 1964," in bound copy of board minutes, Doe Library; second quote from California Senate Permanent Factfinding Committee on Natural Resources, *Third Progress Report to the Legislature* (Sacramento: Senate of the State of California, 1965), 64.

67. *Earthquakes and Geologic Hazards Conference*, 4, reprinted in Hugo Fisher, "Geologic Hazards," *Mineral Information Service* 18 (1965): 45. For support from the Resources Agency, see also "A Study of Resource Policy Directions for California," Dec. 1965, copy in Water Resources Center Archives, Berkeley, 83.

68. Geologic Hazards Advisory Committees for Program and Organization, *Earthquake and Geologic Hazards in California: A Report to the Resources Agency* (n.p., 1967).

69. Ian Campbell, *Fifty-Ninth Report of the State Geologist, July 1963, through June, 1966* (San Francisco: California Division of Mines and Geology, 1966), 18; California State Mining and Geology Board, "Minutes of Meeting, December 2, 1965," in bound copy of board minutes, Doe Library. For the reapportionment, see Jackson K. Putnam, *Modern California Politics*, 2nd ed. (Sacramento: Boyd & Fraser, 1984), 53; for its direct impact on the Division of Mines and Geology's orientation, see California State Mining and Geology Board, "Minutes of Meeting, June 15, 1967," in bound copy of board minutes, Doe Library.

70. Richard M. Stewart to Members of the State Mining and Geology Board, May 20, 1968, RJP, ser. 2, carton 22, file "California Division of Mines and Geology"; California State Mining and Geology Board, "Minutes of Meeting, March 12, 1969," in bound copy of board minutes, Doe Library.

71. State Mining and Geology Board, "Minutes of Meeting, March 12, 1969"; Ian

Campbell, *Sixty-Second Report of the State Geologist, July, 1968 through June, 1969* (San Francisco: California Division of Mines and Geology, 1969); Wesley G. Bruer, *Sixty-Third Report of the State Geologist, July 1, 1969, through June 30, 1970* (Sacramento: California Division of Mines and Geology, 1970), 37. See also California State Mining and Geology Board, "Minutes of Meeting, August 8, 1968," in bound copy of board minutes, Doe Library.

72. Ian Campbell, *Sixty-First Report of the State Geologist, July, 1967 through June, 1968* (San Francisco: California Division of Mines and Geology, 1968), 24–25; Campbell, *Sixty-Second Report,* 24; Bruer, *Sixty-Third Report,* 18–20, 35; Greensfelder and Crice, "Geodimeter Fault Movement Investigations."

73. See, for example, Richard B. Saul, "The Calaveras Fault Zone in Contra Costa County, California," *Mineral Information Service* 20 (1967): 35–37; Thomas H. Rogers, "Where Does the Hayward Fault Go?" ibid., 22 (1969): 54–60; idem, "A Trip to An Active Fault in the City of Hollister," ibid.: 159–164.

74. Greensfelder to Tocher, Aug. 12, 1970, EMLR, box 6, file "200 Agencies."

75. California State Mining and Geology Board, "Minutes, Feb. 3, 1966," in bound copy of board minutes, Doe Library; Campbell to Jahns, June 28, 1968, RJP, ser. 2, carton 25, file "Campbell, Ian"; Campbell, *Sixty-Second Report,* 24.

76. Rice Odell, *The Saving of San Francisco Bay: A Report on Citizen Action and Regional Planning* (Washington, DC: Conservation Foundation, 1972), 5–23.

77. Bruce B. Brugmann, "Redwood Shores Design Unveiled," *RCT,* Dec. 3, 1964, pp. 1, 2.

78. "Conservationists Ask Redwood Shores Halt," *RCT,* May 14, 1964, pp. 1, 2; Ray Spangler, "Filled Marsh Land Isn't the Safest Place," ibid., May 20, 1964, p. 36; Bruce B. Brugmann, "San Francisco Bay Vanishing Wonder," ibid., May 25, 1964, pp. 1, 3; "Speakers Tell Doubts on Shores; City Goes Ahead With District," ibid., May 26, 1964, p. 1.

79. G. Brent Dalrymple and Marvin A. Lanphere, "Potential Earthquake Hazards on Bay-fill and Marshland Adjacent to San Francisco Bay," Dec. 1964, CRP, box 30, folder 6; see also "2 Geologists Testify on Baylands Dangers," *RCT,* Dec. 1, 1964, p. 1; "Geologists Claim Bayfill Peril," ibid., Dec. 2, 1964, p. 3.

80. William W. Moore to Bert W. Levit, Dec. 14, 1964, reproduced in Comptroller General of the United States, "Operation of Federal Housing Programs in Areas of Potential Geologic Instability: Foster City, California," GAO Report B-158554, July 1967, copy in USGS Library, Menlo Park, CA, 101–104, quote on 101; see also "Quake Peril Charges Said 'Half Truths,' " *RCT,* Dec. 3, 1964, pp. 1, 2; Bruce Brugmann, "Foster Issues Attack On Quake Peril Claim," ibid., Dec. 25, 1964, p. 4.

81. Bruce B. Brugmann, "Earthquake Dangers Bring Verbal Combat," *RCT,* Dec. 7, 1964, p. 24; idem, "USGS 'Clarifies' Policy on Quake Statements," ibid., Dec. 7, 1964, pp. 1, 2.

82. Harold Gilliam, "The Case of the Muzzled Geologists," *SFC,* Jan. 24, 1965, "This World" section, pp. 7, 8; Bruce B. Brugmann, "Hazardous Times for Statements on Bay-

fill Quake Hazards," *RCT*, Feb. 1, 1965, p. 3; "Demos Attack Muzzling," ibid., Feb. 15, 1965, p. 2.

83. Jean Travers to Richter, Jan. 9, 1965, and Richter to Travers, Jan. 21, 1965, both in CRP, box 30, folder 6. See also the long but cautiously worded letter of Campbell to Travers, Jan. 17, 1965, ibid.

84. Both quoted in Bruce B. Brugmann, "Two Geologists View Disputed Report on Quake Dangers," *RCT*, Feb. 8, 1965, pp. 13, 14; for a copy of the original letters, see House Committee on Government Operations, "Federal Involvement in Hazardous Geologic Areas: Hearing," 91st Cong., 1st sess., May 7 and 8, 1969, pp. 147-150. See also Stanley N. Davis et al. to Travers, Feb. 10, 1965, RJP, ser. 2, carton 27, file "Redwood Shores."

85. "RC Council Sets Quake Hazard Study," *RCT*, Feb. 9, 1965, p. 1.

86. Degenkolb, "An Engineer's Perspective," quote on 185; see also discussion in Christensen and Bolt, "Earth Movements"; *Earthquake and Geologic Hazards Conference*, 136; "Expert Warns New Buildings May Be Unsafe in Earthquake," *SFC*, Oct. 5, 1966, p. 16.

87. Redwood City Seismic Advisory Board, "Report of Seismic Investigation for Redwood City General Improvement District No. 1-64," Sept. 1965, copy in EERC Library, Richmond, CA, quote on 32; Bruce B. Brugmann, "Quake Experts Approve Shore," *RCT*, Sept. 21, 1965, pp. 1, 2.

88. House Committee on Government Operations, "Federal Involvement in Hazardous Geologic Areas"; "Quake Peril and Peninsula Housing," *SFC*, May 8, 1969, pp. 1, 30; Mike Palmer, "Shores' Future in Doubt," *RCT*, Oct. 10, 1969, pp. 1, 2; Paul N. McCloskey to Redwood City Council, Oct. 13, 1969, RJP, ser. 2, carton 27, file "Redwood Shores."

89. Odell, *Saving of San Francisco Bay*, 24-71.

90. H. Bolton Seed, "Seismic Problems in the Use of Fills in San Francisco Bay," prepared for the San Francisco Bay Conservation and Development Commission, May 1967, copy in Doe Library, quote on 41.

91. Karl V. Steinbrugge, "Seismic Risk to Buildings and Structures on Filled Lands in San Francisco Bay," prepared for the San Francisco Bay Conservation and Development Commission, May 1967, copy in Doe Library, quote on 28.

92. San Francisco Bay Conservation and Development Commission, "Minutes of Meeting of May 19, 1967," in bound copy of Bay Conservation and Development Commission minutes, Doe Library.

93. San Francisco Bay Conservation and Development Commission, "Minutes of Meeting of June 1, 1967," in bound copy of commission minutes, Doe Library; Board of Consultants, "Carrying Out the Bay Plan: The Safety of Fills," prepared for the San Francisco Conservation and Development Commission, Sept. 1968, copy in EERC Library.

94. Wallace, *Earthquakes, Minerals and Me*, sec. 8.

95. Stanley Scott and Robert A. Olson, eds., *California's Earthquake Safety Policy: A Twentieth Anniversary Retrospective, 1969-1989* (Berkeley: Earthquake Engineering Re-

search Center, 1993), 9–10, 13, 16–17; Stanley Scott, "A Model Seismic Safety Program for the Bay Region," in Olson and Wallace, *Geologic Hazards and Public Problems*, 265–282, on 268–269.

96. Steinbrugge, *Earthquake Hazard in the San Francisco Bay Area*, quote on 10; David Perlman, "Quake Threat Ignored," *SFC*, Oct. 24, 1968, p. 2.

97. Stanley Scott, "Preparing For Future Earthquakes: Unfinished Business in the San Francisco Bay Area," *Public Affairs Report: Bulletin of the Institute of Governmental Studies* 9, no. 6 (1968): quotes on 3; "Lifesaving in a Quake," *SFC*, Jan. 7, 1969, p. 2.

98. Scott and Olson, *California's Earthquake Safety Policy*, 1–4; Eugene C. Lee to Jahns, Feb. 21, 1969, and S. Scott, "An Earthquake Commission for the San Francisco Bay Area?" Feb. 20, 1969, both in RJP, ser. 2, carton 25, file "Earthquake Commission"; Stanley Scott, "Earthquake Hazard in the San Francisco Bay Area: Mobilizing for a Model Regional Seismic Safety Program," Mar. 31, 1969, and Lee to Bay Area Legislators, Apr. 1, 1969 (quote), both in ibid., ser. 2, carton 2, no file; "A Call for State Study of Quakes," *SFC*, Apr. 4, 1969, p. 2.

99. Keith Power, "Quake Scare: Predictions of Impending Doom," *SFC*, Mar. 31, 1969, pp. 1, 26 (quotes); Robert Gillette, "Quake-Wise, It's a Very Big Year for Very Bad Predictions," ibid., Jan. 26, 1969, p. 8; Steven V. Roberts, " 'Warning! California Will Fall Into the Ocean in April!' " *NYT Magazine*, Apr. 6, 1969, pp. 12, 14, 96–98; Curt Gentry, *The Last Days of the Late, Great State of California* (New York: Ballantine Books, 1968).

100. Quote from Richter to Robert Iacopi, May 19, 1969, CRP, box 27, folder 1.

101. University of Michigan News Service, "For Immediate Release," Jan. 29, 1968, CRP, box 26, folder 6; R. Rapoport, "Bad Vibrations in the Golden State," *Esquire Magazine*, Dec. 1968, pp. 190–191; Linn Hoover, "Hazards: Fact and Fiction," *Geotimes*, Jan. 1969, p. 9.

102. David Perlman, "Big Fund Asked For Quake Probes," *SFC*, Dec. 31, 1968, pp. 1, 14; Harold M. Schmeck, Jr., "Earthquake Prediction Program Is Called Vital," *NYT*, Dec. 31, 1968, p. 16; "Major Coast Quake by '80 Is Forecast At Senate Hearing," ibid., Mar. 26, 1969, p. 36.

103. Pecora to Hornig, Nov. 15, 1968, FPP, box 9, folder 303; Tocher to EML files, Dec. 11, 1968, EMLR, box 3, file "260 Scientific Activities"; Walter Sullivan, "Coast Quake Area Shows Land Strain," *NYT*, Mar. 16, 1968, p. 33.

104. Walter Sullivan, "In California, the Earthquake Threat is Real," *NYT*, Mar. 23, 1969, IV:9.

105. First lyric from "Day by Day," quoted in Roberts, " 'Warning!' " p. 12; second lyric from "California Earthquake," quoted in ibid., p. 14.

106. "KSAN Broadcasting the Rumblings About California's Next Major Earthquake," press release dated Feb. 3, 1969, EMLR, box 3, file "90 Publicity."

107. Campbell, *Sixty-Second Report*, 35–36; Earl E. Brabb, "Quake Education" (letter to editor), *Geotimes* Apr. 1969, p. 7 (quote).

108. Power, "Quake Scare"; idem, "Science and the Earthquake Hullabaloo"; idem, "A High Quake Risk in Our Buildings," *SFC*, Apr. 2, 1969, p. 13; idem, "The Hidden

Danger of a Quake-Panic," ibid., Apr. 3, 1969, p. 10; Peggy Skarpinski, "When Walls Come Tumbling Down," ibid., Apr. 8, 1969, p. 13; Robert Gillette, "Rumors Create an Insurance Boom," ibid., Mar. 30, 1969, A:20.

109. "Quakes Feared in California," *NYT,* Sept. 8, 1968, p. 26; Kenneth Reich, "Pastor Tells of His 1968 Quake Vision," *LAT,* Feb. 11, 1971, I:1, 3; "The Quake That Didn't Happen," *SFC,* Apr. 5, 1969, pp. 1, 14.

110. Scott and Olson, *California's Earthquake Safety Policy,* 2–3, 5, 7–8; Scott to Stephen Larson, Mar. 24, 1969, SSRG, box 1, folder 107; Scott, "Model Seismic Safety Program," 270–273; Jackson Doyle, "Bill for a Bay Quake Commission," *SFC,* Apr. 9, 1969, p. 8 (quote).

111. Frederick Taylor, "In California, Everyone Talks About the Quakes But Few Do Anything," *Wall Street Journal,* May 5, 1969, clipping in SSRG, box 2, folder 116; E. Jack Schoop to Larson, Apr. 10, 1969, ibid., box 1, folder 107.

112. Odell, *Saving of San Francisco Bay,* 79.

113. Alfred E. Alquist to Members of the Legislature, May 14, 1969, and "Earthquake Film By Alquist Flops," *San Jose Mercury,* May 20, 1969, clipping, both in SSRG, box 2, folder 116; Richard H. Jahns (moderator), "The Role of Geology in Urban Planning and How Best to Ensure Its Use," in Donald R. Nichols and Catherine C. Campbell, eds., *Environmental Planning and Geology* (Washington, DC: GPO, 1971), 90–105, on 93.

114. "And So Much for Those Two Bills," *SFC,* May 28, 1969, p. 11; Alquist to Robert M. White, June 4, 1969, SSRG, box 2, folder 116.

115. "Efforts for Bay Area Quake Study," *SFC,* June 3, 1969, p. 6; "Senate Concurrent Resolution No. 128," as introduced June 9, 1969, CRP, box 24, folder 10; Senate Concurrent Resolution No. 128, as filed with Secretary of State Aug. 25, 1969, in Joint Committee on Seismic Safety, "June 30, 1970 Progress Report," copy in EERC Library, 2–6, quote on 2; Jean Laurin, "Administrative History of the California Seismic Safety Commission," draft of Jan. 18, 1983, RJP, ser. 2, box 2, no file, pp. B-1 to B-2.

116. Alfred E. Alquist, "California's Approach to Seismic Safety," *Proceedings, Structural Engineers Association of California, 39th Convention—1970,* pp. 14–15; "Board of Directors' Action," *EERIN* 4, no. 3 (1970): 1–3; Alquist to George A. Lincoln, May 25, 1970, SSRG, box 2, folder 116.

Chapter 7

1. *Report of the Los Angeles County Earthquake Commission: San Fernando Earthquake, February 9, 1971* (n.p., 1971), 27, 38–39; Dial Torgerson, "Evacuation Order for 80,000 in Valley Extended 2 Days," *LAT,* Feb. 11, 1971, I:1, 34.

2. *Report of the Los Angeles County Earthquake Commission,* 24–28.

3. Joint Panel on the San Fernando Earthquake, *The San Fernando Earthquake of February 9, 1971: Lessons from a Moderate Earthquake on the Fringe of a Densely Populated Region* (Washington, DC: National Academy of Sciences and National Academy of Engineering, 1971); Jim Thompson, "Preliminary Report of the SEASC Ad Hoc Committee on 1971 Earthquake Damage," *Proceedings, 40th Annual Convention, Structural Engineers Association of California,* 1971, pp. 83–86; Clarkson W. Pinkham, "EERI Building Com-

mittee," ibid., 98–101; George W. Housner, "Lessons from the February 9, 1971 San Fernando Earthquake," ibid., 102–109; Paul C. Jennings, ed., *Engineering Features of the San Fernando Earthquake of February 9, 1971,* Earthquake Engineering Research Laboratory Report EERL 71-02 (Pasadena: n.p., 1971), 478–498; *Report of the Los Angeles County Earthquake Commission.*

4. "Comments by Mr. Robert J. Williams," attached to Los Angeles County Earthquake Commission, Summary of Proceedings of Meeting of May 26, 1971, in Kenneth Hahn Papers, Huntington Library, San Marino, CA, preliminary classification 2.4.

5. "The Great Earthquake of '71," *LAT,* Feb. 10, 1971, II:6; "To Cope With Future Quakes," ibid., Feb. 11, 1971, II:6; Ernest Conine, "Shock Waves of Earthquake Will Probably Outlast Its Social Waves," ibid., Feb. 14, 1971, F6; "Time to Reinforce Safety Rules," ibid., Feb. 16, 1971, II:6; "Times Editorial Views of the Week," ibid., Feb. 21, 1971, F6; "The Earthquake," text of KNX Newsradio editorial, Feb. 10, 1971, "Quakes," text of KCBS Newsradio editorial, Feb. 11, 1971, and "Earthquake Planning," text of KPIX TV5 editorial, Feb. 18, 1971, all in SSRG, box 2, folder 130; Russell T. Connors and Thomas L. Wright, "Time Engineers Acted" (letters to editor), *LAT,* Feb. 13, 1971, II:4; "The How of Earthquake Safety," ibid, Aug. 27, 1971, II:6; Adrienne Cole, "Quake Reconstruction" (letter to editor), ibid., Feb. 17, 1972, II:6.

6. Roy Haynes, "Quake Victims to See Agnew on Aid Speedup," *LAT,* Mar. 18, 1971, I:3; "Quake Victims Need More Help," ibid., Mar. 1, 1971, II:6; "Bureaucrats Unshaken by Quake," ibid., June 14, 1971, II:6; Senate Committee on Public Works, *Governmental Response to the California Earthquake Disaster of February 1971,* 92nd Cong., 1st sess., June 10–12, 1971, quote on 6.

7. Senate Committee on Public Works, *Governmental Response,* 682–684; Senate Committee on Commerce, *Earthquakes: Hearings,* 93rd Cong., 1st sess., Apr. 26–27, 1973, pp. 56–59; "Hearing of the Joint Legislative Committee on Seismic Safety Special Subcommittee on the San Fernando Earthquake," Jan. 6, 1972, SSRG, box 4, folder 211, pp. 64–66, 139–147; Joint Legislative Committee on Seismic Safety, "Hearing," Feb. 9, 1972, ibid., folder 213, pp. 81–89; Betty Whirledge to Karl V. Steinbrugge, June 22, 1972, ibid., box 1, folder 111; Charles F. Richter to James A. Hayes, Mar. 30, 1972, ibid., folder 110 (quote).

8. Script for "The City That Waits to Die" in SSRG, box 1, folder 108 (quote on 11); see also John Harris, "British Find Fault With S. F.," *SFE,* Dec. 22, 1970, p. 12.

9. "S. F. Quake Film Big North Beach Attraction," *SFE,* June 15, 1971, p. 16; "Quake-Scare Movie Hottest Underground Item in S. F.," *LAT,* June 15, 1971, I:3; for the background of Duskin's campaign, see also Frederick M. Wirt, *Power in the City: Decision Making in San Francisco* (Berkeley: U of California P, 1974), 177–183, 204–207.

10. Gail Lee Mudgett to Robert Wallace, June 10, 1971, and Mudgett to Alfred Alquist, June 10, 1971, both in SSRG, box 1, folder 108; Mudgett to Alquist, Sept. 19, 1971, and Mudgett to Hans Heilmann, Sept. 20, 1971, both in ibid., box 1, folder 109.

11. "Our Experts Are Unshaken," *SFE,* Dec. 22, 1970, p. 12; Don Tocher to L. R. Alldredge, July 8, 1971, and William W. Moore to KPIX Television, Aug. 9, 1971 (first quote), both in EMLR, box 7, file "90 Public Relations"; "Shake 'Em Up," text of KFRC

editorial, July 29, 1971 (second quote), SSRG, box 2, folder 130; Karl V. Steinbrugge, "Review of 'The City That Waits to Die,'" Dec. 20, 1971 (third quote), ibid., box 1, folder 110.

12. John J. Fried, *Life Along the San Andreas Fault* (New York: Saturday Review Press, 1973), 162–164, 169–172.

13. Henry J. Degenkolb, "Report of Ad Hoc Committee on Direction Study," *Proceedings, Structural Engineers Association of California, 39th Convention—1970*, pp. 16–19; "West Coast High-Rise Buildings May Collapse in Great Earthquake, Expert Warns," NOAA news release for Jan. 29, 1971, EMLR, box 7, file "90 Public Relations"; Jennings, *Engineering Features*, 492–498; Housner, "Lessons"; Robert A. Page, John A. Blume, and William B. Joyner, "Earthquake Shaking and Damage to Buildings," *Science* 189 (1975): 601–608.

14. Leo H. Corning, "New Developments in Seismic Design of Multistory Reinforced Concrete Buildings: The Portland Cement Association Program," *Proceedings, 29th Annual Convention, Structural Engineers Association of California, 1959*, pp. 21–26; Albin Johnson, "Progress Report, SEAOC Seismology Committee," *Proceedings, 34th Annual Convention, Structural Engineers Association of California, 1965*, pp. 38–40; H. S. (Pete) Kellam, "Recommended Lateral Force Requirements By SEAOC, 1966, Part I— Background of Paragraph (j)," *Proceedings, 35th Annual Convention, Structural Engineers Association of California, 1966*, pp. 65–68.

15. "An Earthquake Lesson for S. F.," *SFC*, Aug. 30, 1967, p. 6; Karl V. Steinbrugge and Lloyd S. Cluff, "The Caracas, Venezuela Earthquake of July 29, 1967," *Mineral Information Service* 21 (1968): 3–12; Robert Gilletti, "Experts Worry About Quake Building Peril," *SFE*, Jan. 17, 1971, A26; H. J. Degenkolb, "An Engineer's Perspective on Geologic Hazards," in Robert A. Olson and Mildred M. Wallace, eds., *Geologic Hazards and Public Problems: Conference Proceedings* (Santa Rosa: Office of Emergency Preparedness Region Seven, 1969), 183–195, on 188; Stanley Scott (interviewer), *Henry J. Degenkolb, Connections: EERI Oral History Series*, vol. 1 (Oakland: Earthquake Engineering Research Institute, 1994), 176–179; Fried, *Life Along the San Andreas Fault*, 175–179.

16. Steinmann, "Recommendations"; Pinkham, "EERI Building Committee"; Roland L. Sharpe, "Report of the 1972 Seismology Committee," *Proceedings, Structural Engineers Association of California, 41st Annual Convention, 1972*, pp. 76–87; "Sixteen Los Angeles," *LAT*, Jan. 13, 1972, I:2; "Greater Quake Safety in Building Rules Urged," ibid., Feb. 7, 1972, II:8; Joan Sweeney, "Older Buildings May Have the Flaws That Killed in '71," ibid., Feb. 8, 1981, II:1, 3, 4.

17. J. D. Kaprielian, "Notice," EMLR, box 9, file "210 Committees, Boards, Groups"; Dick Turpin, "Engineers Learn from Earthquake," *LAT*, Oct. 8, 1972, K:2; Erwin Baker, "Code Changes to Cut Quake Damage Passed," ibid., Mar. 14, 1973, II:1; "County OK's Strict Quake Building Code," ibid., July 25, 1973, II:8; "Final Recommendations by SEAOC Seismology Committee on Changes to 1973 UBC," *EERIN* 8, no. 4 (1974): 44; Carl B. Johnson to Henry J. Degenkolb, Dec. 23, 1974, ibid., 9, no. 1 (1974): 30–32; W. J. Hall to Members and Advisors of the SEAOI Seismology Committee, Sept. 20, 1974, Nathan M. Newmark Papers, University of Illinois Archives, Cham-

paign, box 11, folder "SEAOI"; J. M. McLaughlin to Nathan M. Newmark, Dec. 8, 1978, ibid., box 8, folder "ATC 2."

18. Arthur L. Elliott, "Freeway Structures—What the Earthquake Taught Us," *Proceedings, 40th Annual Convention, Structural Engineers Association of California, 1971*, pp. 60-65; Roy Imbsen and James Gates, "Recent Innovations in Seismic Design and Analysis Techniques for Bridge Structures," *Proceedings, 42nd Annual Convention, Structural Engineers Association, 1973*, pp. 81-119; The Governor's Board of Inquiry on the 1989 Loma Prieta Earthquake, *Competing Against Time: Report to Governor George Deukmejian* (North Highlands: State of California Office of Planning and Research, 1990), 65, 122-126; Subcommittee on Highway Structural Safety of the Assembly Transportation Committee, "Transcript of Hearing on Highway Structural Safety," June 8, 1971, Legislative Papers, California State Archives, Sacramento, entry "Legislature-Assembly-Committee on Transportation—Hearing of Subcommittee on Highway Structural Safety."

19. "Hearing of the Joint Legislative Committee," Jan. 6, 1972; John R. Teerink to Alquist, Dec. 18, 1973, SSRG, box 1, folder 113; Ronald B. Robie to Robert A. Olson, June 2, 1976, RJP, ser. 2, carton 11, file "SSC #1"; Gordon W. Dukleth and E. W. Stroppini, "Seismic Safety for California Dams," *California Geology* 29 (1976): 243-248; Joan Sweeney, "1971 San Fernando Disaster Led to Tougher State and Local Laws," *LAT*, Feb. 8, 1981, II:1, 5, 6. On the establishment of the Division of Safety of Dams, see Ian Campbell, "The Influence of Geologic Hazards on Legislation in California," *California Geology* 30 (1977): 219-223.

20. Senate Concurrent Resolution No. 128, filed Aug. 25, 1969, as reprinted in Joint Committee on Seismic Safety, "June 30, 1970 Progress Report," copy in EERC Library, Richmond, CA, 2-7; Stanley Scott and Robert A. Olson, eds., *California's Earthquake Safety Policy: A Twentieth Anniversary Retrospective, 1969-1989* (Berkeley: Earthquake Engineering Research Center, 1993), 24-25, 48-50.

21. "Minutes of the Advisory Groups Coordinating Council," Sept. 3, 1970, SSRG, box 5, folder 242; "Moretti Seeks Quake Study Group," *SB*, Feb. 19, 1971, A5; Special Subcommittee of the Joint Committee on Seismic Safety, "July 31, 1971 Preliminary Report on the San Fernando Earthquake Study," copy in EERC Library, 15-16; Scott and Olson, *California's Earthquake Safety Policy*, 7.

22. Scott and Olson, *California's Earthquake Safety Policy*, 15; Steinbrugge to Alquist, May 3, 1971, SSRG, box 2, folder 117; "Expert Says Capitol Dome Would Topple In Quake," *SB*, May 14, 1971, A1, A20; "If Quake Hits, Dome Would Die," ibid., May 15, 1971, A1, A4; James Dufur, "Experts Advise Lawmakers to Shun Capitol," ibid., June 30, 1972, A1, A24; idem, "Legislators Ban Public Tours In Parts Of 'Unsafe' Capitol," ibid., July 21, 1972, A3; Lee Fremstad, "Reagan: Make Old Capitol Quakeproof," ibid., Dec. 13, 1972, B3; Steve Lawrence, "Sad Solons Move, Sulk," ibid., Aug. 3, 1976, A6.

23. John F. Meehan to R. M. White, June 16, 1967, and J. M. Klaasse to Alldredge, July 5, 1967, both in EMLR, box 1, file "210.20 NAS-NRC Committee Advisory to ESSA"; Keith Power, "The Hidden Danger of a Quake-Panic," *SFC*, Apr. 3, 1969, p. 10;

David Perlman, "Obstacles to Quakeproof Construction," ibid., Jan. 26, 1970, p. 6; Henry Degenkolb, "Rough Draft-Recommendations," undated [shortly before Apr. 23, 1970], RJP, ser. 2, carton 2, envelope "Steinbrugge Task Force Reports."

24. White to Meehan, July 6, 1967, EMLR, box 1, file "210.20 NAS-NRC Committee Advisory to ESSA"; William M. Wells, "Minutes of State Strong Motion Instrumentation Advisory Board Organizational Meeting," Jan. 21, 1972, and National Academy of Sciences–National Academy of Engineering Science and Engineering Committee Advisory to the National Oceanic and Atmospheric Administration, "Report 72-2 of the Panel on Earthquake Engineering," Feb. 1972, both in ibid., box 9, file "210 Committees, Boards, Groups."

25. Erwin Baker, "Code Changes to Cut Quake Damage Given Tentative OK," *LAT*, Feb. 8, 1972, II:1, 5; Don Driggs to Alquist, Nov. 2, 1972, SSRG, box 1, folder 112; "NOAA Facility Analyzes Earthquake's Effect on 22 Modern Buildings," NOAA press release for Dec. 3, 1970, EMLR, box 6, file "90.50 Publicity."

26. "Senate Bill No. 1374," as amended May 28, 1971, Committee Consultant, "Strong-Motion Instrumentation Program," June 10, 1971, and "Amendments to SB 1374," Sept. 23, 1971, all in SSRG, box 4, folder 175; Advisory Group Coordinating Council, minutes of Sept. 2, 1971, ibid., box 5, folder 243; "Chapter 1152," *Statutes of California, 1971 Regular Session*, 2175–2176.

27. Thomas E. Gay, "Second Report on the Strong-Motion Instrumentation Program," California Division of Mines and Geology Sepcial Publication 48, Apr. 1, 1976, excerpted in *EERIN* 10, no. 4 (1976): 88–89; *Sixty-Eighth Report of the State Geologist (1974-1975)* (Sacramento: California Division of Mines and Geology, 1975), 31; California State Mining and Geology Board, "Minutes of Meeting of June 13, 1972," in bound copy of California State Mining and Geology Board minutes, Doe Library, University of California, Berkeley.

28. Advisory Group Coordinating Council, "Minutes of the February 2, 1971 Meeting," SSRG, box 5, folder 243; Sal Bianco to Pete Melnicoe, Feb. 26, 1971, ibid., box 1, folder 108; Bianco to Charles De Maria, Mar. 12, 1971, ibid., box 4, folder 170; Torgerson, "Evacuation Order"; "2 Bills Are Aimed At Earthquake Dangers," *SB*, Feb. 25, 1971, A:7.

29. R. W. Britting to Alquist, Apr. 20, 1971, SSRG, box 1, folder 108; Andrew W. Oppmann, Jr., to Alquist, Apr. 24, 1971, Degenkolb to Steinbrugge, May 5, 1971, and Charles De Maria, "Amendments to S. B. 352," May 20, 1971, all in ibid., box 4, folder 170; Advisory Groups on Engineering Considerations and Earthquake Sciences and Governmental Organization & Performance, "Minutes of Special Review Meeting Apr. 28, 1971," ibid., box 5, folder 243.

30. "Staff Analysis of Senate Bill No. 352 (Alquist) As Amended May 25, 1971," in Senate Committee on Health and Welfare Bill Files, California State Archives, Sacramento, file "1971-SB 352"; John F. Shelley to Alquist, May 28, 1971, and "Amendments to SB 352," undated [ca. June 2, 1971], both in SSRG, box 4, folder 170; Advisory Group Coordinating Council, "Minutes of the July 2, 1971 Meeting," and idem, "Minutes of August 3, 1971 Meeting," both in ibid., box 5, folder 243; "Hospitals Quake Safety Bill

Passes First Test," *SB*, June 3, 1971, D:3; "Senate Finance Unit Kills Quake Standard For Hospitals," ibid., July 16, 1971, A:3.

31. Advisory Group Executive Committee, "Minutes of Meeting of Mar. 7, 1972," and idem, "Minutes of Meeting of June 6, 1972," both in SSRG, box 5, folder 244; Lois C. Lillick to Alquist, Mar. 13, 1972, unsigned notes for revisions of Senate Bill 519, Mar. 24, 1972, Carl B. Johnson to Randolph Collier, June 21, 1972, Steinbrugge to Collier, June 21, 1972, and Degenkolb to Collier, June 22, 1972, all in ibid, box 4, folder 177; "Staff Analysis of Senate Bill No. 519 (Alquist)," May 3, 1972, Senate Committee on Health and Welfare Bill Files, file "1972-SB 519"; James A. Willis to William Campbell, July 12, 1972, Emery B. Dowell to Campbell, July 13, 1972, John P. Sheehan to Campbell, July 18, 1972, and Edward J. Gray to Campbell, July 23, 1972, all in Assembly Committee on Health Bill Files, California State Archives, Sacramento, file "1972-SB 519"; "New Law Makes Hospitals Meet Quake Standards," *SB*, Nov. 24, 1972, C:2; "Chapter 1130," *Statutes of California, 1972 Regular Session*, 2170–2176.

32. James E. Slosson and Carl J. Hauge, "The Public and Geology-Related Legislation in California, 1968–1972," in Douglas E. Moran, James E. Slosson, Richard O. Stone, and Charles A. Yelverton, eds., *Geology, Seismicity, and Environmental Impact* (Los Angeles: Association of Engineering Geologists, 1973), 23–27, on 27; Alquist to Evelle J. Younger, Aug. 10, 1973, James E. Slosson to Jeffrey L. Gunther, Sept. 6, 1973, and "Notes on Building Safety Board Meeting," Oct. 9, 1973, all in SSRG, box 4, folder 187; "Opinion of Evelle J. Younger, Attorney General, Nr. CV73/204," reprinted in *EERIN* 8, no. 5 (1974): 49–53; "Report of Legislative Intent Ad Hoc Subcommittee to Building Safety Board: Summary," ibid., no. 4: 29–31.

33. David J. Leeds, "Report of EERI Nominee to California Hospital Building Safety Board," *EERIN* 8, supplement to no. 3 (1974): 13; John L. Towers to Robert Barnicutt, Aug. 10, 1973, SSRG, box 4, folder 187; J. A. Dobrowolski to Ken Ross, Apr. 19, 1974, ibid., box 8, folder 362; S. B. Barnes to Degenkolb, Apr. 12, 1976, Donald C. Axon to Degenkolb, May 3, 1976, and Degenkolb, "Progress Report of the Task Committee on the Hospital Act of 1972," June 8, 1976, all in RJP, ser. 2, carton 11, file "SSC #1"; Scott, *Henry J. Degenkolb*, 189–199.

34. W. Bryce Connick to Steinbrugge, Olson, and Brynn Kernaghan, shortly after Jan. 24, 1973, Albert A. Marino to Alquist, May 25, 1973, Shelley to Alquist, July 5, 1973, "SB 1372," Aug. 29, 1973 suggested revisions, and Alquist to Advisors to the Joint Committee, Jan. 22, 1974, all in SSRG, box 4, folder 189; Carl Treseder to Steinbrugge, Feb. 16, 1973, Connick to Seismic Safety File, shortly before Mar. 29, 1973, and Connick to Steinbrugge, Olson, and Kernaghan, Apr. 2, 1973, all in ibid., folder 192; Richard Carpenter to Alquist, May 7, 1974, and June 27, 1974, Thomas Van Horne to Alquist, June 26, 1974, Melvin K. Davis to Assembly Committee on Planning, Land Use and Energy, June 26, 1974, and Barbara Hurst to Ed O'Connor et al., July 9, 1974, all in ibid., folder 190; Legislative Analyst, "Analysis of Senate Bill No. 1372 (Alquist)," Dec. 21, 1973, and Committee Consultant, "SB 1372 (Alquist): Analysis," Jan. 8, 1974, both in Senate Committee on Governmental Organization Bill Files, California State Archives, Sacramento, file "1973-74-SB 1372"; "Analysis of SB 1372 (Alquist)," Aug. 11, 1974,

Assembly Committee on Planning and Land Use Bill Files, California State Archives, Sacramento, file "1973-SB 1372."

35. Advisory Group Coordinating Council, "Minutes of October 8, 1970," SSRG, box 5, folder 242; Shelley to Alquist, May 28, 1971, ibid., box 4, folder 170; "Chapter 150," *Statues of California, 1971 Regular Session*, 201–203 (quote on 203).

36. James G. Stearns to Herbert Temple, Jr., and Wesley Bruer, May 22, 1972, EMLR, box 9, file "210.15 Governor's Earthquake Council"; James A. R. Johnson to City and County Administrators, July 17, 1972, ibid., file "210.40 Seismic Investigation and Hazard Survey Adv. Com."

37. *Sixty-Eighth Report of the State Geologist,* 4; John W. Williams, "Seismic Hazards and Urbanization in Santa Clara County," *California Geology* 28 (1975): 224–229; "Study Slated on Impact of Earthquake," *LAT,* Oct. 3, 1971, J8; Nestor Barrett to Tocher, Nov. 8, 1972, EMLR, box 9, file "225 Earthquakes-General"; Larry Liebert, "City Earthquake Study Pinpoints Danger Spots," *SFC,* June 14, 1974, p. 2; idem, "How to Save S. F. From Major Quake Damage," ibid., July 18, 1974, p. 4; John H. Wiggins, Larry T. Lee, Michael R. Ploessel, and William J. Petak, "Seismic Safety Study, City of Los Angeles: Technical Report 74-1199-1," excerpted in *EERIN* 8, no. 5 (1974): 98–101.

38. Tracy Wood, "County's Potential for Disaster," *LAT,* Orange County edition, Mar. 23, 1975, XI:1, 4, 5, reprinted in *EERIN* 9, no. 3 (1975): 154–158; Alan J. Wyner and Dean E. Mann, *Preparing for California's Earthquakes: Local Government and Seismic Safety* (Berkeley: Institute of Governmental Studies, 1986), 10–33; Stuart Cook, "Seismic Hazards and Local Land Use Planning: A Review of California Practice," University of California Center for Environmental Design Research Paper CEDR-WP01-90, 1990, copy in Natural Hazards Research Center Library, Boulder, CO.

39. State of California Resources Agency, *Earthquake and Geologic Hazards Conference* (n.p., 1965), 99–100; Clarence R. Allen, "Earthquakes, Faulting, and Nuclear Reactors," *Engineering and Science* (Caltech alumni magazine), Nov. 1967, pp. 10–16; Lloyd S. Cluff, "Urban Development Within the San Andreas Fault System," *SUPGS* 11 (1968): 55–69; David Perlman, "A Quake Expert's Advice to Builders," *SFC,* Sept. 15, 1967, pp. 1, 10; Ad Hoc Interagency Working Group for Earthquake Research, "Proposal for a Ten-Year National Earthquake Hazards Program: A Partnership of Science and the Community," June 1967, Records of the Office of Science and Technology (RG 359), National Archives II, College Park, MD, entry 2, box 13, file "July 25, 1967," p. 3a; George O. Gates, "Earthquake Hazards," in Olson and Wallace, *Geologic Hazards and Public Problems,* 19–28; "The City That Waits to Die," 6; George Getze, "Quake Risk Factors Explained by Richter," *LAT,* Feb. 14, 1971, A:B, B:3.

40. Clarence Allen to Sal Bianco, Mar. 22, 1971, Sal Bianco, "Construction on Active Faults," undated [ca. Apr. 3, 1971], and Allen to Richard Jahns, Apr. 5, 1971, all in RJP, ser. 2, carton 2, no file; "Senate Bill No. 1206," Apr. 14, 1971, SSRG, box 4, folder 174; Advisory Group Coordinating Council, "Minutes of the July 2, 1971 Meeting," and idem, "Minutes of August 3, 1971 Meeting," both in ibid., box 5, folder 243; Senate Committee on Public Works, *Governmental Response,* 561–562.

41. Paul Priolo, "For Immediate Release," Apr. 15, 1971, Assemblyman Paul Priolo

Bill Files, California State Archives, Sacramento, file "1971-AB 2421"; Jahns to Bianco, Sept. 2, 1971, RJP, ser. 2, carton 2, no file.

42. G. B. Oakeshott, "Proposed Legislation on Hazardous Fault Zones," Feb. 1, 1972, and "Senate Bill No. 520," Mar. 8, 1972, both in SSRG, box 4, folder 178; Advisory Group Executive Committee, "Minutes of Meeting of Mar. 7, 1972," ibid., box 5, folder 244.

43. Priolo to C. Martin Duke, Dec. 20, 1971, and Paul Priolo, "For Immediate Release," Feb. 14, 1972, both in Priolo Bill Files, file "1972-AB 407"; "Assembly Bill No. 407," Feb. 14, 1972, SSRG, box 4, folder 184.

44. Donald R. Nichols to Members, Land Use Planning Advisory Group, Feb. 17, 1972, Nichols to Advisory Group Executive Committee, Mar. 14, 1972, SSRG, box 4, folder 182; Marcella Jacobson to Steinbrugge, Mar. 31, 1972, and Jacobson to GO&P [Governmental Operations and Performance] Subcommittee, Apr. 6, 1972, both in ibid., folder 184; Advisory Group Executive Committee, "Minutes," June 6, 1972, ibid., box 5, folder 244; George G. Mader to Steinbrugge, June 7, 1972, ibid., box 4, folder 178; Nichols to Alquist, June 2, 1972, Senate Committee on Governmental Organization Bill Files, file "1972-SB 520."

45. "Chapter 1354," *Statutes of California, 1972 Regular Session*, 2689–2692; Tim Leslie to Alquist, June 21, 1972, Senate Committee on Governmental Organization Bill Files, file "1972-SB 520"; William G. Holliman, Jr., to Alquist, June 27, 1972, SSRG, box 4, folder 178; "Analysis of SB 520 (Alquist), As Amended 5/20/72," Nov. 14, 1972, Assembly Commmittee on Planning and Land Use Bill Files, California State Archives, Sacramento, file "1972-SB 520"; "Analysis of AB 407 (Priolo)," May 23, 1972, ibid., file "1972-AB 407"; "Support Needed for Land Use Bill," *Sierra Club Capitol Calendar*, Mar. 16, 1972, pp. 3–4, clipping, Frederic A. Lane to Priolo, Apr. 20, 1972, and "Earthquake Zoning Bills Need Some Improving," *Oakland Tribune*, July 26, 1972, clipping, all in Priolo Bill Files, file "1972-AB 407."

46. California State Mining and Geology Board, "Minutes of Meeting, April 26, 1973," idem, "Minutes of Meeting of August 23, 1973," idem, "Minutes of Meeting of October 25, 1973," and idem, "Minutes of Meeting of November 21, 1973," all in bound copy of board minutes, Doe Library.

47. California State Mining and Geology Board, "Minutes of Meeting, February 6, 1973," in bound copy of board minutes, Doe Library; "Alquist-Priolo Geologic Hazard Zones Act," *California Geology* 27 (1974): 7.

48. Edward de Steiguer to State Mining and Geology Board, Mar. 29, 1974, Edgar T. Irvine to State Mining and Geology Board, Apr. 4, 1974 (first quote), Ralph G. Towle to Bruer, Mar. 27, 1974, and W. A. Sidler to John M. Bernard, Mar. 29, 1974, all in RJP, ser. 2, carton 29, file "Special Study Zones"; Alquist to W. W. Rogers, Apr. 1, 1974, ibid., ser. 2, carton 26, file "Ad Hoc Comm. Seismic Safety"; "Suit Questions Validity of Hazard Zones Act," *Hollister Evening Free Lance*, Jan. 17, 1975 (second quote), clipping in *EERIN* 9, no. 2 (1975): 79.

49. Contra Costa County Planning Department, "Response: Preliminary Review Map (per S. B. 520) of Earthquake Fault Special Studies Zones," Mar. 19, 1974, Roy S.

Cameron to Slosson, Mar. 25, 1974, Richard A. Oliver to Slosson, Apr. 1, 1974, and James O. Berkland to State Mining and Geology Board, Apr. 4, 1974, all in RJP, ser. 2, carton 29, file "Special Study Zones."

50. Robert S. Binsacca to State Mining and Geology Board, Mar. 20, 1974, Cameron to Slosson, Mar. 25, 1974, and Ted Tedesco to State Mining and Geology Board, Mar. 28, 1974, all in RJP, ser. 2, carton 29, file "Special Study Zones"; Kernaghan to Jahns et al., Apr. 30, 1974, ibid., carton 26, file "Ad Hoc Comm. Seismic Safety"; California State Mining and Geology Board, "Minutes of Meeting, May 30, 1974," and idem, "Minutes of Meeting, Jan. 23, 1975," both in bound copy of board minutes, Doe Library; Joseph E. Bonadiman to Alquist, Nov. 21, 1974, SSRG, box 1, folder 114; "Alquist-Priolo Special Studies Zones Act," *California Geology* 28 (1975): 212–213.

51. Earl W. Hart, *Fault-Rupture Hazard Zones in California,* California Division of Mines and Geology Special Publication 42, revised 1985 (Sacramento: California Division of Mines and Geology, 1985); Robert Reitherman and David J. Leeds, "A Study of the Effectiveness of the Alquist-Priolo Program," California Division of Mines and Geology Open-File Report 90-18, 1991, copy in EERC Library.

52. D. R. Nichols, "Dam Safety Legislation," Nov. 5, 1971, and C. J. Cortright to Bianco, Dec. 1, 1971, both in SSRG, box 4, folder 181; Advisory Group Executive Committee, "Minutes of Meeting of March 7, 1972," and idem, "Minutes," July 11, 1972, both in ibid., box 5, folder 244; "Chapter 780," *Statutes of California, 1972 Regular Session,* 1394–1396; Everett F. Blizzard, "Report to the Seismic Safety Commission," June 10, 1976, RJP, ser. 2, carton 11, file "SSC #1."

53. Will H. Perry, Jr., to C. A. Hammond, June 19, 1973, SSRG, box 4, folder 189; Temple to Alquist, July 5, 1973, Committee Consultant, "SB 1373 (Alquist): Analysis," Jan. 14, 1974, and Legislative Analyst, "Analysis of Senate Bill No. 1373 (Alquist)," Jan. 16, 1974, all in Senate Committee on Governmental Organization Bill Files, file "1973-74-SB 1373"; "Senate Bill 1373 (Alquist), As Amended Aug. 15: Analysis," Aug. 20, 1974, Assembly Committee on Local Government Bill Files, file "1973-74-SB 1373." On the plans of the Office of Emergency Services and the Office of Emergency Planning, see "Bay Area Plan Coming Up," *SFC,* Jan. 31, 1973, p. 4; Robert W. Neubert, "Preparing For the Big Quake," ibid., Sept. 23, 1973, "California Living" section, pp. 40–45; Robert C. Toth, "California Has More Disasters More Often, U.S. Aide Declares," *LAT,* Mar. 18, 1973, I:3, 24, 25; California Legislature Joint Committee on Seismic Safety, "Public Hearing on Earthquake Prediction," Dec. 13, 1974, SSRG, box 2, folder 143, pp. 26–27.

54. Karl V. Steinbrugge and George O. Gates, "A Unifying Objective," *EERIN* 6, no. 1 (1972), individually paginated attachment (quotes); Joint Committee, "Hearing," Feb. 9, 1972, pp. 72–77.

55. Joint Committee on Seismic Safety, *Meeting the Earthquake Challenge: Final Report to the Legislature* (n.p., 1974), first quote on 11, second quote on 8; see also Alfred E. Alquist, "Government Response to Earthquake Hazards," *EERIN* 8, no. 3 (1974): 64–71; "Sweeping Quake Safety Steps Proposed by Legislative Panel," *LAT,* Jan. 10, 1974, I:3; Steve Lawrence, "Solons Rap Quake Program," *SB,* Jan. 10, 1974, A:7.

56. "Senate Bill No. 1729," Feb. 14, 1974, and anonymous [Steinbrugge?], rough draft of legislation, Jan. 4, 1974, both in SSRG, box 8, folder 364; attribution of rough draft to Steinbrugge is suggested by Kernaghan to Steinbrugge, Jan. 2, 1974, ibid., box 8, folder 363.

57. William Mayer to Alquist, Mar. 18, 1974, SSRG, box 4, folder 196; Lawrence R. Robinson to Alquist, Mar. 19, 1974, and Rodney T. Franz to Kernaghan, Mar. 20, 1974, both in ibid., box 8, folder 362; [Kernaghan], "Proposed Amendments to SB 1729," June 7, 1974, and "Senate Bill No. 1729, As Amended June 20, 1974," both in ibid., folder 364; Kernaghan, "Minutes, Joint Meeting on SB 1729," June 14, 1974, ibid., folder 363.

58. Kernaghan to Ad Hoc Group on Seismic Safety Commission, June 27, 1974, and Olson to Commission Team, July 17, 1974, both in SSRG, box 8, folder 363.

59. Department of Finance, "Enrolled Bill Report—SB 1729," Sept. 13, 1974 (quote), Department of Water Resources, "Enrolled Bill Report—SB 1729," Sept. 12, 1974, Department of General Services, "Enrolled Bill Report—SB 1729," Sept. 18, 1974, Legislative Secretary, "Enrolled Bill Memorandum to Governor, Senate Bill 1729," Sept. 16, 1974, Degenkolb to Ronald Reagan, Sept. 3, 1974, and Henry C. Reyes to Reagan, Sept. 13, 1974, all in Governor's Chaptered Bill Files, California State Archives, Sacramento, file "1974, chap. 1413"; "Chapter 1413," *Statutes of California, 1974 Regular Session*, 3112–3115; "12 Sworn In On New Quake Board," *SFC*, May 28, 1975, p. 15; see also Jean Laurin, "Administrative History of the California Seismic Safety Commission," draft of Jan. 18, 1983, RJP, ser. 2, carton 2, no file, pp. 2-5 to 2-8.

60. C. F. Richter, "Earthquake Disaster—Causes, Effects, Precautions," Nov. 1, 1963 (first and second quotes), CRP, box 31, folder 22; idem, "California Earthquakes—Scare or Real Risk?" Sept. 20, 1966 (fourth quote), ibid., box 32, folder 1; Keith Powers, "A High Quake Risk in Our Buildings" (third quote), *SFC*, Apr. 2, 1969, p. 13.

61. Thomas C. Lynch, "Opinion No. 65/324," May 4, 1966, and Fred W. Cheesebrough, "Statement to Assembly Interim Committee on Education," July 20, 1966, both in Legislative Papers, California State Archives, Sacramento, entry "Committee Hearing on Field Act July 20–21, 1966."

62. California School Boards Association, "Information Bulletin No. 1," June 24, 1966, Legislative Papers, entry "Committee Hearing on Field Act July 20–21, 1966"; Dirk Werkman and Ron Ripley, "Students Suffer As Dollars Saved," *Pasadena Independent*, Mar. 30, 1967, pp. 1, 3; "Assembly Bill 450: Analysis," undated [ca. Apr. 1967], Senate Committee on Education Bill Files, California State Archives, Sacramento, file "1967-AB 450"; "Chapter 214," *Statutes of California, 1967 Regular Session*, 1341–1345.

63. "Legislators Act to Close, Replace Unsafe Schools," *SB*, Mar. 20, 1968, A:6; "State Solons Hit Use of Flag Commercially," ibid., July 5, 1968, A:6; Leroy F. Greene to Reagan, July 10, 1968, Governor's Chaptered Bill Files, file "1968, chap. 692"; "Chapter 692," *Statutes of California, 1968 Regular Session*, 1392; "Leroy F. Greene," biographical data sheet, ca. 1970, SSRG, box 1, folder 36.

64. Charles D. Gibson to Max Rafferty, May 31, 1966, Legislative Papers, entry

"Committee Hearing on Field Act July 20–21, 1966"; Dirk Werkman and Ron Ripley, "Valley Districts Facing Problem," *Pasadena Independent,* Mar. 27, 1967, pp. 1, 4.

65. "Yes Vote Urged on Oakland School Quake Bonds," *SFC,* May 13, 1971, p. 2; "PTA Wants More S. F. Schools Shut," ibid., May 13, 1971, p. 2; "Vote for the School Bonds," ibid., Oct. 22, 1973, p. 38; "Alioto Promises Aid For School Bond Vote," *SFE,* May 20, 1971, p. 3; "Strong Case for Props. A and B," ibid., Oct. 15, 1973, p. 34; "Schools Warned on Election Sloganing," ibid., Oct. 20, 1973, p. 8; "Business and Labor Back Props. A and B," ibid., Oct. 31, 1973, p. 34; "Grassroots Gifts to Promote School Bonds Neared $7,000," *LAT,* June 30, 1971, II:1, 3; "Children's Safety at Stake," ibid., Sept. 23, 1971, II:6; "Record Sum of $165,000 Spent on Proposition A," ibid., Oct. 13, 1971, I:20; "Expert Says Quake Could Rock Capital," *SB,* Oct. 1, 1973, B:3; Stephen C. Bilheimer to Richter, Apr. 30, 1971, CRP, box 34, folder 8; "School Daze," script of KCBS News Radio editorial, Sept. 22, 1969, "Earthquake Proof the Schools (#2)," script of KNBR Radio editorial, "Gambling on School Earthquake Safety," script of KNXT editorial, June 18, 1971, "School Earthquake Safety Bonds," ibid., Sept. 28, 1971, " 'Yes' on School Bonds," script of KNX Newsradio editorial, Oct. 4, 1971, and "New Earthquake Bonds," script of KPIX 5 editorial, Apr. 11, 1973, all in SSRG, box 2, folder 130.

66. Charles Diercksmeyer, "Votes and Bond Issues" (letter to editor), *LAT,* June 4, 1971, II:7; Jack McCurdy, " 'Earthquake Safety' Bond Goes Before Voters on Tuesday," ibid., Oct. 10, 1971, A:B, 19; idem, "L. A. Voters Defeat 'Earthquake Safety' School Bond Issue," ibid., Oct. 13, 1971, I:1, 20; "Ability-to-Pay Tax Needed for Schools" (letters to editor), ibid., Oct. 23, 1971, II:4. On the broader tax protest movement, see Jackson K. Putnam, *Modern California Politics,* 2nd ed. (Sacramento: Boyd & Fraser, 1984), 68–69, 83–85; Mike Davis, *City of Quartz: Excavating the Future in Los Angeles* (New York: Verso, 1990), chap. 3.

67. Warren Erickson to Alquist, Burton and Collier, July 1, 1972 (first quote), SSRG, box 1, folder 111; Frank Hogya, "Open Letter," Mar. 20, 1971 (second quote), and Hogya to Richter, Jan. 29, 1972 (third quote), both in CRP, box 34, folder 8; see also "Action Line," *Pasadena Independent,* July 31, 1968, p. 1, and Aug. 9, 1968, p. 1; Bryce W. Anderson, "Earthquake-Proof Schools Receive a Partial Test," *RCT,* Oct. 8, 1969, p. 11.

68. "Quake-Proof S. F. Schools-Huge Cost," *SFC,* Jan. 14, 1970, pp. 1, 26; Ira M. Rines, "Editorial Rebuttal," undated script for KHJ-AM, and Dolly Swift, "Earthquake Bonds," script for KNX Newsradio editorial reply, Oct. 8, 1971, both in SSRG, box 2, folder 130; Sacramento City Unified School District, "Proposed Building Program for Replacement or Renovation of Older Schools and Growth," June 25, 1971, Assembly Committee on Education Bill Files, California State Archives, Sacramento, file "1971-AB 109"; Jim Ward, "School Votes-Gamble Fails, Dissatisfaction Grows," *SFE,* Nov. 3, 1971, p. A; "Los Angeles County," *LAT,* Dec. 22, 1971, I:2; Richter to Dr. Bonner, undated [ca. Oct. 1968], CRP, box 30, folder 15; see also Fried, *Life Along the San Andreas Fault,* 145–146.

69. Paul Houston, "Quake Bond Issue Facing Legal Snag," *LAT,* May 27, 1971, I:1,

20; Ronald J. Ostrow, "Court Dashes Hope for School Bonds," ibid., June 8, 1971, I:1, 17.

70. Statistics from Wirt, *Power in the City,* 212; "YES on Majority-Vote Bonds," *LAT,* Oct. 26, 1972, II:6; see also Dirk Workman and Ron Ripley, "Lack of Money Key to Tragedy," *Pasadena Independent,* Mar. 28, 1967, pp. 1, 4; Everett R. Holles, "Coast Quake Stirs School Safety Issue," *NYT,* Apr. 4, 1971, p. 41; McCurdy, "L. A. Voters Defeat"; Ward, "School Votes."

71. Leroy F. Greene to Members of the Senate and Assembly, Feb. 11, 1971, SSRG, box 1, folder 108; Assembly Committee on Education, "AB 75 (Fiscal Committee) As Amended 3/3/71," undated bill analysis, Assembly Committee on Education Bill Files, file "1971-AB 75"; Greene to Reagan, May 27, 1971, ibid., file "1971-AB 109"; Jerry Gillam, "Assembly Unit Votes Bond Issue on Schools," *LAT,* Mar. 18, 1971, I:28; idem, "School Quake Bond Proposal Goes to Reagan," ibid., May 12, 1971, I:3, 31; "Reagan OK's Quake Proof School Bill," ibid., May 29, 1971, II:10.

72. "Quake Safety Speed Up Bill Fails to Pass," *SB,* Oct. 19, 1971, A:4; "Senate OKs Bill to Remodel Weak Schools," ibid., Oct. 27, 1971, A:4; "Facts on AB 109 and AB 1966 (1971)," undated [summer 1971], Assembly Committee on Education Bill Files, file "1971-AB 109"; Greene to Reagan, Nov. 8, 1971, ibid., file "1971-AB 1966"; "School Quake Bill Signed," *SFE,* Nov. 13, 1971, p. 3.

73. "YES on Earthquake Safety Bond," *LAT,* Apr. 4, 1972, II:6; "Prop. 2 Can Save Children's Lives," *SFC,* May 30, 1972, p. 30; Californians for Earthquake Safe Schools, "Campaign Prospectus," 1972, Assembly Committee on Education Bill Files, file "1971-AB 75"; Robert Fairbanks, "Narrow Victory for School Bonds Seen as Protest," *LAT,* June 8, 1972, I:3, 31.

74. "Senate Approves Field Act Funds," *SB,* July 13, 1972, A:21; "Governor Gets School Quake Bill," ibid., July 19, 1972, A:4; "Field Act Fund Bill Advances," ibid., Apr. 4, 1973, A:5; "Quake Safety School Money Gets Reagan OK," ibid., B:12; "1,593 Schools Do Not Meet Quake Standards," ibid., July 7, 1972, A:4.

75. Jack McCurdy, "State Board Rejects Riles' Quake Bond Plea," *LAT,* Feb. 12, 1972, II:1; Jerry Gillam, "Bill to Help Passage of School Bonds Filed," ibid., Mar. 22, 1972, II:8; idem, "Senate Votes to Put School Earthquake Measure on Ballot," ibid., May 18, 1972, I:3; "Quake Safety for Schools Makes Ballot," *SB,* July 6, 1972, C:8; "Bond Rules Will Be On Ballot," ibid., July 21, 1972, A:4.

76. "For Safe Schools," *SFE,* Oct. 25, 1972, p. 36; "Minority Rule Vs Safe Schools," *SB,* Oct. 26, 1972, A:22; "YES on Majority-Vote Bonds," *LAT,* Oct. 26, 1972, II:6; Jack McCurdy, "School Safety Bonds Boosted by Prop. 9 OK," ibid., Nov. 9, 1972, I:3.

77. See, for example, "Vote for the School Bonds," *SFC,* Oct. 22, 1973, p. 38; Ron Moskowitz, "School Bond Vote—Morena's Fears," ibid., Oct. 29, 1973, p. 2; "Tour of Potential School Peril Areas," *SFE,* Oct. 31, 1973, p. 34; "School Quake Bonds Win Big," ibid., Nov. 7, 1973, p. 6; Noel Greenwood, "Quake-Safe School Deadline Extended," *LAT,* Apr. 27, 1974, II:1, 10.

78. California State Department of Education, "Structurally Unsafe School Buildings

Progress Report: A Report to the California Legislature," 1976, RJP, ser. 2, carton 2, no file; "Quake Program for Schools Called 'Big Success,'" *LAT,* May 31, 1977, I:21.

79. Robert Rawitch, "Little Done Despite Lesson of '33," *LAT,* Aug. 29, 1971, C:1; "Policy OK'd For City's Quake-Hazard Buildings," ibid., Mar. 18, 1975, I:22; Ellen Hume, "Unsafe Buildings Endanger 200,000," ibid., Apr. 7, 1975, II:1, 2, 3; Donald Canter, "Grim Quake Forecast for Downtown S. F.," *SFE,* June 14, 1974, p. 3; Joint Committee on Seismic Safety, "Public Hearing on Earthquake Prediction," 56; Henrik Bull, "Potential Seismic Hazard versus Certain Personal Disaster," *Proceedings, Workshop on the Seismic Upgrading of Existing Buildings, September 9-10, 1982,* pp. 33-41, copy in EERC Library.

80. Daniel J. Alesch and William J. Petak, *The Politics and Economics of Earthquake Hazard Mitigation: Unreinforced Masonry Buildings in Southern California* ([Boulder]: Institute of Behavioral Sciences, University of Colorado, 1986), chap. 4; Edward M. O'Connor, "Pre-Field Act Buildings in the City of Long Beach," *Proceedings, Structural Engineers Association of California, 39th Convention—1970,* pp. 20-24; "Affidavit of Edward M. O'Connor in Support of Defendants' Motion to Dismiss," July 9, 1970, CRP, box 34, folder 7; E. M. O'Connor, "Existing Earthquake Hazardous Buildings," May 8, 1972, RJP, ser. 2, carton 2, no file; Fried, *Life Along the San Andreas Fault,* 149-151, 155-160.

81. Wyner and Mann, *Preparing for California's Earthquakes,* 47-49; Fried, *Life Along the San Andreas Fault,* 152-154. A similar ordinance was rescinded soon after being enacted in Oroville after a destructive earthquake in 1975; see Robert A. Olson, Richard S. Olson, and David L. Messinger, *"Standing Rubble": The 1975-1976 Oroville, California Experience with Earthquake-Damaged Buildings* (Sacramento: VSP Associates, 1988).

82. Alesch and Petak, *Politics and Economics,* 57-68; Erwin Baker, "Panel OKs Shift in Quake Safety Policy," *LAT,* Aug. 28, 1976, II:1, 10; C. W. Carson, "'The Cyclone Didn't Hear Us'" (letter to editor), ibid., Sept. 21, 1976, II:6; Baker, "Stronger Rules on Quake Safety Urged," ibid., Oct. 23, 1976, II:1; John F. Schwarzer, "'A Fair Shake for Safety'" (letter to editor), ibid., Nov. 4, 1976, II:6; Apartment Association of Los Angeles County, "Evicted" (advertisement), ibid., Dec. 2, 1976, IV:34; "Quake Safety Proposal Gets Setback in Council," ibid., Dec. 10, 1976, II:8; Baker, "Council Shifts Emphasis on Quake Safety of Buildings," ibid., Jan. 25, 1977, II:1, 6. See also Howard Jarvis with Robert Pack, *I'm Mad As Hell: The Exclusive Story of the Tax Revolt and Its Leader* (New York: Times Books, 1979), on 68, where Jarvis takes credit for killing "a proposed earthquake ordinance that would have required the demolition of 14,000 brick buildings and put 100,000 tenants out of their apartments."

Chapter 8

1. Victory Cohn, "Earth Scientists Again Call For Serious Quake Studies," *Miami Herald,* Feb. 12, 1971 (first quote), clipping in FPP, box 24, folder 745; "Earthquake Hearings," KNX Newsradio editorial, June 11, 1971 (second quote), transcript in SSRG, box 2, folder 130; see also Jay Sharbutt, "Are Earthquakes Predictable? Even the Experts

Can't Agree," *LAT,* Feb. 16, 1971, I:3; Keith Power, "Earthquake's Hard Lesson," *SFC,* Feb. 10, 1971, p. 6.

2. "Acceleration of the National Earthquake Hazards Reduction Program," undated, and W. H. Radlinski and Robert M. White to Edward E. David, Jr., May 28, 1971, both in EMLR, box 6, file "10 FY72 Budget Amendment"; Morton J. Rubin to Dr. Townsend, Nov. 11, 1971, ibid., box 7, file "200.12 Geological Survey"; Don Tocher, handwritten notes, Jan. 19, 1972, ibid., box 8, file "10 Budget"; Dallas L. Peck to Members, Advisory Panel to the National Center for Earthquake Research, Mar. 16, 1972, RJP, ser. 2, carton 22, file "U.S.G.S."

3. Frank Press to Vincent McKelvey, Apr. 28, 1972, FPP, box 32, folder 981; C. B. Raleigh to Press, May 5, 1972, ibid., box 31, folder 973.

4. Jan Foley to Richard H. Jahns, Oct. 19, 1971, RJP, ser. 2, carton 25, folder "Allan [*sic*] Cranston"; Bill Boyarsky, "$120 Million Sought for Quake Warning, Safe Building Plan," *LAT,* Dec. 21, 1971, II:1, 3; Senate Committee on Commerce, "Earthquake Hazards Act: Hearing," 92nd Cong., 2nd sess., May 16, 1972, pp. 3–11; Alan Cranston to Henry J. Degenkolb, July 14, 1975, *EERIN* 9, no. 5A (1975): 7; J. P. Eaton, *Microearthquake Seismology in USGS Volcano and Earthquake Hazards Studies: 1953-1995,* USGS Open-File Report 96-54 (Washington, DC: GPO, 1996), 8.

5. Senate Committee on Commerce, "Earthquake Hazards Act," 24–34.

6. M. A. Sadovsky et al., "The Processes Preceding Strong Earthquakes in Some Regions of Middle Asia," *Tectonophysics,* 14 (1972): 295–307, on 301; Amos Nur, "Dilatancy, Pore Fluids, and Premonitory Variations of ts/tp Travel Times," *BSSA* 62 (1972): 1217–1222, on 1221; Yash P. Aggarwal, Lynn R. Sykes, John Armbruster, and Marc L. Sbar, "Premonitory Changes in Seismic Velocities and Prediction of Earthquakes," *Nature* 241 (1973): 101–104, on 101.

7. Geomagnetism Correspondent, "Precursory V_P/V_s Changes," *Nature* 243 (1973): 380; T. Rikitake, "Editorial: Symposium on Earthquake Mechanics," *Tectonophysics* 9 (1970): 97–99, on 99; [David Davies], "Predicting Earthquakes," *Nature* 245 (1973): 121–122; F. Gilman Blake to L. A. Goldmuntz, Oct. 1, 1971, Edward E. David Papers, Richard M. Nixon Project, National Archives II, College Park, MD, box 82, folder "Natural Disasters (3 of 4)"; Walter Sullivan, "Scientists Envision Quake Prediction and Prevention," *NYT,* Apr. 25, 1972, p. 85.

8. Aggarwal et al., "Premonitory Changes."

9. James H. Whitcomb, Jan D. Garmany, and Don L. Anderson, "Earthquake Prediction: Variation of Seismic Velocities Before the San Francisco [*sic*—for 'San Fernando'] Earthquake," *Science,* 180 (1973): 632–635; Geomagnetism Correspondent, "Precursory V_P/V_s Changes."

10. Nur, "Dilatancy"; T. A. Heppenheimer, *The Coming Quake: Science and Trembling on the California Earthquake Frontier* (New York: Times Books, 1988), 140–141.

11. Christopher H. Scholz, Lynn R. Sykes, and Yash P. Aggarwal, "Earthquake Prediction: A Physical Basis," *Science* 180 (1973): 803–810.

12. Scholz et al., "Earthquake Prediction," 803 (quote); Allen L. Hammond, "Earthquake Predictions: Breakthrough in Theoretical Insight?" *Science* 180 (1973): 851–853;

"Quake Forecast Viewed as Near," *NYT,* Apr. 18, 1973, p. 94; [Davies], "Predicting Earthquakes"; Heppenheimer, *The Coming Quake,* 139, 142–143.

13. "Minutes, Committee on Seismology, May 11, 1973" (first quote), National Academy of Sciences Archives, Washington, DC, Divisional Series, "DIV ES: Com on Seismology: Meetings: Agendas and Minutes, 1973 May"; Carl Greenberg, "Cranston Tells of Dangers in Future Quakes," *LAT,* Apr. 27, 1973, I:26 (second quote).

14. Yash P. Aggarwal, Lynn R. Sykes, David W. Simpson, and Paul G. Richards, "Spatial and Temporal Variations in ts/tp and in P Wave Residuals at Blue Mountain Lake, New York: Application to Earthquake Prediction," *Journal of Geophysical Research* 80 (1975): 718–732; Peter J. Smith, "Successful Quake Prediction Made," *Nature* 255 (1975): 282; Robert Reinhold, "Clue to Quake Forecasting Sought in Tiny Temblors," *NYT,* Aug. 11, 1973, pp. 21, 38; "Predicting the Quake," *Time,* Aug. 27, 1973, p. 36; "Forecast: Earthquake," ibid., Sept. 1, 1975, pp. 36–41; Harvey Ardman, "The Case of the Earthquake Detectives," *American Legion Magazine,* May 1974, pp. 6–9, 34–36, 38, quote on 6 (clipping courtesy of Yash P. Aggarwal, New City, NY). For exuberant media coverage of earthquake prediction, see also Gordon D. Friedlander, "Earthquake Prediction: A New Art," *IEEE Spectrum,* Sept. 1973, pp. 46–57; Steve Aaronson, "The Shape of Quakes to Come," *The Sciences,* Apr. 1974, pp. 15–20; Christopher H. Scholtz [*sic*], "Toward Infallible Earthquake Prediction," *Natural History,* May 1974, pp. 54–59; Carl Kisslinger, "Earthquake Prediction," *Physics Today,* Sept. 1974, pp. 36–42;

15. T. V. McEvilly and L. R. Johnson, "Earthquakes of Strike-Slip Type in Central California: Evidence on the Question of Dilatancy," *Science* 182 (1973): 581–584; Clarence R. Allen and Donald V. Helmberger, "Search for Temporal Changes in Seismic Velocities Using Large Explosions in Southern California," *SUPGS* 13 (1973): 436–445; William H. Bakun, Roger M. Stewart, and Don Tocher, "Variations in V_P/V_s in Bear Valley in 1972," ibid., 13 (1973): 453–462; Chris H. Cramer and Robert L. Kovach, "A Search for Teleseismic Travel-Time Anomalies Along the San Andreas Fault Zone," *Geophysical Research Letters* 1 (1974): 90–92; see also Reinhold, "Clue to Quake Forecasting."

16. Geomagnetism Correspondent, "No Dilatancy," *Nature* 246 (1973): 332–333; "Minutes of the December 7 and 8, 1973 Meeting of the USGS Earthquake Studies Advisory Panel," RJP, ser. 2, carton 22, file "U.S.G.S.," p. 20; Chi-yuen Wang, "Earthquake Prediction and Oriented Microcracks in Rocks," *Nature* 251 (1974): 405–406.

17. D. J. Sutton, "A Fall in P-Wave Velocity Before the Gisborne, New Zealand, Earthquake of 1966," *BSSA* 64 (1974): 1501–1508; M. Wyss and A. C. Johnston, "A Search for Teleseismic P Residual Changes Before Large Earthquakes in New Zealand," *Journal of Geophysical Research* 79 (1974): 3283–3290; Gordon S. Stewart, "Prediction of the Pt. Mugu Earthquake By Two Methods," *SUPGS* 13 (1973): 473–478.

18. M. Wyss and D. J. Holcomb, "Earthquake Prediction Based on Station Residuals," *Nature* 245 (1973): 139–140; Amos Nur, "Matsushiro, Japan, Earthquake Swarm: Confirmation of the Dilatancy–Fluid Diffusion Model," *Geology* 2 (1974): 217–221; Walter Sullivan, "Quake Prediction Gets New Impetus," *NYT,* Oct. 14, 1973, p. 80; see also Raymon Brown, "Precursory Changes in V_P/V_s Before Strike-Slip Events," *SUPGS*

13 (1973): 463–472; Max Wyss, "A Search for Precursors to the Sitka, 1972, Earthquake: Sea Level, Magnetic Field, and P-Residuals," *Pure and Applied Geophysics* 113 (1975): 297–309.

19. Russell Robinson, Robert L. Wesson, and William L. Ellsworth, "Variation of P-Wave Velocity Before the Bear Valley, California, Earthquake of 24 February 1972," *Science* 184 (1974): 1281–1283; C. H. Cramer and R. L. Kovach, "Time Variations in Tele-seismic Residuals Prior to the Magnitude 5.1 Bear Valley Earthquake of February 24, 1972," *Pure and Applied Geophysics* 113 (1975): 281–292; Peter J. Smith, "Earthquakes Predicted," *Nature* 252 (1974): 9–11.

20. Allen L. Hammond, "Dilatancy: Growing Acceptance as an Earthquake Mecha-nism," *Science* 184 (1974): 551–552.

21. Marvin Miles, "Possible Quake Link to Seismic Waves Reported," *LAT*, Aug. 21, 1973, II:1, 8; "Earthquakes: Pattern For Prediction," *Engineering and Science* [Caltech alumni magazine], Oct. 1973, pp. 19–21; "Seismic Wave Velocities in Riverside," *California Geology* 26 (1973): 247; Hiroo Kanamori and Wai-Ying Chang, "Temporal Changes in P-Wave Velocity in Southern California," *Tectonophysics* 23 (1974): 67–78.

22. George Alexander, "Scientist's Prediction of Quake Comes True," *LAT*, Apr. 11, 1974, I:1, 36; "An Earthquake—On Schedule," *Engineering and Science*, May 1974, p. 13; "Earthquake Successfully Predicted," *California Geology* 27 (1974): 257.

23. Allen L. Hammond, "Earthquake Prediction: Progress in California, Hesitation in Washington," *Science* 187 (175): 419–420, both quotes on 419; Peter J. Smith, "Get-ting Closer to Prediction," *Nature* 253 (1975): 593; B. E. Smith and M. J. S. Johnston, "A Tectonomagnetic Effect Observed Before A Magnitude 5.2 Earthquake Near Hollister, California," *Journal of Geophysical Research* 81 (1976): 3556–3560; C. E. Mortensen and M. J. S. Johnston, "Anomalous Tilt Preceding the Hollister Earthquake of November 28, 1974," ibid.: 3561–3566.

24. *United States Geological Survey Annual Report, Fiscal Year 1975* (Washington, DC: GPO, 1975), 54–59, quote on 58; see also California Legislature Joint Committee on Seismic Safety, "Public Hearing on Earthquake Prediction," Dec. 13, 1974, SSRG, box 2, folder 143, p. 6; "Successful Prediction of Earthquake Reported," *LAT*, Jan. 9, 1975, I:30; "A Landscape Tilt May Mean Earthquake," *NYT*, Jan. 19, 1975, IV:8; Robert Strand, "In Two Years Scientists May Be Able to Predict Coming Quake," *SB*, Mar. 1, 1975, C:14.

25. Hammond, "Earthquake Prediction"; Deborah Shapley, "Chinese Earthquakes: The Maoist Approach to Seismology," *Science* 193 (1976): 656–657; David Davies, "Earthquake Prediction in China," *Nature* 258 (1975): 286–287; C. H. Scholz, "A Physi-cal Interpretation of the Haicheng Earthquake Prediction," ibid., 267 (1977): 121–124; "China Says She Predicted Big Quake," *NYT*, Mar. 1, 1975, p. 8; George Alexander, "Professor Hails China's Quake Prediction Efforts," *LAT*, Apr. 27, 1975, II:1, 3; Peter Gwynne, "Thunder Out of China," *Newsweek*, June 16, 1975, p. 45.

26. National Research Council Panel on Earthquake Prediction, *Predicting Earth-quakes: A Scientific and Technical Evaluation with Implications for Society* (Washington, DC: National Academy of Sciences, 1976), first quote on 2; Robert M. Hamilton, "The

Future of Earthquake Prediction," in *Geological Survey Annual Report, Fiscal Year 1975*, 7-10, second quote on 9; "Minutes of the December 10 and 11, 1974 Meeting of the USGS Earthquake Studies Advisory Panel" (third quote), RJP, ser. 2, carton 22, file "U.S.G.S."; Robert M. Hamilton to Members of the USGS Earthquake Studies Advisory Panel, June 16, 1975, and Hamilton, "U.S. Geological Survey Earthquake Prediction Council," draft, Apr. 21, 1976, both in FPP, box 32, folder 991; *Earthquake Prediction—Opportunity to Avert Disaster*, USGS Circular 729 (Washington, DC: GPO, 1976); "Earthquake Prediction Raises Issues," *NYT*, Nov. 9, 1975, p. 42; USGS, "Warning and Preparedness for Geologic-Related Hazards: Proposed Procedures," *Federal Register* 42 (1977): 19292-19296.

27. "Draft Minutes of the NOAA Earthquake Research Committee (ERC) Meeting," Dec. 11, 1970, EMLR, box 6, file "210.15 Earthquake Research Committee"; Comptroller General of the United States, "Need for a National Earthquake Research Program," GAO Report B-176621, Sept. 11, 1972, quote on 1; Nicholas Wade, "Earthquake Research: A Consequence of the Pluralistic System," *Nature* 178 (1972): 39-43.

28. "Advisory Panel, National Center for Earthquake Research: Meeting, Sept. 14-15, 1972," RJP, ser. 2, carton 22, file "U.S.G.S."; Geologic Division, "Earthquake Hazard Reduction Program of the U.S. Geological Survey," draft of Sept. 1972, EMLR, box 9, folder "200.25 U.S. Geological Survey," pp. 72-84; Dwight A. Ink to William Morrill, Oct. 24, 1972, and Clifford Berg to Pat Dinneen, Oct. 26, 1972 (quote), both in Records of the Office of Management and Budget (RG 51), National Archives II, College Park, MD, entry 175, box 35, folder "R 1-6 Disasters and Emergencies (Other than Military)."

29. Tocher, handwritten notes, Dec. 15, 1972, EMLR, box 8, file "10 Budget"; White, "Message from the Administrator," Jan. 29, 1973, ibid., box 15, file "Peck Committee"; Department of the Interior, "NOAA Earthquake Programs to USGS, NSF," press release, June 5, 1973, and Hamilton to Ann Wray, May 13, 1974, both in Records of the U.S. Senate (RG 46), National Archives, Washington, DC, series "Committee on Commerce, Science, and Transportation," subseries "94th Congress Legislative Files: S. 1174," box 7, loose material.

30. Senate Committee on Commerce, "Earthquakes: Hearings," 93rd Cong., 1st sess., Apr. 26 and 27, 1973, first quote on 114, second quote on 115; Joint Committee on Seismic Safety, "Public Hearing," third quote on 56.

31. Robert E. Wallace, *Goals, Strategy, and Tasks of the Earthquake Hazard Reduction Program*, USGS Circular 701 (Washington, DC: GPO, 1974); see also "Minutes of the June 6 and 7, 1974 Meeting of the USGS Earthquake Studies Advisory Panel," RJP, ser. 2, carton 22, file "U.S.G.S."

32. "New Funds for Quake Research Are Proposed," *SFC*, Mar. 14, 1975, p. 16; Cranston to C. Martin Duke, Apr. 23, 1975, *EERIN* 9, no. 4 (1975): 41-49; Senate Committee on Commerce, "Earthquake Disaster Mitigation Act of 1975: Hearing," 94th Cong., 2nd sess., Feb. 19, 1976.

33. Ernest F. Hollings and Cranston to Dear Colleague, Feb. 20, 1975, U.S. Senate Records, series "Committee on Commerce, Science, and Transportation," subseries "Earthquake Legislation 1972-1975: Miscellaneous Materials," box 10, folder "Earth-

quake Legislation Corres. Pre-1976"; Senate Committee on Aeronautical and Space Sciences, "Earthquake Research and Knowledge: Hearing," 94th Cong., 1st sess., Apr. 26, 1975, p. 184; "Earthquake Hazards Reduction Act," *Congressional Record,* May 12, 1977, pp. S7471–S7478, on S7476.

34. "Advisory Committee on Emergency Planning," Oct. 20, 1973, Appendix A to "Agenda for Panel on Earthquake Prediction Meeting of December 12, 1973," National Academy of Sciences Archives, Divisional Series, folder "DIV ES: Com on Seismology: Panel on Earthquake Prediction: Mtg: Agenda & Minutes, Dec/1973"; National Research Council Panel on the Public Policy Implications of Earthquake Prediction, *Earthquake Prediction and Public Policy* (Washington, DC: National Academy of Sciences, 1975), quotes on 3; Ralph H. Turner, "Earthquake Predictions: Potential Blessing," *California Geology* 29 (1976): 208–209.

35. *Earthquake Prediction—Opportunity to Avert Disaster,* 17–19; George Alexander, "Quake Forecasts Could Be Harmful, Experts Say," *LAT,* Aug. 26, 1975, II:1, 18; "Forecasting Earthquakes," ibid., Aug. 27, 1975, II:6; George Alexander, "Quake Forecasts: How Will Community React?" ibid., Oct. 12, 1975, II:1, 5.

36. H. Guyford Stever to Degenkolb, Aug. 12, 1975, *EERIN* 9, no. 6A (1975): 54; "Minutes of the December 5 and 6, 1975, Meeting of the USGS Earthquake Studes Advisory Panel," FPP, box 32, folder 985.

37. Robert O. Castle, John N. Alt, James C. Savage, and Emery I. Balazs, "Elevation Changes Preceding the San Fernando Earthquake of February 9, 1971," *Geology* 2 (1974): 61–66; Robert O. Castle, Jack P. Church, and Michael R. Elliott, "Aseismic Uplift in Southern California," *Science* 192 (1976): 251–253; Charles R. Real and John H. Bennett, "Palmdale Bulge," *California Geology* 29 (1976): 171–173; "Minutes of the December 5 and 6, 1975, Meeting."

38. Press, handwritten notes taken at USGS advisory panel meeting, Dec. 5, 1975, FPP, box 32, folder 994.

39. Hamilton to the Record, Dec. 30, 1975, FPP, box 22, folder 681; "Minutes of the Advisory Group on Anticipated Advances in Science and Technology," Jan. 14, 1976, ibid., box 22, folder 677; "The U.S. National Program in Earthquake Prediction: Summary of the U.S. Geological Survey Presentation to the President's Advisory Panel on Anticipated Advances in Science and Technology," Jan. 14, 1976, Glenn R. Schleede Files, Gerald R. Ford Presidential Library, Ann Arbor, MI, box 9, folder "Earthquake Prediction, 1976: Background Information."

40. Press to Nelson A. Rockefeller, Jan. 21, 1976 (quotes), Schleede Files, box 9, folder "Earthquake Prediction, 1976: Correspondence"; see also Rockefeller to Gerald R. Ford, Feb. 5, 1976, ibid., folder "Earthquake Prediction, 1976: Memoranda for the President"; Press to Stever, Jan. 21, 1976, FPP, box 22, folder 681.

41. Dick Allison to Glenn Schleede, Feb. 5, 1976, and James T. Lynn and James Cannon, "Memorandum for the Vice President," Feb. 10, 1976, both in Schleede Files, box 9, folder "Earthquake Prediction, 1976: General."

42. Press to Rockefeller, Jan. 21, and Feb. 5, 1976, and Stever to Rockefeller, Feb. 9,

1976 (two letters), all in Schleede Files, box 9, folder "Earthquake Prediction, 1976: Correspondence"; McKelvey to Secretary of the Interior, Feb. 9, 1976, and Stanley D. Doremus to James L. Mitchell, Feb. 9, 1976, both in ibid., folder "Earthquake Prediction, 1976: Background Information."

43. Lynn and Cannon, "Memorandum for the Vice President"; Rockefeller, "Memorandum for the President," Feb. 11, 1976, and Schleede to Mitchell, Feb. 18, 1976, both in Schleede Files, box 9, folder "Earthquake Prediction, 1976: Memoranda for the President"; Rockefeller, "Memorandum for the President," Feb. 19, 1976, Dennis Barnes Files, Gerald R. Ford Library, Ann Arbor, MI, box 1, file "Earthquakes (1)."

44. Press, handwritten notes, undated [ca. spring 1976], FPP, box 22, folder 675; see also Dorothy M. Rolleri, "Request for Proposal RFP 287W," June 10, 1976, RJP, ser. 2, carton 22, file "U.S.G.S."; Jim McCullough to Mitchell, July 29, 1976, Schleede Files, box 9, folder "Earthquake Prediction, 1976: General."

45. Charles Petit, "A 6-Inch Bulge in the Fault," *SFC,* Feb. 13, 1976, p. 7.

46. "U.S. Geological Survey—State of California Briefing Concerning Recent Land Uplift in Southern California," Mar. 17, 1976 (quote), RJP, ser. 2, carton 11, file "SSC #1"; Paul H. O'Neill, "Memorandum for the President," Mar. 23, 1976, White House Central Files, Gerald R. Ford Library, Ann Arbor, MI, Subject Files, box DI1, folder "DI2 1/1/76–12/14/76"; George Alexander, "Get Ready for Giant Quake, Panel Warns," *LAT,* Apr. 18, 1976, I:3, 18; Hamilton, "Chronology of Events Relating to Accelerated Earthquake Studies," June 3, 1976, FPP, box 32, folder 991.

47. George Alexander, "Concern Voiced Over Bulge in Earth Along San Andreas Fault," *LAT,* Mar. 12, 1976, II:1, 2; State of California Seismic Safety Commission, "Resolution No. 1-76," Apr. 8, 1976 (quote), *EERIN* 10, no. 3 (1976): 40–41; idem, "Resolution No. 2-76," undated [Apr. 8, 1976], ibid.: 42–43; "The Bubble in Our Midst," *LAT,* Apr. 8, 1976, II:4.

48. Alexander, "Get Ready for Giant Quake" (quote). For a brief history of the council, see California Earthquake Prediction Evaluation Council, "Earthquake Prediction Evaluation Guidelines," *California Geology* 30 (1977): 158–160.

49. Charles Manfred to Chairmen, Boards of Supervisors, et al., Mar. 30, 1976 (quote), H. R. Pulley, "OES Report to the Seismic Safety Commission Concerning its Activities with Respect to Earthquake Prediction," May 13, 1976, and Ronald B. Robie to Robert A. Olson, June 2, 1976, all in RJP, ser. 2, carton 11, file "SSC #1"; Seismic Safety Commission, " 'Palmdale Bulge' Discussions," July 8, 1976, ibid., file "SSC #2"; C. E. Forbes to R. J. Datel, June 2, 1976, *EERIN* 10, no. 4 (1976): 57–59; Seismic Safety Commission, "Minutes, Aug. 12, 1976," ibid., no. 5 (1976): 22–25. See also Ralph H. Turner, Joanne M. Nigg, and Denise Heller Paz, *Waiting for Disaster: Earthquake Watch in California* (Berkeley: U of California P, 1986).

50. "Palmdale Bulge May Foreshadow a Major Southern California Earthquake," *Congressional Record,* Mar. 16, 1976, p. S3512; Richard C. Paddock, "Study of China's Success in Quake Warning Urged," *LAT,* Apr. 15, 1976, I:1, 20.

51. George Alexander, "Caltech Scientist Offers Cautious Quake Prediction," *LAT,*

Apr. 21, 1976, I:1, 3, 29; idem, "Experts Not Convinced by Data on Quake Prediction," ibid., May 1, 1976, I:1, 10; James H. Whitcomb, "Quake Prediction: An Emerging Science," ibid., May 2, 1976, VIII:5; Real and Bennett, "Palmdale Bulge" (reproduces a graph of V_P/V_s data from a preprint of Whitcomb's never-published paper); "Testing a Hypothesis," *Engineering and Science,* May 1976, pp. 8–9.

52. George Alexander, "Don't Panic, Stay Put and Avoid Windows," *LAT,* Apr. 22, 1976, II:1, 4; idem, "Like the Boy Scouts, We Should Be Prepared," ibid., May 2, 1976, VIII:5; Betty Liddick, "Quake Predictions on Firm Ground?" ibid., Apr. 29, 1976, IV:1, 8, 9; Deborah Shapley, "Earthquakes: Los Angeles Prediction Suggests Faults in Federal Policy," *Science* 192 (1976): 535–537; Turner et al., *Waiting for Disaster,* chap. 2.

53. Ellen Goodman, "Don't Kill the Messenger—Sue Him," *Washington Post,* [ca. May 1976] clipping in U.S. Senate Records, series "Committee on Commerce, Science, and Transportation," subseries "Earthquake Hazards Reduction Act: General Information and Miscellaneous Publications, 1976," box 8, loose material; George Alexander, " 'You . . . Are There' as Scientist Tests Quake Forecast Theory," *LAT,* May 16, 1976, IX:1, 4; "An Investment in Safety," ibid., June 1, 1976, II:4; "Spending for a Big Payoff," ibid., June 29, 1976, II:4; "A Calamitous Reminder," ibid., July 30, 1976, II:4; "Spending That's No Waste," ibid., Sept. 22, 1976, II:6; "Finding Ways to Predict Quakes," *SFC,* Feb. 17, 1976, p. 34; "The Science of Quake Warnings," ibid., Apr. 16, 1976, p. 38; "The Predicted Chinese Quakes," ibid., July 30, 1976, p. 34.

54. Senate Committee on Commerce, "Earthquake Disaster Mitigation Act of 1975"; Staff to Senator Magnuson, Apr. 23, 1976, U.S. Senate Records, series "Committee on Commerce, Science, and Transportation," subseries "94th Congress Legislative Files: S. 1174," box 7, loose material; "Quake Warning Bill Advances," *LAT,* May 5, 1976, I:29; "Earthquake Hazard Reduction Act," *Congressional Record,* May 24, 1976, S7795–S7799; Richard C. Paddock, "Quake Forecast Funds OK'd" (quote), *LAT,* May 25, 1976, I:1, 8.

55. "An Effort to Reduce Earthquake Hazards," *Congressional Record,* June 4, 1976, pp. E3122–E3123; "Earthquake Hazards Reduction Legislation," ibid., June 4, 1976, pp. E3130–E3131; House Committee on Science and Technology, "Earthquake: Hearings," 94th Cong., 2nd sess., June 22, 23, and 24, 1976.

56. "Report of the Subcommittee on Science, Research and Technology: To Accompany HR 14876 as Amended by the Subcommittee," Aug. 12, 1976, copy in Barnes Files, box 2, folder "Earthquakes-Legislation (2)"; House of Representatives, "Earthquake Hazards Reduction Act of 1976: Report," 94th Cong., 2nd sess., Rept. 94-1440, Aug. 26, 1976; Barnes to Lynn May and Schleede, Sept. 3, 1976, Schleede Files, box 9, folder "Earthquake Prediction, 1976: Legislation"; "Earthquake Hazards Reduction Act of 1976," *Congressional Record,* Sept. 20, 1976, pp. H10530–H10544; "Reducing the Hazards of Earthquakes," ibid., pp. H10563–H10564; Paul Huston, "Funds for Quake Research Killed," *LAT,* Sept. 21, 1976, I:1, 9; William M. Ketchum, " 'Spending That's No Waste' " (letter to editor), ibid., Oct. 8, 1976, II:6.

57. McCullough, "Memorandum for Jim Mitchell," July 12, 1976, and Cannon to Ford, undated [Aug. 25, 1976], both in Barnes Files, box 1, folder "Earthquakes (2)";

Schleede et al. to Cannon, Aug. 20, 1976, and Schleede to Cannon, Aug. 30, and Aug. 31, 1976, all in ibid., box 2, folder "Earthquakes-Legislation (2)."

58. Rockefeller, "Memorandum for the President," Feb. 11, 1976 (quote); Phil Smith to Stever, May 17, 1976, Records of the National Science Foundation (RG 307), National Archives II, College Park, MD, series "Office of the Director-Science Advisor G. Stever, Subject Files 1973-76," box 12, file "Advisory Group/Earthquake Prediction and Hazard Mitigation"; "Summary Minutes of the Advisory Gruop on Earthquake Prediction and Hazard Mitigation," June 14, 1976, Barnes Files, box 1, folder "Earthquakes-Advisory Group . . . (1)"; "Earthquake Prediction and Hazard Mitigation—Options for USGS and NSF Programs," Sept. 15, 1976, copy in Nathan M. Newmark Papers, University of Illinois Archives, Champaign, box 10, folder "NSF Adv." The penultimate draft of the report had called for a Federal Earthquake Prediction and Hazards Reduction Advisory Committee to periodically review the program, but this was dropped in the final version: see "Earthquake Prediction and Hazard Mitigation—Options for USGS and NSF Programs," draft of Sept. 1, 1976, copy in ibid.

59. Stever to Lynn, Oct. 18, and Nov. 19, 1976, both in H. Guyford Stever Papers, Gerald R. Ford Library, Ann Arbor, MI, box 89, folder "Earthquake Prediction and Hazard Mitigation"; Office of Management and Budget, "1978 Presidential Review: Certain Interior and Related Programs," Dec. 8, 1976, James Cannon Papers, Gerald R. Ford Library, Ann Arbor, MI, box 17, folder "Pres. Rev. Energy Related Issues + Earthquake Research"; "Ford Urges Doubling of Quake Research Funds With Goal of Reliable Predictions in 10 Years," LAT, Jan. 18, 1977, I:14.

60. Philip M. Boffey, "Frank Press, Long-Shot Candidate, May Become Science Advisor," Science 195 (1977): 763-766; Hamilton, "Summary of FY78 Budget Situation, USGS Earthquake Hazards Reduction Program," undated (before Sept. 12, 1977), Newmark Papers, box 12, folder "USGS 2"; House of Representatives, "Department of the Interior and Related Agencies Appropriation Bill, 1978," 95th Cong., 1st sess., Rept. 95-392, quote on 36.

61. "Cranston Seeks $220 Million for Quake Study, Preparation," LAT, Jan. 12, 1977, II:6; Press to George E. Brown, Apr. 15, 1977, and Press to Jimmy Carter, Apr. 19, 1977, both in FPP, box 36, folder "Chron File April 1977"; Magnuson to Members of the Senate Commerce Committee, May 3, 1977, U.S. Senate Records, series "Committee on Commerce, Science, and Transportation," subseries "95th Congress Legislative Files, S. 126," box 9, folder "Correspondence 1977"; "Earthquake Hazards Reduction Act," Congressional Record; Paul Houston, "House Passes Bill to Speed Study of Quake Forecasts," LAT, Sept. 10, 1977, I:1, 6.

62. Chen Yong, ed., The Great Tangshan Earthquake of 1976: An Anatomy of Disaster (New York: Pergamon Press, 1988); Schleede to Cannon, Aug. 31, 1976, Barnes Files, box 2, folder "Earthquakes-Legislation (2)."

63. Jack Bennett, "Palmdale 'Bulge' Update," California Geology 30 (1977): 187-189; Robert Lindsey, "San Andreas 'Bulge' Puzzles Scientists," NYT, Mar. 13, 1977, p. 15; "Minutes of the May 25-26, 1978 Meeting of the USGS Earthquake Studies Advisory Panel," Newmark Papers, box 12, folder "USGS"; George Alexander, "Palmdale Bulge—

It's Sinking in One Place," *LAT,* Feb. 17, 1977, II:1, 3; idem, "U.S. Scientists Doubt Soviet Prediction of Palmdale Quake," ibid., Apr. 22, 1978, I:24; Robert C. Toth, "The Bulge—Study Fails to Find Answer," ibid., June 21, 1979, I:3, 30; George Alexander, "Palmdale Bulge—A Statistical Mirage?" ibid., Dec. 6, 1979, I:3, 25; Richard A. Kerr, "Does California Bulge or Does It Jiggle?" *Science* 219 (1983): 1205–1206; Robert J. Geller, "Earthquake Prediction: A Critical Review," *Geophysical Journal International* 131 (1997): 425–450, on 433.

64. Art Seidenbaum, "All Shook Up by Doomsday Seers," *LAT,* Dec. 31, 1976, IV:1; James H. Whitcomb, "Earthquake Prediction–Related Research at the Seismological Laboratory, California Institute of Technology, 1974–1976," in C. Kisslinger and Z. Suzuki, eds., *Earthquake Precursors: Proceedings of the U.S.-Japan Seminar on Theoretical and Experimental Investigations of Earthquake Precursors* (Tokyo: Center for Academic Publications Japan, 1978), 1–11.

65. A. G. Lindh, D. A. Lockner, and W. H. K. Lee, "Velocity Anomalies: An Alternative Explanation," *BSSA* 68 (1978): 721–734; William D. Stuart and Malcolm J. S. Johnston, "Intrusive Origin of the Matsushiro Earthquake Swarm," *Geology* 3 (1975): 63–67; J. B. Walsh, "An Analysis of Local Changes in Gravity Due to Deformation," *Pure and Applied Geophysics* 113 (1975): 97–106.

66. Hiroo Kanamori, "Earthquake Prediction," *Engineering and Science,* Oct. 1974, pp. 18–21; Hiroo Kanamori and David Hadley, "Crustal Structure and Temporal Velocity Change in Southern California," *Pure and Applied Geophysics* 113 (1975): 257–280; Hiroo Kanamori and Gary Fuis, "Variation of P-Wave Velocity Before and After the Galway Lake Earthquake (ML-5.2) and the Goat Mountain Earthquakes (ML-4.7, 4.7), 1975, in the Mojave Desert, California," *BSSA* 66 (1976): 2017–2037; Susan A. Raikes, "The Temporal Variation of Teleseismic P-Residuals for Stations in Southern California," ibid., 68 (1978): 711–720; Chris H. Cramer, Charles G. Bufe, and Paul W. Morrison, "P-Wave Travel-Time Variations Before the August 1, 1975 Oroville, California Earthquake," ibid., 67 (177): 9–26; Bruce A. Bolt, "Constancy of P Travel Times From Nevada Explosions to Oroville Dam Station, 1970–1976," ibid., 67 (1977): 27–32; see also David M. Boore, Thomas V. McEvilly, and Allan Lindh, "Quarry Blast Sources and Earthquake Prediction: The Parkfield, California, Earthquake of June 28, 1966," *Pure and Applied Geophysics* 113 (1975): 293–296.

67. Richard A. Kerr, "Earthquakes: Prediction Proving Elusive," *Science* 200 (1978): 419–421 (both quotes); idem, "Prospects for Earthquake Prediction Wane," ibid., 206 (1979): 542–545; George Alexander, "Quake That Hit Without Warning Puzzles Scientists," *LAT,* Oct. 27, 1979, I:1, 30; for the waning of enthusiasm, see also Council of State Governments, "National Conference on Earthquakes and Related Hazards," Nov. 16–18, 1977, copy of proceedings in U.S. Senate Records, series "Committee on Commerce, Science, and Transportation," subseries "Earthquake Hazards Reduction Act: General Information and Miscellaneous Publications, 1976," box 8, folder "News Clippings, Articles, Reports, etc."

68. Quotation from "Earthquake Hazards Reduction Act of 1977," Public Law 95-124.

Chapter 9

1. VSP Associates, "To Save Lives and Protect Property: A Policy Assessment of Federal Earthquake Activities, 1964-1987," final report prepared for the Federal Emergency Management Agency, Nov. 1, 1988, copy in EERC Library, Richmond, CA, 32-34. For an early example of the belief in information as the critical variable in mitigation, see [USGS] Geologic Division, "Earthquake Hazard Reduction Program of the U.S. Geological Survey," draft of Sept. 1972, in EMLR, box 9, folder "200.25 U.S. Geological Survey."

2. VSP Associates, "To Save Lives," 63-64, 69-80; Frank Press, "Memorandum for the President," May 11, 1978, reprinted in ibid., 227-235; Jim McIntyre and Press to Howard W. Cannon and Olin E. Teague, Aug. 25, 1978, Records of the U.S. Senate (RG 46), National Archives, Washington, DC, series "Committee on Commerce, Science, and Transportation," subseries "95th Congress Legislative Files," box 7, folder "Earthquake Hazards Reduction."

3. VSP Associates, "To Save Lives," 78-89, 130-131; U.S. Congress Office of Technology Assessment, *Reducing Earthquake Losses,* Report OTA-ETI-623 (Washington, DC: GPO, 1995), 18, 25-27.

4. VSP Associates, "To Save Lives," chap. 6; Office of Technology Assessment, *Reducing Earthquake Losses,* 12-17; Robert C. Toth, "Budget Cuts Jolt Project to Reduce Quake Hazards," *LAT,* Oct. 15, 1978, I:1, 5.

5. T. A. Heppenheimer, *The Coming Quake: Science and Trembling on the California Earthquake Frontier* (New York: Times Books, 1988), 122-136, 156-164; George Alexander, "Rock-Layer Changes May Bring on Quake Advisories," *LAT,* Jan. 27, 1980, II:1, 5.

6. Heppenheimer, *The Coming Quake,* 165-188; Robert E. Wallace, *Earthquakes, Minerals and Me: With the USGS, 1942-1995,* USGS Open-File Report 96-260 (Washington, DC: GPO, 1996), available online at http://quake.wr.usgs.gov/study/history/OFR96-260-wallace/wallace_contents.html, sec. 7; Allan G. Lindh, "Earthquake Prediction Comes of Age," *Technology Review* [MIT alumni magazine], 93, no. 2 (1990): 43-51; "Once Again Parkfield Quake Is a No-Show," *Science* 262 (1993): 1369; J. Savage, "The Parkfield Prediction Fallacy," *BSSA* 83 (1993): 1-6; Robert J. Geller, "Earthquake Prediction: A Critical Review," *Geophysical Journal International* 131 (1997): 425-450, on 438-440.

7. Robert L. Wesson and Robert E. Wallace, "Predicting the Next Great Earthquake in California," *Scientific American,* Feb. 1985, pp. 35-43; Lindh, "Earthquake Prediction," 45; *83rd Report of the State Geologist, 1990-1991* (Sacramento: California Division of Mines and Geology, 1991), 10-11. For a criticism of long-term probabilistic forecasts, see J. Savage, "Criticism of Some Forecasts of the National Earthquake Prediction Council," *BSSA* 81 (1991): 862-881.

8. Seismic Safety Commission, *Northridge Earthquake: Turning Loss to Gain,* report SSC 95-01 (Sacramento: Seismic Safety Commission, 1995), 114; Mark B. Roman, "Finding Fault," *Discover,* Aug. 1988, pp. 56-63.

9. W. Henry Lambright, "The Role of States in Earthquake and Natural Hazard

Innovation at the Local Level: A Decision-Making Study," final report on NSF grant PFR-8018710, Dec. 1984, copy in Natural Hazards Research Center Library, Boulder, CO, sec. 7; VSP Associates, "To Save Lives," 57–58; see also Kristine Moe, "Businesses Urged to Plan to Survive Quake," *San Diego Union-Tribune,* Apr. 6, 1987, A:4.

10. Office of Technology Assessment, *Reducing Earthquake Losses,* 16; Stanley Scott (interviewer), *Henry J. Degenkolb,* Connections: The EERI Oral History Series, vol. 1 (Oakland: Earthquake Engineering Research Institute, 1994), 144–149.

11. Susan R. Schrepfer, "The Nuclear Crucible: Diablo Canyon and the Transformation of the Sierra Club, 1965–1985," *California History* 71 (1992): 213–237; William J. Lanouette, "Unsettling Decision," *National Journal,* Mar. 27, 1982, p. 558; C. A. Hall, Jr., "San Simeon-Hosgri Fault System, Coastal California: Economic and Environmental Implications," *Science* 190 (1975): 1291–1294; Larry Pryor, "Fate of Atom Plant in Doubt," *LAT,* Jan. 18, 1976, I:3, 28; "Nader Urges A-Plant Closings," ibid., Apr. 9, 1976, I:26; George Alexander, "Scientists Study Possibility of Link in 3 Offshore Faults," ibid., Apr. 25, 1976, II:1; idem, "Quakes and A-Plants: How Great Is the Threat?" ibid., May 30, 1976, II:1, 3, 4.

12. Schrepfer, "Nuclear Crucible"; Bryce Nelson, "Diablo Canyon A-Plants Safe, Panel Declares," *LAT,* July 12, 1978, I:3, 12; "A-Plant Judged Safe From Quake," ibid., Oct. 2, 1979, I:3, 24; Judith Cummings, "Blueprint Switch at Coast A-Plant Widens U.S. Inquiry on Its Safety," *NYT,* Oct. 1, 1981, p. 1; idem, "Reactor License Voted, But Coast Fight Goes On," ibid., Sept. 9, 1984, p. 70; Thomas C. Hayes, "Pacific Gas's Next Challenge," ibid., Nov. 20, 1984, D:1. On the 1976 nuclear moratorium initiative, see Thomas Raymond Wellock, *Critical Masses: Opposition to Nuclear Power in California, 1958–1978* (U of Wisconsin P, 1998), chap. 4.

13. John Kendall, "U.S. May Have to Redesign Auburn Dam for Quakes," *LAT,* Sept. 29, 1976, I:1, 3, 32, 33; "State Supports Dam Despite Quake Danger," ibid., Mar. 22, 1977, I:3; "Potential for Big Auburn Dam Quake Reported," ibid., June 29, 1977, I:3, 31; John Kendall "U.S. Sets Auburn Dam Quake Standards," ibid., Sept. 15, 1978, I:24; idem, "Auburn Dam to Be Redesigned," ibid., Jan. 27, 1979, I:1, 24; Gladwin Hill, "California Aides Veto Federal Plan for Dam, Citing Quake Damages," *NYT,* Jan. 5, 1979, p. 12; Kimberly A. Moy, "Auburn Dam Foes Cite Quake Dangers," *SB,* Sept. 26, 1995, B:1.

14. "Quakes: Shaking Up Our Thinking," *LAT,* July 8, 1980, II:4; "Vital First Step," ibid., Dec. 16, 1980, II:6; "Facts, Not Flyers," ibid., Jan. 1, 1981, II:6; Hugo Martin, "Seismic Safety Puts Benson on Firm Footing on Council," *LAT,* May 22, 1994, A:1; Stanley Scott and Robert A. Olson, eds., *California's Earthquake Safety Policy: A Twentieth Anniversary Retrospective, 1969–1989* (Berkeley: Earthquake Engineering Research Center, 1993), 6.

15. Erwin Baker, "Council OKs Earthquake Safety Study," *LAT,* Oct. 25, 1979, II:1; Claire Spiegel, "7,876 Brick Buildings in L. A. Predate Quake Codes," ibid., Nov. 25, 1979, II:1, 3; Erwin Baker, "Cost of Reinforcing Old Buildings Told," ibid., June 18, 1980, II:1, 8; Paul Manuele, "Proposed Ordinance Will Require Upgrading of Unsafe Structures," ibid., July 27, 1980, IX:37, 39; Erwin Baker, "Quake Standards for Old

Buildings Urged," ibid., Dec. 5, 1980, II:9; idem, "Quake Safety Ordinance for Masonry Buildings Gains," ibid., Dec. 17, 1980, I:1, 22; idem, "400 Renters Assail Quake Safety Ordinance, Tell Fears of Eviction," ibid., Dec. 24, 1980, II:1, 2; Joan Sweeney, "Council Will Act on Tougher Quake Standards," ibid., Jan. 5, 1981, II:1, 2, 8; Harry Weinstein, "New Quake Safety Standards Enacted," ibid., Jan. 8, 1981, II:1, 5. See also Daniel J. Alesch and William J. Petak, *The Politics and Economics of Earthquake Hazard Mitigation: Unreinforced Masonry Buildings in Southern California* ([Boulder:] Institute of Behavioral Sciences, University of Colorado, 1986), chap. 5.

16. Alesch and Petak, *Politics and Economics of Earthquake Hazard Mitigation,* chap. 6; Philip R. Berke and Timothy Beatley, *Planning for Earthquakes: Risk, Politics, and Policy* (Baltimore: Johns Hopkins UP, 1992), chap. 4; Erin Kelly, "Risky Buildings: Which Ones Are They?" *Orange County Register,* Oct. 14, 1990, L:10; Barbara Metzler, "Still Shaky," *Riverside Press-Enterprise,* Oct. 11, 1992, B:1; Seismic Safety Commission, *Status of California's Unreinforced Masonry Building Law: 1997 Biannual Report to the Legislature,* Report SSC 97-03 (Sacramento: Seismic Safety Commission, 1997); idem, *Northridge Earthquake,* 48-50.

17. "Quake Protection Plans: They're on Shaky Ground," *San Diego Union-Tribune,* July 11, 1986, A:1; Thor Kamban Biberman, "New State Law Requires Hospitals to be Seismically Fit," *San Diego Daily Transcript,* Sept. 3, 1997, p. 1C; Seismic Safety Commission, *Northridge Earthquake,* 68.

18. Scott Harris, "Bond Issue for Quake Funds Proposed," *LAT,* Oct. 5, 1985, II:3; Carl Ingram, "$800-Million Plan to Boost Quake Safety in State Buildings Proposed," ibid., Nov. 15, 1986, I:34; Erin McCormick, "Billions Unspent From '89 Temblor," *SFE,* Feb. 13, 1994, A:1; Cynthia H. Craft, "Quake Retrofit Funds About to be Unearthed," *LAT,* June 21, 1995, A:3; Kenneth R. Weiss, "UC Berkeley Seeks $700 Million for Seismic Work," ibid., Oct. 25, 1997, A:18; Stanley Scott (interviewer), *George W. Housner, Connections: The EERI Oral History Series,* vol. 4 (Oakland: Earthquake Engineering Research Institute, 1997), 181.

19. The Governor's Board of Inquiry on the 1989 Loma Prieta Earthquake, *Competing Against Time: Report to Governor George Deukmejian* (North Highlands: State of California Office of Planning and Research, 1990); William P. McGowan, "Fault-Lines: Seismic Safety and the Changing Political Economy of California's Transportation System," *California History* 72 (1993): 171-192.

20. Virginia Ellis, "State to Retrofit Bridges Faster," *LAT,* Mar. 10, 1994, A:1; Dan Bernstein, "Bonds' Defeat Strains State Budget," *SB,* June 9, 1994, A:29; Greg Lucas, "Bridge Tolls May Double," *SFC,* June 14, 1996, A:1; Max Vanzi, "State Funding of Bridge Work Splits North and South," *LAT,* May 26, 1997, A:3.

21. "Chapter 1521," *Statutes of California,* 1985.

22. Earl W. Hart, *Fault-Rupture Hazard Zones in California,* California Division of Geology Special Publication 42, revised 1985 edition (Sacramento: California Division of Mines and Geology, 1985); Alan J. Wyner and Dean E. Mann, *Preparing for California's Earthquakes: Local Government and Seismic Safety* (Berkeley: Institute of Governmental Studies, 1986), chap. 2; Robert Reitherman and David J. Leeds, "A Study of the

Effectiveness of the Alquist-Priolo Program," California Division of Mines and Geology Open-File Report 90-18, copy in EERC Library; James E. Slosson, Jeffrey R. Keaton, and Jeffrey A. Johnson, "Fault-Rupture Hazards, the Alquist-Priolo Fault Hazard Act, and Siting Decisions in California," *Bulletin of the Association of Engineering Geologists* 31 (1994): 183–189; *83rd Report of the State Geologist,* 5–9; Kenneth Reich, "State Releases Maps Showing Quake Hazard Zones in East Valley," *LAT,* Feb. 9, 1996, B:1.

23. E. L. Jackson, "Public Response to Earthquake Hazard," *California Geology* 30 (1977): 278–280. On the character of the regulatory-state apparatus for earthquake hazard mitigation, see also the excellent analysis in Robert Stallings, *Promoting Risk: Establishing the Earthquake Threat* (New York: Aldine de Gruyter, 1995).

24. See, for example, Kendrick A. Clements, "Engineers and Conservationists in the Progressive Era," *California History* 58 (1979): 282–303; Susan R. Schrepfer, *The Fight to Save the Redwoods: A History of Environmental Reform, 1917–1978* (Madison: U of Wisconsin P, 1983), chaps. 1–5; Michael L. Smith, *Pacific Visions: California Scientists and the Environment, 1850–1915* (New Haven: Yale UP, 1987); Anne F. Hyde, "William Kent: The Puzzle of Progressive Conservationists," in William Deverell and Tom Sitton, eds., *California Progressivism Revisited* (Berkeley: U of California P, 1994), 34–56.

25. See, for example, Schrepfer, "The Nuclear Crucible"; Wellock, *Critical Masses,* chap. 1.

Essay on Sources

Historiography

The history of seismology and of earthquake hazard mitigation in California has been subjected to few serious historical studies. The only related topic that has received adequate attention so far has been the fight against the Bodega Head and Malibu Canyon nuclear reactors, for which see David Okrent, *Nuclear Reactor Safety: On the History of the Regulatory Process* (Madison: University of Wisconsin Press, 1981), chap. 17; Brian Balogh, *Chain Reaction: Expert Debate and Public Participation in American Commercial Nuclear Power, 1945–1975* (Cambridge: Cambridge University Press, 1991), 240–258; J. Samuel Walker, *Containing the Atom: Nuclear Regulation in a Changing Environment, 1963–1971* (Berkeley: University of California Press, 1992), chap. 5; and Thomas Raymond Wellock, *Critical Masses: Opposition to Nuclear Power in California, 1958–1978* (Madison: University of Wisconsin Press, 1998), chap. 1.

The 1906 San Francisco earthquake and the reactions it produced among scientists and engineers have been analyzed in Arnold J. Meltsner, "The Communication of Scientific Information to the Wider Public: The Case of Seismology in California," *Minerva* 17 (1979): 331–354; Gladys Hansen and Emmet Condon, *Denial of Disaster* (San Francisco: Cameron, 1989); Dennis R. Dean, "The San Francisco Earthquake of 1906," *Annals of Science* 50 (1993): 501–521; and James C. Williams, "Earthquake Engineering: Designing Unseen Technology against Invisible Forces," *ICON: Journal of the International Committee for the History of Technology* 1 (1995): 172–194. Reactions to the 1868 San Francisco Bay area earthquake are discussed in William H. Prescott, "Circumstances Surrounding the Preparation and Suppression of a Report on the 1868 California Earthquake," *Bulletin of the Seismological Society of America* 72 (1982): 2389–2393; Michele L. Aldrich, Bruce A. Bolt, Alan E. Leviton, and Peter U. Rodda, "The 'Report' of the 1868 Haywards Earthquake," ibid., 76 (1986): 71–76; and Charles Wollenberg, "Life on the Seismic Frontier: The Great San Francisco Earthquake (of 1868)," *California History* 71 (1992): 494–509.

Judith R. Goodstein provides an excellent discussion of the early years of the Caltech Seismological Laboratory in "Waves in the Earth: Seismology Comes to Southern

California," *Historical Studies in the Physical Sciences* 14 (1984): 201–230. The efforts to induce retrofitting of unreinforced masonry buildings are covered in Daniel J. Alesch and William J. Petak, *The Politics and Economics of Earthquake Hazard Mitigation: Unreinforced Masonry Buildings in Southern California* ([Boulder:] Institute of Behavioral Sciences, University of Colorado, 1986). Much information on the Palmdale Bulge and the Whitcomb prediction can be found in Ralph H. Turner, Joanne M. Nigg, and Denise Heller Paz, *Waiting for Disaster: Earthquake Watch in California* (Berkeley: University of California Press, 1986).

For a fascinating sociological analysis of the "earthquake establishment" that has dominated seismic safety advocacy in the past thirty years, see Robert Stallings, *Promoting Risk: Establishing the Earthquake Threat* (New York: Aldine de Gruyter, 1995). Two valuable, albeit mostly undocumented journalistic accounts are John J. Fried, *Life along the San Andreas Fault* (New York: Saturday Review Press, 1973), which details issues surrounding building-code and land-use measures in the 1960s and early 1970s, and T. A. Heppenheimer, *The Coming Quake: Science and Trembling on the California Earthquake Frontier* (New York: Times Books, 1988), which concentrates on earthquake prediction in the 1970s and 1980s.

For the broader history of seismology, I have relied on Charles Davison, *The Founders of Seismology* (Cambridge: Cambridge University Press, 1927); Dennis R. Dean, "Robert Mallet and the Founding of Seismology," *Annals of Science* 48 (1991): 39–67; James Dewey and Perry Byerly, "The Early History of Seismometry (to 1900)," *Bulletin of the Seismological Society of America* 59 (1969): 183–227; A. L. Herbert-Gustar and P. A. Nott, *John Milne: Father of Modern Seismology* (Tenterden, Kent: Paul Norbury Publications, 1980); Stephen Brush, "Discovery of the Earth's Core," *American Journal of Physics* 48 (1980): 705–724; Benjamin F. Howell, *An Introduction to Seismological Research: History and Development* (Cambridge: Cambridge University Press, 1990); Bruce Bolt, *Nuclear Explosions and Earthquakes: The Parted Veil* (San Francisco: W. H. Freeman and Co., 1976); Charles C. Bates, Thomas F. Gaskell, and Robert B. Rice, *Geophysics in the Affairs of Man: A Personalized History of Exploration Geophysics and Its Allied Sciences of Seismology and Oceanography* (New York: Pergamon Press, 1982); and Kai-Henrik Barth, "Science and Politics in Early Nuclear Test Ban Negotiations," *Physics Today* 51 (March 1998): 34–39.

My overall historical framework owes much to the so-called organizational synthesis of American history, for which see in particular Brian Balogh, "Reorganizing the Organizational Synthesis: Federal-Professional Relations in Modern America," *Studies in American Political Development* 5 (1991): 119–172; idem, ed., *Integrating the Sixties: The Origins, Structures, and Legitimacy of Public Policy in a Turbulent Decade* (University Park: Pennsylvania State University Press, 1996); David Vogel, "The 'New' Social Regulation in Historical and Comparative Perspective," in Thomas K. McCraw, ed., *Regulation in Perspective: Historical Essays* (Cambridge: Harvard University Press, 1981), 155–185; and Richard A. Harris and Sidney M. Milkis, *The Politics of Regulatory Change: A Tale of Two Agencies,* 2nd ed. (New York: Oxford University Press, 1996).

My interpretation of the history of environmentalism in California and elsewhere derives in particular from Samuel P. Hays, *Conservation and the Gospel of Efficiency: The Progressive Conservation Movement, 1890-1920* (Cambridge: Harvard University Press, 1959); Kendrick A. Clements, "Engineers and Conservationists in the Progressive Era," *California History* 58 (1979): 282-303; Susan R. Schrepfer, *The Fight to Save the Redwoods: A History of Environmental Reform, 1917-1978* (Madison: University of Wisconsin Press, 1983); Michael L. Smith, *Pacific Visions: California Scientists and the Environment, 1850-1915* (New Haven: Yale University Press, 1987); Anne F. Hyde, "William Kent: The Puzzle of Progressive Conservationists," in William Deverell and Tom Sitton, eds., *California Progressivism Revisited* (Berkeley: University of California Press, 1994), 34-56; Samuel P. Hays, *Beauty, Health, and Permanence: Environmental Politics in the United States, 1955-1985* (Cambridge: Cambridge University Press, 1987); Michael J. Lacey, ed., *Government and Environmental Politics: Essays on Historical Developments since World War Two* (Washington: Woodrow Wilson Center Press, 1989); and George Hoberg, *Pluralism by Design: Environmental Policy and the American Regulatory State* (New York: Praeger, 1992).

In approaching the problem of how a natural hazard becomes a publicly recognized problem, I have also been much influenced by the constructionist sociology of social problems; see, for example, John R. Gusfield, *The Culture of Public Problems: Drinking-Driving and the Symbolic Order* (Chicago: University of Chicago Press, 1981); Mary Douglas and Aaron Wildavsky, *Risk and Culture: An Essay on the Selection of Technical and Environmental Dangers* (Berkeley: University of California Press, 1982); Malcolm Spector and John I. Kitsuse, *Constructing Social Problems* (New York: Aldine de Gruyter, 1987); and John A. Hannigan, *Environmental Sociology: A Social Constructivist Perspective* (New York: Routledge, 1995).

Archival Sources

California seismologists, particularly those active in the early years of the Seismological Society of America and those working in the 1960s and '70s, have left behind abundant records. Archival documentation is much more difficult to find for the 1940s and '50s and for the activities of engineers.

Those papers of John C. Branner in the Stanford University archives, which contain a nearly complete set of both his incoming and outgoing correspondence, are the preeminent source for the history of seismology in California from 1906 to 1919. Branner's correspondence for 1920 and 1921, as well as drafts for an autobiography, can be found in the Huntington Library in San Marino, California. For the work of the State Earthquake Investigation Commission, the Branner papers can be supplemented with the sparser records in the George C. Davidson and Andrew C. Lawson papers at the Bancroft Library, on the campus of the University of California at Berkeley, and a few items in the records of the U.S. Geological Survey at the National Archives II in College Park, Maryland. The papers of commission member A. O. Leuschner at the Bancroft Library contain nothing of interest. The correspondence of the Seismological Society of America at the Bancroft covers the years when Sidney Townley and Perry Byerly were the society's secretaries, that is, 1911 to about 1950. There is no additional information in the papers of Townley at Stanford. The records of the U.S. Weather Bureau at the National Archives II give a very detailed picture of instrumental seismology in the United States in the 1910s.

For California seismology in the 1920s and early '30s, two important sources are the papers of Bailey Willis at the Huntington Library (his voluminous correspondence has been only incompletely preserved in this collection) and the papers of Harry O. Wood at the California Institute of Technology archives (his correspondence after 1921, both incoming and outgoing, appears to have been completely preserved). The latter collection can be usefully supplemented with the records of the Carnegie Institution, at its headquarters in Washington, D.C., which discuss the institution's support for seismology in southern California; the papers of Robert A. Millikan at Caltech (available on microfilm as well), which cover some of the political activities of Caltech seismologists; and the papers of Ralph Arnold at the Huntington, which describe the activities of one of Wood's allies in the southern California geological community. The Henry Pritchett papers at the Library of Congress contain important documents giving insight into the California boosters' efforts to suppress information after the Santa Barbara earthquake. The Robert Hill papers at Southern Methodist University in Dallas provide a fascinating look into the efforts of real estate speculators to shoot down Willis's earthquake prediction. Insights into the work of engineers at this time can be gleaned from the papers of John Freeman at the Massachusetts Institute of Technology archives and the records of the School of Engineering at Stanford. For the early history of the structural engineers' associations, I have relied on the office files of both the northern California association in San Francisco and the southern California one in Whittier, both of which I visited in 1994. When I returned to San Francisco in 1998, the northern California association had moved its office, and its records were no longer accessible.

Records for the immediate postwar period are sparse; the papers of engineer Lydik Jacobsen, which are housed at Stanford, are quite fragmentary. For the history of seismology in the 1960s and '70s, five collections are eminent sources: the papers of Charles F. Richter at Caltech, which concentrate on his public relations efforts on behalf of seismic safety; the papers of Frank Press at MIT, which document in detail the campaign for an earthquake prediction research program in the United States and also provide great insight into Geological Survey activities during this period; the papers of Richard H. Jahns (not yet completely sorted), at Stanford, which offer a very eclectic mix of documents on hazard mitigation at both the state and the federal level; the records of the Joint Committee on Seismic Safety at the California State Archives in Sacramento (supplemented by the legislative bill and committee files also kept there), which provide a complete record of that important body's activities; and the records of the Earthquake Mechanism Laboratory preserved at the National Archives regional center in San Bruno, California, which document the work of the Geological Survey's main rival in earthquake research.

For the Bodega Head nuclear reactor controversy, I located one resource not yet used by other historians: the papers of Elmer Marliave at the Water Resources Research Center archives on the Berkeley campus, which tell the story from the point of view of a consulting geologist for Pacific Gas and Electric. The papers of Donald Hornig at the Lyndon B. Johnson Presidential Library in Austin, Texas, add little to what is in the Press papers, because they preserve only Hornig's outgoing correspondence. A bound copy of the minutes of the State Mining and Geology Board, which documents the evolution of the California Division of Mines and Geology, can be found in the stacks of the Doe Library on the Berkeley campus. Much insight on the final establishment of the National Earthquake Hazards Reduction Program can be gleaned from the papers of presidential staff members at the Gerald Ford Library in Ann Arbor, Michigan, and the papers of engineer Nathan Newmark at the University of Illinois archives; conversely, I was not able to find anything of value in the as-yet-unprocessed papers of Alan Cranston at the Bancroft.

Other Primary Sources

Much information on public reactions to earthquakes in California can be gleaned from the state's newspapers. Published indexes to the major newspapers are available beginning in 1970. The Los Angeles Public Library has a very useful index for the *Los Angeles Times* covering the 1930s and '40s. The *Los Angeles Examiner* morgue, containing clippings filed by subject, is in the Regional History Collection at the University of Southern California. The pre-1970 index to San Francisco newspapers that is available on microform proved unhelpful. For the most part, I found myself scanning issues of the major newspapers that appeared in the months after a significant earthquake or other event (a most useful collection of newspapers on microfilm is at the California State Library in Sacramento; microfilm copies of local newspapers can also be found in most public libraries); I also relied heavily on newspaper clippings found in various archival collections.

In the absence of archival collections, the best evidence for engineers' attitudes toward seismic hazards comes from trade publications. For the period before World War II, I leaned most heavily on the *Engineering Supplement to the American Builders' Review* (which contains papers discussed at the biweekly meetings of the Structural Association of San Francisco), the *Architect and Engineer of California,* the weekly *Engineering News-Record,* and the *Southwest Builder and Contractor* (available on microfilm in the Los Angeles Public Library). For the postwar period, I relied on the proceedings of the annual meeting of the Structural Engineers Association of California and on the newsletter of the Earthquake Engineering Research Institute (which beginning about 1973 reprinted numerous meeting minutes, letters, and other archival matter concerning earthquake engineering); also of great value were the oral histories of earthquake engineers recorded by Stan Scott and published by the EERI.

I found much valuable material by systematically scanning the shelves of special libraries for rare reports, unpublished draft manuscripts, and other such "grey" literature. Among the libraries I searched with profit were: the Earthquake Engineering Research Center library at the University of California's field station in Richmond, which contains an absolute wealth of hard-to-find material; the Natural Hazards Library at the University of Colorado, which provides a valuable collection of social science studies on hazard mitigation; the geology libraries at Caltech, Stanford, Berkeley, and the Lamont-Doherty Observatory of Columbia University; the library of the U.S. Geological Survey in Menlo Park; the earthquake engineering and general engineering libraries at Caltech; the engineering library at the University of Illinois; and the Los Angeles, Pasadena, and Long Beach public libraries. The law library on the Berkeley campus contains a complete collection of all bills introduced in the California legislature and all amendments thereto; this is particularly valuable for tracking the legislative history of earthquake laws.

Index